Ionic Channels in Cells and Model Systems

Series of the Centro de Estudios Científicos de Santiago

Series Editor: Claudio Teitelboim
Centro de Estudios Científicos de Santiago
Santiago, Chile
and University of Texas at Austin
Austin, Texas, USA

IONIC CHANNELS IN CELLS AND MODEL SYSTEMS
Edited by Ramon Latorre

Ionic Channels in Cells and Model Systems

Edited by
Ramon Latorre
*Centro de Estudios Científicos de Santiago
and Facultad de Ciencias
Universidad de Chile
Santiago, Chile*

PLENUM PRESS • NEW YORK AND LONDON

Library of Congress Cataloging in Publication Data

Ionic channels in cells and model systems.

(Series of the Centro de Estudios Científicos de Santiago)
Includes bibliographies and index.
1. Ion channels. 2. Cell membranes. 3. Biological models. I. Latorre, Ramon. II. Series.
QH601.I67 1986   574.87′5   86-12265
ISBN 0-306-42194-1

© 1986 Plenum Press, New York
A Division of Plenum Publishing Corporation
233 Spring Street, New York, N.Y. 10013

All rights reserved

No part of this book may be reproduced, stored in a retrieval system, or transmitted in any form or by any means, electronic, mechanical, photocopying, microfilming, recording, or otherwise, without written permission from the Publisher

Printed in the United States of America

Contributors

Osvaldo Alvarez, Departamento de Biología, Facultad de Ciencias, Universidad de Chile, Santiago, Chile; and Centro de Estudios Científicos de Santiago, Santiago, Chile

Clay M. Armstrong, Department of Physiology, School of Medicine, University of Pennsylvania, Philadelphia, Pennsylvania 19104

Illani Atwater, Laboratory of Cell Biology and Genetics, NIADDK, National Institutes of Health, Bethesda, Maryland 20205

Juan Bacigalupo, Departamento de Biología, Facultad de Ciencias, Universidad de Chile, Santiago, Chile

Charles Bare, Agricultural Research Service, Weed Science Laboratory, Agricultural Environmental Quality Institute, Beltsville, Maryland 20705

Francisco J. Barrantes, Instituto de Investigaciones Bioquimicas, Universidad Nacional del Sur, 8000 Bahia Blanca, Argentina

Dale J. Benos, Department of Physiology and Biophysics, Laboratory of Human Reproduction and Reproductive Biology, Harvard Medical School, Boston, Massachusetts 02115; *present address:* Department of Physiology and Biophysics, University of Alabama at Birmingham, Birmingham, Alabama 35294

Francisco Bezanilla, Department of Physiology, Ahmanson Laboratory of Neurobiology; and Jerry Lewis Neuromuscular Research Center, University of California at Los Angeles, Los Angeles, California 90024

Ximena Cecchi, Departamento de Biología, Facultad de Ciencias, Universidad de Santiago, Chile; and Centro de Estudios Cientifícos de Santiago, Santiago, Chile

John A. Dani, Department of Physiology, University of California Medical School, Los Angeles, California 90024; *present address:* Section of Molecular Neurobiology, Yale University, New Haven, Connecticut 06510

A. Darszon, Departamento de Bioquímica, Centro de Investigación y de Estudios Avanzados del I.P.N., México, D.F. 07000, Mexico

Gerald Ehrenstein, Laboratory of Biophysics, NINCDS, National Institutes of Health, Bethesda, Maryland 20205

George Eisenman, Department of Physiology, University of California Medical School, Los Angeles, California 90024

Richard FitzHugh, Laboratory of Biophysics, NINCDS, National Institutes of Health, Bethesda, Maryland 20205

Robert J. French, University of Maryland School of Medicine, Baltimore, Maryland 21201

Harold Gainer, National Institutes of Health, Bethesda, Maryland 20205

Ana Maria Garcia, Boston Biomedical Research Institute, Department of Muscle Research, Boston, Massachusetts 02114; *present address:* Whitehead Institute, Cambridge, Massachusetts 02142

J. García-Soto, Departamento de Bioquímica, Centro de Investigación y de Estudios Avanzados de I.P.N., México D.F. 07000, México

Sandra Guggino, Laboratory of Molecular Aging, Gerontology Research Center, National Institute on Aging, National Institutes of Health, Baltimore, Maryland 21224

Cecilia Hidalgo, Department of Muscle Research, Boston Biomedical Research Institute, and the Department of Neurology, Harvard Medical School, Boston, Massachusetts 02114; *present address:* Departamento de Fisiología y Biofísica, Facultad de Medicina, Universidad de Chile; and Centro de Estudios Científicos de Santiago, Santiago, Chile.

A. D. Islas-Trejo, Departamento de Bioquímica, Centro de Investigación y de Estudios Avanzados del I.P.N., México, D.F. 07000, México

Kinihiko Iwasa, Laboratory of Biophysics, NINCDS, National Institutes of Health, Bethesda, Maryland 20205

Contributors

Enrique Jaimovich, Departamento de Fisiología y Biofísica, Facultad de Medicina, Universidad de Chile, Santiago, Chile

Bruce K. Krueger, University of Maryland School of Medicine, Baltimore, Maryland 21201

Ramon Latorre, Departmento de Biología, Facultad de Ciencias, Universidad de Chile, Santiago, Chile; and Centro de Estudios Científicos de Santiago, Santiago, Chile

Harold Lecar, Laboratory of Biophysics, National Institute of Neurological and Communicative Disorders and Stroke, National Institutes of Health, Bethesda, Maryland 20205

Irwin B. Levitan, Graduate Department of Biochemistry, Brandeis University, Waltham, Massachusetts 02254

A. Liévano, Departamento de Bioquímica, Centro de Investigación y de Estudios Avanzados del I.P.N., México, D.F. 07000, México

Werner R. Loewenstein, Department of Physiology and Biophysics, University of Miami School of Medicine, Miami, Florida 33101

M. Luxoro, Laboratorio de Fisiología Celular, Facultades de Medicina y Ciencias, Universidad de Chile, Santiago, Chile

Christopher Miller, Graduate Department of Biochemistry, Brandeis University, Waltham, Massachusetts 02254

Charles Mischke, Agricultural Research Service, Weed Science Laboratory, Agricultural Environmental Quality Institute, Beltsville, Maryland 20705

Nava Moran, Laboratory of Biophysics, NINCDS, National Institutes of Health, Bethesda, Maryland 20205; *present address:* Department of Neurobiology, The Weizmann Institute of Science, Center for Neurosciences and Behavioural Research, Rehovot, Israel 76100

V. Nassar-Gentina, Laboratorio de Fisiología Celular, Facultades de Medicina y Ciencias, Universidad de Chile, Santiago, Chile

Ana Lia Obaid, University of Pennsylvania, Philadelphia, Pennsylvania 19104

Harvey B. Pollard, Laboratory of Cell Biology and Genetics, NIADDK, National Institutes of Health, Bethesda, Maryland 20205

Stanley I. Rapoport, Laboratory of Neurosciences, National Institute on Aging, National Institutes of Health, Bethesda, Maryland 20892

John Rinzel, Mathematical Research Branch, NIADDK, National Institutes of Health, Bethesda, Maryland 20205

Eduardo Rojas, Laboratory of Cell Biology and Genetics, NIADDK, National Institutes of Health, Bethesda, Maryland 20205

Frederick Sachs, Department of Biophysics, State University of New York at Buffalo School of Medicine, Buffalo, New York 14214

Brian M. Salzberg, University of Pennsylvania, Philadelphia, Pennsylvania 19104

J. Sánchez, Departamentos de Bioquímica y Farmacología, Centro de Investigación y de Estudios Avanzados del I.P.N., México, D.F. 07000, México

Rosa Maria Santos, Laboratory of Cell Biology and Genetics, NIADDK, National Institutes of Health, Bethesda, Maryland 20205

Andres Stutzin, Laboratory of Cell Biology and Genetics, NIADDK, National Institutes of Health, Bethesda, Maryland 20205

Benjamin A. Suarez-Isla, Laboratory of Neurosciences, National Institute on Aging, National Institutes of Health, Bethesda, Maryland 20892; *present address:* Centro de Estudios Científicos de Santiago, Santiago, Chile

Robert E. Taylor, Laboratory of Biophysics, NINCDS, National Institutes of Health, Bethesda, Maryland 20205

Cecilia Vergara, Departmento de Biología, Facultad de Ciencias, Universidad de Chile, Santiago, Chile; and Centro de Estudios Científicos de Santiagos, Santiago, Chile

Daniel Wolff, Departmento de Biología, Facultad de Ciencias, Universidad de Chile, Santiago, Chile; and Centro de Estudios Científicos de Santiago, Santiago, Chile

Jennings F. Worley, III, University of Maryland School of Medicine, Baltimore, Maryland 21201

Preface

This book is based on a series of lectures for a course on ionic channels held in Santiago, Chile, on November 17–20, 1984. It is intended as a tutorial guide on the properties, function, modulation, and reconstitution of ionic channels, and it should be accessible to graduate students taking their first steps in this field.

In the presentation there has been a deliberate emphasis on the specific methodologies used toward the understanding of the workings and function of channels. Thus, in the first section, we learn to "read" single-channel records: how to interpret them in the theoretical frame of kinetic models, which information can be extracted from gating currents in relation to the closing and opening processes, and how ion transport through an open channel can be explained in terms of fluctuating energy barriers. The importance of assessing unequivocally the origin and purity of membrane preparations and the use of membrane vesicles and optical techniques in the study of ionic channels are also discussed in this section.

The patch-clamp technique has made it possible to study ion channels in a variety of different cells and tissues not amenable to more conventional electrophysiological methods. The second section, therefore, deals with the use of this technique in the characterization of ionic channels in different types of cells, ranging from plant protoplasts to photoreceptors. Several chapters on ionic channel reconstitution form part of the third section. Here we learn how this methodology has made it possible to study the channels contained in membrane systems inaccessible to patch electrodes such as sarcoplasmic reticulum or muscle transverse tubules. Furthermore, single-channel recording in cells, in cell-free membrane patches, and in planar lipid bilayers allows us to reveal how the metabolic machinery of the cell and the cell-to-cell interaction regulates the behavior of these proteins. These subjects are covered in the fourth section. Finally, chapters on models, channel structure, and function form part of the last section.

I would like to thank Osvaldo Alvarez, Cecilia Hidalgo, Enrique Jaimovich, and Maria Luisa Valdovinos for their invaluable help in the organization of the course. The enormous enthusiasm put forward by the assisting students and professors provided a very pleasant and exciting environment that resulted in an excellent meeting. I would also like to thank the Tinker Foundation and the PNUD, UNESCO program for funding.

Ramon Latorre

Santiago, Chile

Foreword

Robert E. Taylor

In the United States it is often asked why there are so many Chileans working on channels and on active transport. There are probably a number of reasons for this, but surely prominent among them is the existence of a small laboratory in Montemar, near Viña del Mar, Chile. How did it start, what happened, and who did it?

It started because Mario Luxoro got his Ph.D. with Francis O. Schmitt at MIT in Cambridge, Massachusetts in 1957. Luxoro and Schmitt were interested in the axoplasm of the giant axon of the squid then available in Massachusetts. In 1955, Schmitt and friends had caught specimens of the large squid, *Dosidicus gigas,* off Iquique, in northern Chile. With the cooperation of Dr. P. Yanez, a unit was set up in the Estación de Biología Marina in Montemar for the procuring, processing, and shipping of chilled as well as frozen and dried axoplasm to Cambridge. On his return to Chile, Luxoro was involved in this arrangement. That year Eduardo Rojas, as a medical student, spent some time in Montemar, and Pancho Hunneus-Cox went to MIT for 2 years.

The idea of the facilities of the Estacion in Montemar being used as merely a source of axoplasm for someone at MIT was not too attractive to the Chilean scientists. Luxoro worked with Eduardo Rojas in the sum-

---

ROBERT E. TAYLOR • Laboratory of Biophysics, NINCDS, National Institutes of Health, Bethesda, Maryland 20205.

mer of 1959–60 on the microinjection of trypsin into squid axons, Hunneus-Cox returned in 1962 and worked in Montemar on squid, studying S–S bonds, and in 1963 Mitzy Canessa returned from postdoctoral work in the United States to become involved in studies of the biochemistry of the axon membrane with Sigmund Fischer. Fernando Vargas was there, and Ichichi Tasaki went to Montemar in 1964 and introduced internal perfusion with Luxoro.

In November of 1963, a group consisting of Clay Armstrong and Daniel Gilbert, from the Laboratory of Biophysics at the NIH in Bethesda, along with Clara Franzini-Armstrong, Rita Guttman, and Werner Loewenstein joined Luxoro in Montemar for a few months. This arrangement grew out of discussions that Luxoro had had with Dr. Kenneth S. Cole, who wanted to take advantage of the availability of the large squid in Chile. Cole was not able to go, and the administration devolved on me. I went to Chile in February of 1966 and continued to spend part of each Chilean summer there until 1972. That is basically the reason why I am the one who is writing this.

By the summer of 1964–65, Eduardo Rojas had completed his Ph.D. in Chicago and was working at the NIH biophysics laboratory, and with Gerald Ehrenstein of that laboratory, he went to Montemar to perfuse the axon of the giant squid. About this time various problems arose between the workers and the administration of the Estación, and the Chancellor of the University of Chile provided some money to buy an old house across the street, and the Laboratorio de Fisiología Celular was born; it was in operation when I first went there in 1966.

The many students who appeared in the lab in Montemar from time to time make an impressive list. The ones I got to know well include Ramon Latorre, Cecilia Hidalgo, Eugenia Yanez, Francisco Bezanilla, Julio Vergara, and Veronica Nassar. There were others, like F. Zambrano, Cristian Bennett, and many more after 1972.

It is important to remember that the axon of the giant squid, available only in Chile during the period of about 1957 to 1971, was of great importance to our understanding of channels. Not only was it a superb preparation, mainly because of the absence of branches, but it served to gather people together with common interests. Some people worked on the voltage-dependent ion movements, some on the biochemistry of the membrane, and some were, and still are, interested in the role of the axoplasm. We might recall that in the 1960s it was not known for sure that there were individual ionic channels and that they were composed of protein (or lipid–protein–carbohydrate complexes). Many people thought so, but the pioneering work of Luxoro and Hunneus-Cox and then Mitzy Canessa with Sigmund Fischer and later the work of Rojas, Armstrong,

Foreword

and Atwater with internally perfused pronase was all important in focusing attention on the proteins.

These are only a few comments about this small but important laboratory. Perhaps someone will write a proper history that would include the work of Mario Luxoro, Veronica Nassar, Francisco Bezanilla, Julio Vergara, Juan Bacigalupo, Cecilia Vergara, Elizabeth Bosch, Rafael Torres, and Victor Corvalan since there have been no giant squid. Some of these people are still associated with this laboratory, and many are spread about the world, but the work goes on. Most of the great ideas are probably wrong, but by training students and fighting about the ideas, progress occurs.

Contents

Introduction 1

I. Methodologies

Chapter 1
Kinetic Models and Channel Fluctuations
Osvaldo Alvarez

1. Introduction 5
2. Two-State Channel 5
3. Two Two-State Channels 11
4. Three-State Channel with Three Conductances 13
5. Three-State Channel with Only Two Conductances 13
   References 15

Chapter 2
Single-Channel Currents and Postsynaptic Drug Actions
Harold Lecar

1. Introduction 17
2. Channel Gating as a Stochastic Process 18

3. Postsynaptic Channels in the Presence of Drugs 24
4. Reconstructing the Postsynaptic Current 30
5. Macroscopic and Molecular Consequences 33
   References 34

Chapter 3
Voltage-Dependent Gating: Gating Current Measurement and Interpretation
Francisco Bezanilla

1. Introduction 37
2. Voltage Gating 37
3. Gating Current Is a Capacitive Current 38
4. Measurement of Gating Currents 39
5. Gating of the Sodium Channel 45
   References 51

Chapter 4
Characterizing the Electrical Behavior of an Open Channel via the Energy Profile for Ion Permeation: A Prototype Using a Fluctuating Barrier Model for the Acetylcholine Receptor Channel
George Eisenman and John A. Dani

1. Introduction 53
2. Theory 54
3. Confrontation with Experimental Data for the AChR Channel .. 63
4. Discussion 81
   Appendix 83
   References 85

Chapter 5
The Use of Specific Ligands to Study Sodium Channels in Muscle
Enrique Jaimovich

1. Introduction 89
2. Molecular Pharmacology of the Sodium Channel in Muscle 90
3. Sodium Channel in Cardiac Muscle: Are All Sodium Channels Alike? 93
4. Surface and Tubular Sodium Channels in Skeletal Muscle 94

5. Models for Sodium Channels in Muscle Membranes 96
References 98

Chapter 6
Isolation of Muscle Membranes Containing Functional Ionic Channels
Cecilia Hidalgo

1. Introduction 101
2. Excitation–Contraction Coupling 101
3. Ionic Channels and E–C Coupling 102
4. Isolation of Muscle Membranes 105
5. Concluding Remarks 120
References 120

Chapter 7
Methodologies to Study Channel-Mediated Ion Fluxes in Membrane Vesicles
Ana Maria Garcia

1. Introduction 127
2. Channel-Mediated Tl^+ Flux Measured by Fluorescence Quenching 129
3. Channel-Mediated Ion Fluxes Measured by Light Scattering ... 136
References 139

Chapter 8
Optical Studies on Ionic Channels in Intact Vertebrate Nerve Terminals
Brian M. Salzberg, Ana Lia Obaid, and Harold Gainer

1. Introduction 141
2. Equivalence of Optical and Electrical Measurements of Membrane Potential 142
3. Optical Recording of Action Potentials from Nerve Terminals of the Frog *Xenopus* 147
4. Properties of the Action Potential in the Nerve Terminals 151
5. Ionic Basis of the Depolarizing Phase of the Action Potential .. 154
6. Concluding Remarks 159
References 160

Chapter 9
Optical Detection of ATP Release from Stimulated Endocrine Cells: A Universal Marker of Exocytotic Secretion of Hormones
Eduardo Rojas, Rosa Maria Santos, Andres Stutzin, and Harvey B. Pollard

1. Introduction 163
2. Methodological Considerations 164
3. Acetylcholine-Induced ATP Release from Chromaffin Cells: Calcium Dependence 168
4. Nicotinic Receptor Desensitization 169
5. Granular Nature of the Secreted ATP 170
6. ATP Release Evoked by Membrane Depolarization Is Mediated by Activation of Voltage-Gated Calcium Channels 173
7. ATP Release from Collagenase-Isolated Islets of Langerhans ... 174
8. Conclusion 175
9. Summary 176
   References 177

II. Channels in Biological Membranes

Chapter 10
Mechanotransducing Ion Channels
Frederick Sachs

1. Introduction 181
2. Recording SA Channels 183
3. General Characteristics 184
4. Conductance Properties 185
5. Kinetic Properties 185
6. The Model 188
7. Comparing the Model to the Data 189
8. Future Prospects 192
   References 192

Chapter 11
Ionic Channels in Plant Protoplasts
Nava Moran, Gerald Ehrenstein, Kunihiko Iwasa, Charles Baré, and Charles Mischke

1. Introduction 195
2. Some Methodological Considerations 197
3. Voltage-Dependent Channels Opened by Hyperpolarization 199

4. Channels Affected by TEA	201
5. Conclusions	203
References	204

Chapter 12
Channels in Kidney Epithelial Cells
Sandra Guggino

1. Introduction	207
2. Cell Culture	209
3. Patch-Clamp Methodology	209
4. Potassium Channel Characteristics	211
5. Channel Modulation	215
6. Conclusions	215
References	218

Chapter 13
Channels in Photoreceptors
Juan Bacigalupo

1. Introduction	221
2. Vertebrate Photoreceptors	223
3. Invertebrate Photoreceptors	225
References	232

Chapter 14
Inactivation of Calcium Currents in Muscle Fibers from Balanus
M. Luxoro and V. Nassar-Gentina

1. Introduction	235
2. Methodological Considerations	236
3. Characteristics of Inward Currents	237
4. Mechanism of Inactivation	239
References	241

Chapter 15
Electrophysiological Studies in Endocrine Cells
Clay M. Armstrong

| 1. Introduction | 243 |

2. Whole-Cell Patch-Clamp Methodology 245
3. Cell Culture 247
4. Ionic Currents in GH3 Cells 247
5. Characteristics of Calcium Channels 249
6. Conclusions 252
   References 254

III. Ionic Channel Reconstitution

Chapter 16

Ion Channel Reconstitution: Why Bother?

Christopher Miller

1. Introduction and Background 257
2. Unexpected Surprises 263
3. Unconstrained Variables 264
4. Unrealized Hopes 267
   References 269

Chapter 17

From Brain to Bilayer: Sodium Channels from Rat Neurons Incorporated into Planar Lipid Membranes

Robert J. French, Bruce K. Krueger, and Jennings F. Worley III

1. Perspectives and Background 273
2. Electrophysiology without Cells 277
3. A Closer Look at Batrachotoxin-Activated Sodium Channels in Bilayer Membranes 280
4. Looking Ahead 287
   References 288

Chapter 18

Ionic Channels in the Plasma Membrane of Sea Urchin Sperm

A. Darszon, J. García-Soto, A. Liévano, J. A. Sánchez, and A. D. Islas-Trejo

1. Introduction 291
2. Are There Channels in Sea Urchin Sperm? 293
3. Reconstitution Studies with Isolated Sea Urchin Sperm Plasma Membrane 294

4. Channels in the Plasma Membrane of Sea Urchin Sperm:
   Implications for the Acrosome Reaction 300
5. Are There Receptors to the Egg Jelly in the Sea Urchin Sperm
   Plasma Membranes? 300
6. Perspectives 302
   References 302

Chapter 19

Characterization of Large-Unitary-Conductance Calcium-Activated Potassium Channels in Planar Lipid Bilayers

Daniel Wolff, Cecilia Vergara, Ximena Cecchi, and Ramon Latorre

1. Introduction 307
2. Channel Gating 307
3. Channel Conductance and Selectivity 311
4. Conductance of the Calcium-Activated K^+ Channels 312
5. Selectivity of the Ca-K Channels 313
6. Blockade of the Ca-K Channels 314
7. Conclusions 320
   References 321

IV. Ionic Channel Modulation

Chapter 20

Metabolic Regulation of Ion Channels

Irwin B. Levitan

1. Introduction 325
2. Second Messengers 327
3. Protein Phosphorylation 329
4. Summary 331
   References 332

Chapter 21

The Cell-to-Cell Membrane Channel: Its Regulation by Cellular Phosphorylation

Werner R. Loewenstein

1. Introduction 335

2. The Cell-to-Cell Channels Are Up-Regulated by
   cAMP-Dependent Phosphorylation 338
3. The Cell-to-Cell Channels are Down-Regulated by Tyrosine
   Phosphorylation 344
   References 349

Chapter 22

The β-Cell Bursting Pattern and Intracellular Calcium

Illani Atwater and John Rinzel

1. Introduction 353
2. Role of $[Ca^{2+}]_i$: Dependence on Glucose 354
3. A Biophysical/Mathematical Model 356
4. Burst Frequency Depends on the Ratio
   [free $Ca^{2+}]_i$/[total Ca]$_i$ 359
5. Summary 361
   References 361

Chapter 23

Neurotrophic Effects of in Vitro Innervation of Cultured Muscle Cells. Modulation of Ca^{2+}-Activated K^+ Conductances

Benjamin A. Suarez-Isla and Stanley I. Rapoport

1. Introduction 363
2. Methodological Considerations 364
3. Innervation and Muscle Cell Electrical Activity 367
4. Conclusions 380
   References 381

V. Ionic Channel Structure, Functions, and Models

Chapter 24

Correlation of the Molecular Structure with Functional Properties of the Acetylcholine Receptor Protein

Francisco J. Barrantes

1. Introduction 385
2. The AChR Macromolecule 386
3. Arrangement of Subunits in the AChR Macromolecule 387
4. The AChR Primary Structure, cDNA Recombinant Techniques,
   and Modeling Receptor Structure 389

5. Immunochemistry of AChR and the Testing of Models 392
6. Voltage-Gated and Agonist-Gated Channels: A Comparison 392
7. Dynamics of AChR and Lipids in the Membrane 393
8. Acetylcholine-Receptor-Controlled Channel Properties 395
   References 397

Chapter 25
Amiloride-Sensitive Epithelial Sodium Channels
Dale J. Benos

1. Introduction 401
2. Amiloride-Sensitive Na^+ Transport Processes 403
3. Characterization of Amiloride-Sensitive Na^+ Channels in
   Intact Epithelia 405
4. Incorporation of Amiloride-Sensitive Na^+ Channels into
   Planar Bilayers 412
5. Concluding Remarks 416
   References 417

Chapter 26
A Channel Model for Development of the Fertilization Membrane in Sea Urchin Eggs
Gerald Ehrenstein and Richard FitzHugh

1. Introduction 421
2. Processes Following Fertilization 421
3. Experimental Basis for Model 422
4. Description of Model 423
5. Equations of Model 425
6. Solutions of Model Equations 426
   References 429

Index 431

Introduction

Since the work of Ringer (1883), it has been clear that cells modify their behavior according to the ionic composition of the external medium. Thus, Ringer showed that the excised frog heart requires sodium, potassium, and calcium to continue beating. On this basis, Berstein (1902) proposed a mechanism for the generation of the resting potential. He suggested that the resting membrane of a nerve or a muscle is only permeant to potassium ions. He further proposed that during activity the membrane permeability became transiently nonselective. Thus, we have known for a long time that membranes are permeable to ions. Since a lipid bilayer presents an enormous energy barrier to the movement of small ions like potassium (60 kcal/mol), a class of membrane protein has evolved to catalyze ion movement. Ion channels can be defined as integral membrane proteins spanning the bilayer and able to connect the internal with the external medium of the cell.

Proteins of this type are widely distributed throughout the biological world, and they are involved in an immense variety of physiological processes including the propagation of the electrical impulse in nerve and muscle, hormone secretion, visual transduction, the immune response, transepithelial electrolyte transport, cell volume regulation, activation of contraction, fertilization, and cell-to-cell communication.

Ion channels play the role of transmitter–amplifiers, and some of them are receptor–transmitter–amplifiers. The last class are receptors because they are able to respond to an external stimulus, and in this sense they capture signals coming from the external environment of the cell. A typical receptor–transmitter–amplifier channel is the acetylcholine receptor channel. In this case, the channel conduction is modulated by the presence in the external medium of acetylcholine. In other words, channel opening is controlled by this chemical transmitter. In the case of the neuromuscular junction, the opening of ACh channels leads to a depolarization of the cell membrane that turns on channels that sense the electrical potential: the sodium channels responsible for the action potential, the delayed rectifier (a potassium channel), and calcium channels. The important thing to keep in mind is that in all the cases mentioned, the channel, in response to the stimulus, goes from a nonconducting configuration (closed) to a conducting one (open). This process is known as channel gating.

We have said that channels are transmitters. Here we referred to their capacity, once they are open, to allow the passage of a current carried by ions. This process is known as channel conduction. Many channels let only one type of ionic species pass through them; that is to say, they are selective. A large channel selectivity demands a specific interaction between some of the channel protein groups and the ion that is allowed to pass through the channel. The ensemble of protein groups able to interact with the ion is known as the selectivity filter. Gating, conduction, and channel selectivity define a given channel.

Inasmuch as we are postulating a site in the channel able to interact with the ion, this process can be viewed in the same way as the interaction of a substrate with an enzyme. Channels are, then, enzymes that catalyze the flow of ions through biological membranes. Actually, they reduce the energies for transmembrane ionic diffusion from the 60 kcal/mol, the energy cost for the transfer of an ion across the lipid bilayer, to values in the range of 5 kcal/mol. The rates at which ion channels catalyze ion flow are extremely high (10^6–10^8 ions/sec) compared with "classical" enzymes. Ion transport channels are therefore able to amplify the electric current that passes through membranes by several orders of magnitude, allowing nerve conduction and other physiological processes to occur.

Part I

Methodologies

Chapter 1

Kinetic Models and Channel Fluctuations

Osvaldo Alvarez

1. INTRODUCTION

In this chapter the essentials of the kinetics of opening and closing of ionic channels will be discussed. I present several kinetic models and show the type of single channel records to be expected for each model and how to calculate kinetic constant from these records. The models I discuss are two state channels, three-state channels with three conductances, and three-state channels with only two conductances. The basic ideas for this chapter were taken from Dionne and Leibowitz (1982) and Ehrenstein *et al.* (1974).

2. TWO-STATE CHANNEL

A record of the current passing a membrane with only one two-state channel presents two levels of current corresponding to either the closed or the open state of the channel (see Fig. 1A). The channel fluctuates

---

OSVALDO ALVAREZ • Departamento de Biología, Facultad de Ciencias, Universidad de Chile, Santiago, Chile; and Centro de Estudios Cientificos de Santiago, Santiago, Chile.

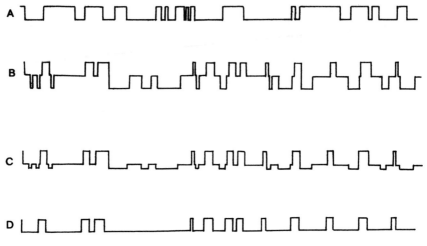

Figure 1. Samples of single-channel records simulated by the computer. A: Single two-state channel. B: Two two-state channels. C: A three-state channel with two closed states with different conductances. D: Three-state channel with two closed states with the same conductance. For the simulation, the values for all the kinetic constants were equal. Note that the sequence of changes of state in records B, C, and D is the same, but the conductances of the states differ.

between the closed and open states, and this process can be described by the following kinetic scheme:

$$\text{CLOSED} \underset{\beta}{\overset{\alpha}{\rightleftarrows}} \text{OPEN}$$

In this scheme, the constants α and β represent the rate constants connecting the two states of the channel. These constants have a clear meaning when we are describing a membrane with many channels. For the case of a membrane with only one channel, it is more convenient to define these constants as the probability per unit time that a channel changes state.

α = Probability per unit time that a closed channel opens

β = Probability per unit time that an open channel closes

With these two rate constants defined, the expected fraction of time a channel spends in the open state can be calculated. This fraction of time, P_o in a stationary current record, is a constant independent of time and defined by the kinetic constants only.

Kinetic Models and Channel Fluctuations

For the analysis of single-channel current records, it is convenient to have an expression for the probability of finding the channel open after some time interval measured from the moment when a closed-to-open transition has occurred. This probability, $P_o(t)$ is defined as:

$P_o(t)$ = Probability of finding an open channel when the channel opened at time 0

In a membrane with many channels, $P_o(t)$ would be the number of channels left open after a time t beginning with all the channels open at time zero. $P_o(t)$ clearly decreases as time passes, and it is related to α and β by the following differential equation:

$$dP_o(t)/dt = -\beta P_o(t) + \alpha[1 - P_o(t)]$$

Variables can be separated for integration

$$dP_o(t)/[P_o(t) - \alpha/(\alpha + \beta)] = -(\alpha + \beta) \, dt$$

The numerator of the left-side fraction contains the differential of the denominator, and therefore the integral is a natural logarithm function of the denominator. The initial condition is that for $t = 0$, $P_o(0) = 1$, and the integral expression is:

$$\ln[P_o(t) - \alpha/(\alpha + \beta)] - \ln[1 - \alpha/(\alpha + \beta)] = -(\alpha + \beta) t$$

By taking the exponential function of this equation, $P_o(t)$ can be solved for:

$$P_o(t) = \alpha/(\alpha + \beta) + [1 - \alpha/(\alpha + \beta)] \exp[-(\alpha + \beta) t]$$

After a long time, $P_o(t)$ reaches a constant value, since the exponential function vanishes, and the probability (P_o) of finding an open channel in a stationary record is

$$P_o = \alpha/(\alpha + \beta)$$

The function describing $P_o(t)$ is a function with an initial value of 1 and a final value P_o. The time constant of the exponential decay, τ, is:

$$\tau = 1/(\alpha + \beta)$$

From the values of P_o and τ, the constants α and β can be calculated. This can be done once we have the experimentally obtained function for

$P_o(t)$. This is constructed from single-channel current records as shown in Fig. 2. The current record is divided into segments of duration Δt, and a histogram with the same bin width is constructed. The histogram initially contains zero in all the bins. The histogram is constructed by maintaining two pointers, one pointing to the current record and the other to the histogram. Initially, the record pointer is at a closed-to-open transition and histogram pointer is at bin 0. The histogram construction procedure is: if the point at the record corresponds to an open state, then add 1 to the count at the current bin; increment both pointers and repeat the operation until the last histogram bin has been done. Then reset the record pointer to the next closed-to-open transition and the histogram pointer back to bin 0. The procedure continues over the whole current

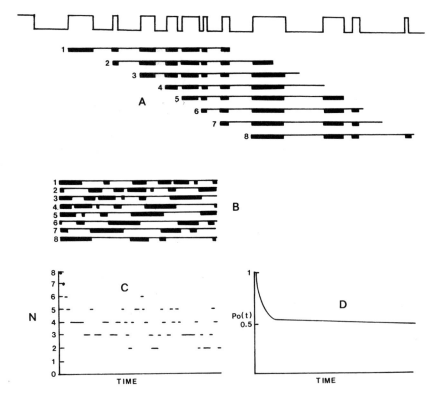

Figure 2. Construction of the probability of finding an open channel as a function of time. A: Samples of the record were taken starting from a closed-to-open transition, and the periods of open channels are marked. B: All the samples are collected, placed in register, and the marked periods added. In the figure, eight samples were taken producing, a very scattered histogram. C: When many samples are analyzed, the smooth function for $P(t)$ is obtained (D).

record. When done, the contents of all the histogram bins are divided by the number of closed-to-open transitions examined, i.e., the contents of bin 0.

The constants α and β can be calculated from the single-channel current records by examining the time the channel dwells on each state between two transitions. I here define the probability $p_o(t)$:

$p_o(t)$ = Probability that a channel stays open at a time t after a closed-to-open transition at time 0

The probability $p_o(t)$ is different from $P_o(t)$. The probability $P_o(t)$ does not consider that the channel may have closed several times before we looked at it. When we constructed $P_o(t)$ we were only interested in whether the channel was open or closed. The probability $p_o(t)$ is the probability that the channel remains open and has never closed until we look at it. This probability decreases as time passes after the closed-to-open transition. The rate of decrease is proportional to the probability of the channel still being open, $p_o(t)$, and to the probability that this open channel closes in a time unit, β.

$$dp_o(t)/dt = -\beta\, p_o(t)$$

The integration is a logarithmic function, and the initial conditions are that $p_o = 1$ for $t = 0$, so the integral, after taking the exponential function, is:

$$p_o(t) = \exp(-\beta t)$$

The same reasoning can be used for the closed channel, defining a probability of duration $p_c(t)$:

$p_c(t)$ = Probability that a closed channel stays closed after a time t measured from an open-to-closed transition

$$p_c(t) = \exp(-\alpha t)$$

Now the constants α and β can be calculated from the experimental functions $p_o(t)$ and $p_c(t)$. These can be constructed from the single-channel records as shown in Fig. 3. The current record is divided into segments of duration Δt, and a histogram with the same bin width is constructed. The histogram initially contains zero in all the bins. The histogram is constructed by maintaining two pointers, one pointing to the current record and the other to the histogram. Initially, the record pointer is at a

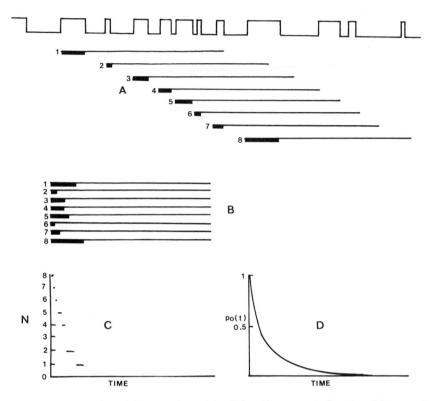

Figure 3. Construction of the open-channel dwell-time histogram. A: Samples of the record were taken starting from a closed-to-open transition, marking the sample up to the first open-to-closed transition. B: All the samples are collected, placed in register, and the marked periods added. The result is the histogram in C. When many samples are analyzed, the smooth function for $p(t)$ is obtained (D).

closed-to-open transition, and the histogram pointer is at bin 0. The histogram construction procedure is: if the point on the record corresponds to an open state, then add 1 to the count at the current bin, increment both pointers, and repeat until a closed state is found. When an open-to-closed transition is detected, reset the record pointer to the next closed-to-open transition and the histogram pointer back to bin 0. The procedure continues over the whole current record. When done, the contents of all the histogram bins are divided by the number of closed-to-open transitions examined, i.e., the contents of bin 0. The procedure is basically the same for $p_c(t)$.

There is still a third method to calculate the constants α and β from single-channel records. This involves the measurement of the mean open and mean closed times. Since we already know the distribution of the channel durations, we can calculate the mean open or closed time by computing a weighted average in which the weighting factors are the probability of each open time:

$$\text{mean open time} = \int t \exp(-\beta t) / \int \exp(-\beta t)$$

When the integrals are evaluated from 0 to infinity, the mean open time is $1/\beta$, and by the same reasoning the mean closed time is $1/\alpha$.

3. TWO TWO-STATE CHANNELS

We can now examine the case of a membrane with two identical two-state channels. When these two channels open and close, they generate a current record with three current states: two closed, one open and one closed, and two open channels (see Fig. 1B). The expected fractions of time the current record is found in each of the three states are P_0, P_1, and P_2. The probabilities of remaining in one state after arriving that state at time zero are p_0, p_1, and p_2.

The P probabilities can be calculated by recalling that for a single two-state channel the probability of finding it open is $P_o = \alpha/(\alpha + \beta)$; then:

$$P_0 = (1 - P_o)^2 = [\beta/(\alpha + \beta)]^2$$

$$P_1 = 2P_o(1 - P_o) = 2\alpha\beta/(\alpha + \beta)^2$$

$$P_2 = P_o^2 = [\alpha/(\alpha + \beta)]^2$$

Experimental values of the Ps are calculated by measuring the total time the membrane spent at each conductance level and dividing it by the total observation time.

The p probabilities are functions of time. For example, the probability of remaining in state 0 after arriving at state 0 at time zero decreases at a rate proportional to p_0 and the probability per unit time of leaving state 0. In this case, we have two channels, so the probability per unit time of leaving state 0 is 2α. This leads to an exponential distribution of the state-

0 dwell times, and the mean duration in state 0 is $1/2\alpha$. Following the same reasoning, all p_i can be found, and they are:

$$p_0 = 1/2\alpha$$

$$p_1 = 1/(\alpha + \beta)$$

$$p_2 = 1/2\beta$$

The experimental values for the ps are obtained from three dwell-time histograms, using a procedure similar to the one used for the single-channel case.

Estimations of α and β can be obtained by combining the experimental values of p and P for each state.

From state 1 additional information on the constants can be obtained by computing the ratio of the number of transitions from state 1 to state 2 divided by the number of transitions from state 1 to state 0. This ratio gives the ratio α/β, which are the probabilities of opening and of closing, respectively.

No. of upward transitions/No. of downward transitions = α/β

On a membrane with two two-state channels, three estimations of α and β are obtained, one for each state. Some of these measurements may have little significance; for instance, if $\alpha/(\alpha + \beta)$ is a small number, then the probability P_2 will be very small and difficult to measure accurately because the number of time intervals the membrane has the conductance corresponding to state P_2 will be a small number, and the error of the count is the square root of the count. The significance of each measurement must be evaluated before all the estimates of the constants are combined.

This analysis can be extended for membranes with any number of two-state channels. The limitations are of two kinds: experimental and theoretical. The experimental limitation arises from the difficulty of resolving the different conductance states and of knowing the number of channels in the membrane. The other problem is that the probability of having two transitions occurring simultaneously on a time interval increases with the number of channels. In the above reasoning it is implicit that the time intervals used in the analysis are very small and simultaneous transitions are not considered.

4. THREE-STATE CHANNEL WITH THREE CONDUCTANCES

We can now consider a channel that has three states, all with different conductances. Let us assume that the first state does not conduct, the second state has a very small conductance, and the third state has a large conductance. I will call the states: closed 1 (C_1), closed 2 (C_2), and open (O). The conductances need not be so defined for the present discussion, but this will be useful to treat the next model, in which the conductance of both closed states is zero. A membrane with one channel of this type presents three levels of conductance, one for each state, as seen in Fig. 1C. The kinetic scheme for this channel is:

$$C_1 \underset{b}{\overset{a}{\rightleftarrows}} C_2 \underset{\beta}{\overset{\alpha}{\rightleftarrows}} O$$

The calculation of the four constants a, b, α, and β is based on the same reasoning used for the case of a membrane with two two-state channels. In this case, the dwell-time distributions are exponential functions, and the mean dwell times for each of the three states are:

Mean time in state O = $1/\beta$

Mean time in state C_2 = $1/(\alpha + b)$

Mean time in state C_1 = $1/a$

To separate the constants α and b, the ratio α/b is obtained from the ratio of the number of upward and downward transitions from state C_1.

5. THREE-STATE CHANNEL WITH ONLY TWO CONDUCTANCES

In this model the two closed states do not conduct, and therefore it is not possible to determine from the current level in which of the two closed states the channel is. The current record for a channel of this type presents only two levels of current, but short and long closed times are observed (see Fig. 1D). If long and short closed times are very different, the record appears as bursts of activity separated by silent periods. If long and short closed times are not very different, they may not be apparent on the current record, but the closed time distribution will not be a single exponential function. The kinetic sheme for this channel is the same as that of the previous case.

The open time distribution is still an exponential function, and the constant β can be calculated from the mean open time.

$$\text{Mean time in state O} = 1/\beta$$

The closed-time distribution is not an exponential function. The probability $p_{1,2}$ is

$p_{1,2}$ = Probability of remaining in state C_1 or C_2 after a transition from open to closed at time 0.

This probability decreases with time, and the rate of decrease is proportional to $p_{1,2}$ and to the probability per unit time of leaving the closed state. Since this state can only be exited from state C_2, the probability of leaving the closed state is the probability of being in state C_2, $P_2(t)$, times the probability per unit time of an upward transition, α.

$$dp_{1,2}/dt = -\alpha\, p_{1,2}\, P_2(t)$$

The probability $P_2(t)$ is a function of time, since immediately after the open-to-closed transition $P_2(0) = 1$, and it relaxes to a steady value as time passes. We can now think of the interconversion of the two closed states as a two-state channel, and, from the equations derived for that case, we can write an expression for $P_2(t)$

$$P_2(t) = a/(a + b) + [1 - 1/(a + b)]\exp[-(a + b)t]$$

Combining these two equations and integrating, the probability $p_{1,2}$ can be found:

$$p_{1,2} = \exp\{-\alpha[a/(a + b)]t - [\alpha b/(a + b)^2](1 - \exp[-(a + b)t]\}$$

The constants a, b, and α can be obtained by fitting this function to the experimental closed-time distribution histogram.

For the case that a and b are large, i.e., the exchange between the two closed states is very fast, the right-hand exponential function vanishes, the equation reduces to a single exponential, and the mean closed time is now $(a + b)/\alpha a$. However, since the closed distribution is a single exponential, there is no way to know that there are actually two closed states.

For the case that a and b are small constants, i.e., the exchange between the two closed states is very slow, the current record will present "bursts" of activity, and it will be possible to distinguish the long closed times from the short closed times. If this distinction is done correctly, two histograms can be constructed for long and short closed times, and the mean closed times are then $1/\alpha$ for the short times and $1/a$ for the long times. The constant b can now be determined by the ratio of transitions from state C_1 to state O divided by the number of transitions from state C_1 to state C_2. This is α/b and is measured from the number of opening transitions into a "burst."

ACKNOWLEDGMENTS. I thank Mr. Juan Espinoza for the preparation of the figures. This work was supported by Fondo Nacional de Investigación (Chile) Grant 1922 and Universidad de Chile Grant B-1985/8413 from Departamento de Investigación y Bibliotecas.

REFERENCES

Dionne, V. E., and Leibowitz, M. D., 1982, Acetylcholine receptor kinetics, a description from single-channel current at snake neuromuscular junctions, *Biophys. J.* **39**:253–261.

Ehrenstein, G., Blumenthal, R., Latorre, R., and Lecar, H., 1974, Kinetics of the opening and closing of individual excitability-inducing material, *J. Gen. Physiol.* **63**:707–721.

Chapter 2

Single-Channel Currents and Postsynaptic Drug Actions

Harold Lecar

1. INTRODUCTION

In single-channel recordings, the gating transitions of individual ionic channels appear as a stochastic process. The description of the random transitions of a channel between its discrete conductance states complements the kinetic picture of the gating process derived from macroscopic experiments such as voltage clamp, noise analysis, or gating current. The goals of kinetic measurements are to derive unique reaction schemes for modeling macroscopic excitation phenomena and perhaps to obtain some insights about the molecular motions involved in the conformation changes attendant to gating. In this chapter, I show how the rates of a kinetic process are inferred from single-channel data and how such analysis can be applied to the study of drug action on postsynaptic channels.

Single-channel studies suggest a description of gating as a finite-state Markov process. That is, the channel has reasonably small numbers of kinetically distinguishable open and closed states, and the rate of transition between any two states depends only on the present condition of the system and not on past history. For many physical systems, the Markov-process description at a coarse time scale, such as the millisecond

---

HAROLD LECAR • Laboratory of Biophysics, National Institute of Neurological and Communicative Disorders and Stroke, National Institutes of Health, Bethesda, Maryland 20205.

time scale of conductance transitions, can coexist with another Markov-process description at a finer time scale, such as the much faster dynamic fluctuations of the channel protein (van Kampen, 1981). Thus, we might question whether the fit of the stochastic data to a finite-state model constitutes evidence for the physical reality of the hypothetical states or whether such a fit is merely a construct of the data-fitting procedure.

For electrically excitable channels, these questions are implicit in voltage-clamp analysis. The Hodgkin–Huxley equations interpreted literally would suggest that channel opening requires the concerted activation of several independent subunits. Structural data on sodium channels (Noda *et al.*, 1984) show some evidence for protein domains that might be the gating subunits, but they are not likely to move independently. The best electrical evidence for subunits is probably the gating current, which is a result of transitions involving charge displacements and tends to show less time delay than channel opening (Armstrong and Bezanilla, 1974). Single-channel studies would provide an independent measurement to be fit to a kinetic model of the transitions among the various closed states of the channel (Horn and Vandenburg, 1984). A definitive kinetic model would somehow have to accommodate all the data.

For chemically excitable channels, the fit to finite-state models should be less conjectural than for voltage-dependent channels, because part of the process involves definite bound and unbound states of the agonist and receptor. In the Markov-process description, transitions involving binding will have concentration-dependent rates and therefore should be identifiable. A recent review by Horn (1984) surveys the use of single-channel data to verify different kinetic models of the gating process. To see how the actions of different postsynaptic pharmacological agents can be sorted out from single-channel kinetics, we can calculate probability distributions of open- and closed-state dwell times for a postsynaptic channel in the presence of various prototypical drugs. Single-channel measurements can distinguish between different molecular mechanisms that might be confused in macroscopic dose–response experiments. When the drug-altered gating-kinetic schemes have been specified, the model can be employed to predict the form of the macroscopic postsynaptic response. Predicting the postsynaptic current from the single-channel kinetic model provides a correspondence between the microscopic steady-state fluctuations and the macroscopic relaxation process.

2. CHANNEL GATING AS A STOCHASTIC PROCESS

A channel macromolecule in steady state fluctuates among several open and closed states. However, only transitions between states of different ionic conductance are observable. The basic statistical information is then the distribution of dwell times in the observable aggregates of states. From the dwell-time distributions, we can obtain rate constants to fit the gating process to a finite-state Markov process.

For a simple two-state channel (Ehrenstein et al., 1970), the durations are distributed exponentially. The probabilities of open and closed duration being greater than some time t are given by the distributions

$$F_o(t) = \exp(-at)$$
$$F_c(t) = \exp(-bt)$$
(1)

where the parameter for the open-time distribution, b, is the closing rate, and the parameter for the closed-time distribution, a, is the opening rate. These two parameters suffice to describe the macroscopic behavior of the system. For example, if a and b are known as functions of membrane potential, the relaxation of the macroscopic conductance in response to a step of potential would obey the equation

$$dg/dt = (a + b)(g_{max} a/(a + b) - g) \qquad (2)$$

where g is the macroscopic conductance and g_{max} is the value of g when all the channels are open. Similarly, the steady-state membrane current noise would have the spectrum

$$S(f) = S(O)/\{1 + [2\pi f/(a + b)]^2\} \qquad (3)$$

where f is frequency.

For channels with multiple open or closed states, the distributions of dwell times are multiexponential, representing sojourns in complexes of states. A complex of N states gives rise to a distribution of dwell times requiring N exponential terms. This is analogous to the description of voltage-clamp transients by sums of exponentials or the description of membrane noise by a power spectrum composed of several Debye relaxation spectra. The theory of channel gating as a multistate stochastic process has been discussed extensively (Colquhoun and Hawkes, 1977, 1981, 1983; Horn, 1984; Fredkin et al., 1983).

For a system having two observable conditions, conducting and nonconducting, there will be two dwell-time distributions,

$$F_c(t) = \sum_{i}^{N_c} A_i \exp(R_i t)$$
$$F_o(t) = \sum_{i}^{N_o} B_i \exp(R_i' t)$$
(4)

Here, $F_c(t)$ and $F_o(t)$ are called the open- and closed-time distributions. N_c and N_o are the numbers of closed and open states, and A_i, B_i, R_i, and R_i' are parameters obtained from the Markov model of the gating process.

A Markov process of N states can be described by N time-dependent probabilities, P_i, which obey the set of first-order equations

$$dP_i/dt = \sum_{i=1}^{N} Q_{ij} P_j \tag{5}$$

where N is the number of states. The system is governed by the stochastic matrix $Q = (Q_{ij})$. In order to derive the distributions of open- and closed-state dwell times, a special process can be set up to represent the sojourn in a particular group of states. For example, a system with two groups of states can be represented by the partitioned matrix

$$Q = \left(\begin{array}{c|c} Q_{oo} & Q_{oc} \\ \hline Q_{co} & Q_{cc} \end{array} \right) \tag{6}$$

To calculate the coefficients of equation 4 for the distribution of closed times, for example, we would solve the equation for the probability of being in one of the closed states following a particular initial condition. The equation for the probability of being in the ith closed state is

$$dP_i(t)/dt = \sum_{j=1}^{N_c} (Q_{cc})_{ij} P_j(t) \tag{7}$$

The appropriate initial condition is that $P_i(0)$ equals the probability that the sojourn begins in state i immediately on closing. $P_i(0)$ is obtained by taking the product of the equilibrium probability of being in a particular open state connecting to i and the probability of transition into i and summing over all such terms. Thus, $P_i(0)$ can be written as

$$P_i(0) = \sum_{j=1}^{N_o} \left[Q_{ij} \Big/ \left(\sum_{i=N_o+1}^{N} \sum_{j=1}^{N_o} Q_{ij} \right) \right] \bar{P}_j \tag{8}$$

The solution of equation 8 can be written as

$$P_i(t) \sum_{j=1}^{N_c} a_{ij} \exp(R_j t) \tag{9}$$

where the R_j are the eigenvalues of Q_{cc} and the N_c^2 coefficients a_{ij} come from matching the initial conditions and are given by the ratio of determinants

$$a_{ik} = D^{-1} \begin{vmatrix} 1 & \cdots & P_i(0) & \cdots & 1 \\ R_1 & \cdots & \dot{P}_i(0) & \cdots & R_{N_c} \\ R_1^2 & \cdots & \ddot{P}_i(0) & \cdots & R_{N_c}^2 \\ \cdots & \cdots & \cdots & \cdots & \cdots \\ R_1^{N_c-1} & \cdots & P_c^{(N-1)}(0) & \cdots & R_{N_c}^{N_c-1} \end{vmatrix} \quad (10)$$

$$D = \begin{vmatrix} 1 & \cdots & 1 & \cdots & 1 \\ R_1 & \cdots & R_i & \cdots & R_{N_c} \\ R_1^2 & \cdots & R_i^2 & \cdots & R_{N_c}^2 \\ \cdots & \cdots & \cdots & \cdots & \cdots \\ R_1^{N_c-1} & \cdots & R_i^{N_c-1} & \cdots & R_{N_c}^{N_c-1} \end{vmatrix}$$

where $\dot{P}_i(0), \ddot{P}_i(0),\ldots,P_i^{(k)}(0)$ are the initial conditions of the 1,2,... k derivatives of $P_i(t)$.

The distribution of closed dwell times is just the probability of the channel being in any of the closed states,

$$F_c(t) = \sum_{i=1}^{N_c} P_i(t) = \sum_{j=1}^{N_c} \left(\sum_{i=1}^{N_c} a_{ij} \right) \exp(R_j t) \quad (11)$$

In general, the distribution of dwell times can be a complicated function of the rates in the stochastic matrix. Only for relatively simple kinetic models can we write an explicit expression for $F_c(t)$ or $F_o(t)$ in terms of the rates Q_{ij}. For a complex of N_c closed states, the dwell-time distribution yields $2N_c - 1$ parameters, N_c eigenvalues, and $N_c - 1$ amplitudes. Thus, for a model with N_o open states and N_c closed states, the two dwell-time distributions give $2(N - 1)$ relations among the rates.

Figure 1 shows a number of kinetic schemes used to analyze channel gating. We take the general three-state model of Fig. 1B as an example for the calculation of the dwell times. Figure 2 outlines the calculation. The open-time distribution is just

$$F_o(t) = \exp[-(q_{10} + q_{20})t] \quad (12)$$

and the closed-time distribution is

$$F_c(t) = [(R_2 + A)/(R_2 - R_1)] \exp(R_1 t) \\ + [(R_1 + A)/(R_1 - R_2)] \exp(R_2 t) \quad (13)$$

A. Two State

$$C \underset{b}{\overset{a}{\rightleftarrows}} O \qquad Q = \begin{pmatrix} -a & b \\ \hline a & -b \end{pmatrix}$$

B. Three State

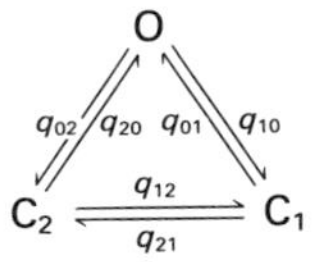

$$\begin{array}{c} O \\ q_{02} \Big\Updownarrow q_{20} \quad q_{01} \Big\Updownarrow q_{10} \\ C_2 \underset{q_{21}}{\overset{q_{12}}{\rightleftarrows}} C_1 \end{array} \qquad Q = \begin{bmatrix} -(q_{10}+q_{20}) & q_{01} & q_{02} \\ \hline q_{10} & -(q_{01}+q_{21}) & q_{12} \\ q_{20} & q_{21} & -(q_{02}+q_{12}) \end{bmatrix}$$

C. del Castillo–Katz Agonist Scheme

$$C_2 \underset{d}{\overset{k_x X}{\rightleftarrows}} C_1 \underset{b}{\overset{a}{\rightleftarrows}} O \qquad Q = \begin{bmatrix} -b & a & 0 \\ \hline b & -(a+d) & k_x X \\ 0 & d & k_x X \end{bmatrix}$$

D. Hodgkin–Huxley m^3h Scheme for Na$^+$

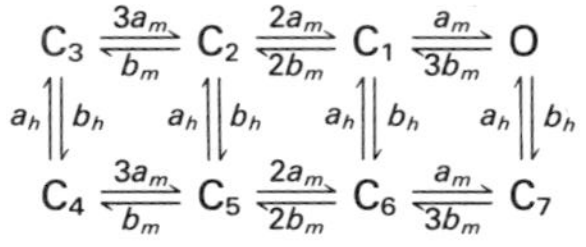

$$\begin{array}{ccccccc} C_3 & \underset{b_m}{\overset{3a_m}{\rightleftarrows}} & C_2 & \underset{2b_m}{\overset{2a_m}{\rightleftarrows}} & C_1 & \underset{3b_m}{\overset{a_m}{\rightleftarrows}} & O \\ a_h \Big\Updownarrow b_h & & a_h \Big\Updownarrow b_h & & a_h \Big\Updownarrow b_h & & a_h \Big\Updownarrow b_h \\ C_4 & \underset{b_m}{\overset{3a_m}{\rightleftarrows}} & C_5 & \underset{2b_m}{\overset{2a_m}{\rightleftarrows}} & C_6 & \underset{3b_m}{\overset{a_m}{\rightleftarrows}} & C_7 \end{array}$$

$$Q = \begin{bmatrix} -(3b_m+b_h) & a_m & 0 & 0 & 0 & 0 & 0 & a_h \\ 3b_m & -(2b_m+b_h+a_m) & 2a_m & 0 & 0 & 0 & a_h & 0 \\ 0 & 2b_m & -(2a_m+b_m+b_h) & 3a_m & 0 & a_h & 0 & 0 \\ 0 & 0 & b_m & -(3a_m+b_h) & a_h & 0 & 0 & 0 \\ \hline 0 & 0 & 0 & b_h & -(3a_m+a_h) & b_m & 0 & 0 \\ 0 & 0 & b_h & 0 & 3a_m & -(2a_m+b_m+a_h) & 2b_m & 0 \\ 0 & b_h & 0 & 0 & 0 & 2a_m & -(a_m+b_m+a_h) & b_h \\ b_h & 0 & 0 & 0 & 0 & 0 & a_m & -(b_m+a_h) \end{bmatrix}$$

Figure 1. Some transition schemes and their stochastic matrices. The dashed lines show the partitioning into Q_{oo}, Q_{cc}, Q_{oc}, and Q_{co} as in equation 6.

Postsynaptic Drug Actions

Three-State Model

A. Open Time

$$\xleftarrow{q_{20}} O \xrightarrow{q_{10}} \qquad Q_{oo} = -(q_{10} + q_{20})$$

$$dP_o/dt = -(q_{10} + q_{20}) P_o \qquad P_o(0) = 1$$

B. Closed Time

$$\xleftarrow{q_{02}} C_2 \underset{q_{21}}{\overset{q_{12}}{\rightleftarrows}} C_1 \xrightarrow{q_{01}}$$

$$Q_{cc} = \begin{pmatrix} -(q_{01} + q_{21}) & | & q_{12} \\ \text{\textemdash\textemdash\textemdash\textemdash\textemdash\textemdash} & | & \text{\textemdash\textemdash\textemdash\textemdash\textemdash\textemdash} \\ q_{21} & | & -(q_{02} + q_{12}) \end{pmatrix}$$

$$dP_1/dt = -(q_{01} + q_{21})P_1 + q_{12} P_2; \; P_1(0) = q_{10}/(q_{10} + q_{20})$$
$$dP_2/dt = q_{21} P_1 - (q_{02} + q_{12})P_2; \; P_2(0) = q_{20}/(q_{10} + q_{20})$$
$$R^2 + (q_{01} + q_{02} + q_{12} + q_{21})R + (q_{01}q_{02} + q_{01}q_{12} + q_{02}q_{21}) = 0$$

Figure 2. General three-state model and the stochastic processes for obtaining the open and closed dwell times. The schemes imply the differential equations shown with the indicated initial conditions.

where

$$R_1, R_2 = -s/2 \pm (\tfrac{1}{2})(s^2 - 4m)^{\tfrac{1}{2}}$$

$$s = q_{01} + q_{02} + q_{12} + q_{21}$$

$$m = q_{01}q_{02} + q_{01}q_{12} + q_{02}q_{21} \qquad (14)$$

$$A = (q_{01}q_{10} + q_{02}q_{20})/(q_{10} + q_{20})$$

In Fig. 1C, we see that the del Castillo–Katz model of agonist action is a three-state scheme with one open state and two closed states, which can be derived from the general three-state model by letting $q_{02} = q_{20} = 0$. Similarly, the blocked two-state channel is obtained by letting $q_{12} = q_{21} = 0$. We use these results in Section 3.

For these two simple models, the data of two dwell-time histograms are sufficient to specify all the transition rates. However, for the general three-state model there are six rates and only four parameters obtainable from the histograms. For a more complex model, such as the Hodgkin–Huxley sodium scheme, shown in Fig. 1D, the histogram data are insufficient unless other constraints are imposed. In general, specification of the model will require some knowledge of how some of the rates might change when experimental parameters such as membrane potential or agonist concentration are varied.

Since the single-channel experiments monitor the stochastic process itself, there should be more information than just that contained in the histograms. For example, there can be correlational information in the sequence of dwell times. Consider the two schemes shown in Fig. 3. Let us say the two classes of openings are distinguishable, either by having different conductances or by having very different closing rates. In Fig. 3A, the different openings come from a common precursor state so that the probability of pairs of like events is given in terms of the *a priori* probability of the individual events. In Fig. 3B, however, if the rate connecting the two closed-state precursors is slow, there can be a bias to having like pairs, and they will be overrepresented. Jackson *et al.* (1983) and Labarca *et al.* (1985) have applied such correlation analyses to the AChR channel. A theoretical analysis by Fredkin *et al.* (1983) shows the conditions in which the two-time distributions give additional information for discriminating between models. There is no additional information in higher-order correlations. Moreover, there will be no correlation data for any system that does not have at least two open and two closed states; for example, systems with many closed states but only a single open state contain no information beyond that in the dwell-time histograms.

3. POSTSYNAPTIC CHANNELS IN THE PRESENCE OF DRUGS

To look at the application of single-channel recording to postsynaptic pharmacology, we start with the usual scheme for chemically activated channels and inquire how the kinetic scheme is altered in the presence

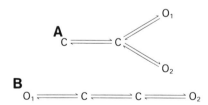

Figure 3. Two schemes that can be distinguished by evaluating two-dimensional distributions. In the branching scheme, O_1 and O_2 have a common precursor state and therefore occur independently. In the double-ended scheme, one of the open states could undergo much faster reversible transitions than the other and give correlations.

of different postsynaptic drugs. We want to examine situations in which we can distinguish the different pharmacological actions even when the kinetic model is not specified completely, for specifying all the details of even a relatively simple agonist interaction can be a formidable task. Single-channel experiments at high time resolution often exhibit such complexities as multiple open states, conductance sublevels, and very fast flickering, which have no parallel in macroscopic observations.

The basic scheme for agonist activation was shown in Fig. 1. The simplest possible model would have two states, but the usual paradigm, first proposed by del Castillo and Katz (1957), shows three states, unliganded and liganded closed states and an open state that is reached by a thermal transition from the agonist-bound state. The need for a three-state model was not always apparent in noise and relaxation measurements (Steinbach, 1980). Single-channel experiments often show reopenings of channels when the rates connecting the activated state to the open state are faster than the rate of agonist binding. Because such reopening phenomena are manifest in single-channel records, the three states must be the minimal model for agonist binding. The basic model really has at least four states, since dose–response experiments (Steinbach, 1978) and structural evidence (Conti-Tronconi and Raftery, 1982) both suggest that two bound agonist molecules are needed for the major channel opening.

Let us consider some hypothetical drug interactions. Blockers and antagonists are both drugs that decrease the postsynaptic current; potentiators and agonists both enhance the postsynaptic response. We define a blocker as a molecule that binds to the lumen of the channel when it is open and an antagonist as a molecule that competes with the endogenous transmitter for access to the receptor site. Agonists are substitutes for the endogenous transmitter, whereas potentiators cause transmitter action to be more effective but do not act as agonists themselves. All of these drugs alter the postsynaptic response in ways that can be predicted from the drug-altered single-channel kinetics. Channel-kinetic schemes in the presence of the drugs may often be too complicated to solve in general, but a study of the single-channel kinetics suggests special experimental conditions for which the different drug effects will be salient.

Consider the difference between an open-channel blocker and an antagonist for the simplified model in which only one agonist binding is needed to activate a channel. Both drugs cause channels to close in a dose-dependent manner, but the kinetic schemes are topologically distinct, as shown in Fig. 4. The sequential scheme for the antagonist has the open state at the end of a sequence of closed states, whereas the scheme for a blocker has the open state between two disjoint groups of closed states. These differences lead to observable differences in the

A. Open-Channel Blocker

$$C \underset{d}{\overset{k_xX}{\rightleftarrows}} C \underset{b}{\overset{a}{\rightleftarrows}} O \underset{d_b}{\overset{k_bB}{\rightleftarrows}} B$$

$$Q = \begin{bmatrix} -(b + k_bB) & b & 0 & k_bB \\ a & -(a + d) & d & 0 \\ 0 & k_xX & -k_xX & 0 \\ d_b & 0 & 0 & -d_b \end{bmatrix}$$

B. Antagonist

$$A \underset{k_aA}{\overset{d_a}{\rightleftarrows}} C \underset{d}{\overset{k_xX}{\rightleftarrows}} C \underset{b}{\overset{a}{\rightleftarrows}} O$$

$$Q = \begin{bmatrix} -b & b & 0 & 0 \\ a & -(a + d) & d & 0 \\ 0 & k_xX & -(k_xX + k_aA) & k_aA \\ 0 & 0 & d_a & -d_a \end{bmatrix}$$

C. Potentiator

$$\begin{array}{ccc} C_2 & \underset{}{\overset{k_xX}{\rightleftarrows}} C_1 & \underset{b}{\overset{a}{\rightleftarrows}} O \\ k_pP \updownarrow e_p & k_p'P \updownarrow e_p' & \\ C_{P1} & \rightleftarrows C_{P2} \rightleftarrows & O_P \end{array}$$

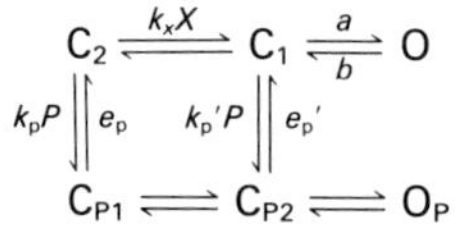

At high potentiator concentration,

$$C_{P2} \underset{d_p}{\overset{k_xX}{\rightleftarrows}} C_{P1} \underset{b}{\overset{a}{\rightleftarrows}} O$$

$$Q = \begin{bmatrix} -b & b & 0 \\ a & -(a + d_p) & k_xK \\ 0 & d_p & k_xX \end{bmatrix}$$

D. Two Agonists

$$O_2 \underset{a_2}{\overset{b_2}{\rightleftarrows}} C_2 \underset{d_2}{\overset{k_2X_2}{\rightleftarrows}} C_3 \underset{d_1}{\overset{k_1X_1}{\rightleftarrows}} C_1 \underset{a_1}{\overset{b_1}{\rightleftarrows}} O_1$$

$$Q = \begin{bmatrix} -b_1 & 0 & b_1 & 0 & 0 \\ 0 & -b_2 & 0 & b_2 & 0 \\ a_1 & 0 & -(a_1 + d_1) & k_1X_1 & 0 \\ 0 & a_2 & k_2X_2 & -(a_2 + d_2) & d_2 \\ 0 & 0 & 0 & d_2 & -(k_1X_1 + k_2X_2) \end{bmatrix}$$

Figure 4. Transition schemes and stochastic matrices for a chemically activated channel in the presence of various drugs.

channel fluctuations. The stochastic matrices for the two kinetic schemes are shown in Fig. 4. The open- and closed-state dwell-time distributions for the blocker are

$$F_o(t) = \exp[-(b + k_b B)t] \qquad (12a)$$

$$F_c(t) = [b/(b + k_b B)] \{[(R_2 + a)/(R_2 - R_1)] \exp(R_1 t) \\ + [(R_1 + a)/(R_1 - R_2)] \exp(R_2 t)\} \\ + [k_b B/(b + k_b B)] \exp(R_3 t) \qquad (12b)$$

where the closed-time rates are

$$R_1, R_2 = -\tfrac{1}{2}(a + d + k_x X) \pm \tfrac{1}{2}[(a + d + k_x X)^2 - 4ak_x X]^{\frac{1}{2}} \qquad (12c)$$

$$R_3 = -d_b$$

The transition rates a, d, k_x, X, and d_b are defined in Fig. 4A. For the antagonist, the dwell-time distributions are

$$F_o(t) = \exp(-at) \qquad (13a)$$

$$F_c(t) = D^{-1} [S_2 S_3 (S_3 - S_2) + a(S_3^2 - S_2^2) \\ + a(a + b)(S_3 - S_2)] \exp(S_1 t) \\ + [S_1 S_3 (S_1 - S_3) + a(S_1^2 - S_2^2) + a(a + b)(S_1 - S_3)] \exp(S_2 t) \\ + (S_1 S_2 (S_2 - S_1) + a(S_2^2 - S_1^2) + a(a + b)(S_2 - S_1)] \exp(S_3 t) \qquad (13b)$$

where S_1, S_2, and S_3 are solutions of the cubic equation

$$S^3 + (d_a + k_x X + k_a A) S^2 + (d_a b + d d_a + a k_x X \\ + 2 d k_x X - k_a k_x X + a k_a A + d_a k_a A) S + d b k_a A = 0 \qquad (13c)$$

and D is a determinant of the form given in equation 10,

$$D = S_2 S_3 (S_3 - S_2) + S_1 S_3 (S_1 - S_3) + S_1 S_2 (S_2 - S_1) \qquad (13d)$$

The open-state lifetime distribution for both cases is a single exponential with mean lifetime given by the reciprocal of the sum of all the rates leading out of the open state. The blocker has an open-state lifetime that is blocker-concentration dependent, but the antagonist does not alter the open-state lifetime. This criterion has been used to show that the classical antagonist curare can also block the acetylcholine receptor channel (Trautmann, 1982; Morris *et al.*, 1983).

The closed-time distributions are a bit more complicated but are simplified in the limit of low agonist concentration. In this limit, the closed-time distribution for the open-channel blocker becomes

$$\begin{aligned}F_c(t) = &\ [b/(b + k_bB)] \{[b/(b + d)] \exp[-(b + d)t] \\&+ [d/(b + d)] \exp[-bk_xXt)/(b + d)]\} \\&+ [k_bB/(b + k_bB)] \exp(-d_bt)\end{aligned} \quad (14)$$

By comparing equations 14 and 13b, we see that there is a blocker-concentration range for which the shortened "blocked" openings may appear in bursts. Such blocker-concentration dependence has been demonstrated with the local anesthetic QX-222 (Neher and Steinbach, 1978; Neher, 1983). Bursting phenomena have been analyzed theoretically (Colquhoun and Hawkes, 1982), but when the bursts are not evident to the eye, the same information comes from studying the closed-time histograms.

In a similar limiting case, antagonists do not exhibit bursting but rather show a weaker drug-concentration dependence for the closed-time distribution. The distribution for the antagonist case is

$$\begin{aligned}F_c(t) = &\ \{[dS_2 - a(a + d)]/S_1(S_2 - S_1)\} \exp(S_1 t) \\&+ \{[aS_1 + a(a + d)]/S_1(S_2 - S_1)\} \exp(S_2 t) \\&+ \{[1 + a(S_2 + S_1) + a(a + d)]/S_1 S_2\} \exp(S_3 t)\end{aligned} \quad (15a)$$

with

$$S_1 = -(k_aA + d_a + a + d)$$

$$S_2 = -(a + d)(k_aA + d_a)/(k_aA + d_a + a + d) \quad (15b)$$

$$S_3 = -(2d + a) d_a k_x X/(a + d)(k_aA + d_a)$$

We note here that the amplitudes depend differently on agonist concentration from the forms obtained for blockers; the slowest component, S_3, dominates in this limit.

These differences in the single-channel behavior are reflected in the macroscopic postsynaptic currents as shown in Section 4. Blockers will affect the time course as well as the amplitude of the postsynaptic response, but antagonists only diminish the amplitude of the response without altering the duration, as is shown in Section 4.

We next contrast two drugs that enhance the postsynaptic response. Agonists, of course, are ideal candidates for single-channel study because we can detect agonists that open channels infrequently with the same sensitivity as the more effective agonists. Figure 5 shows for two different

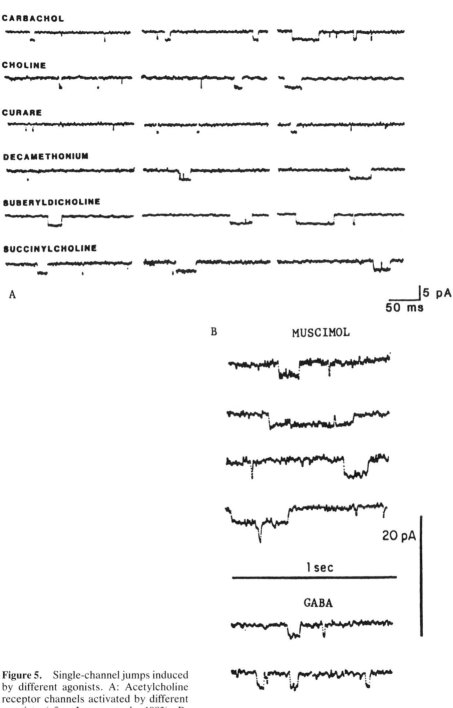

Figure 5. Single-channel jumps induced by different agonists. A: Acetylcholine receptor channels activated by different agonists (after Lecar et al., 1983). B: GABA channels excited by different agonists (after Jackson et al., 1983).

channels that various agonists differ in kinetics and not in the current that flows through the open channel.

Our example of a potentiator is the anticonvulsant and anxiolytic drug diazepam (Valium®), which may act by potentiating the GABA response of the inhibitory postsynaptic channels in CNS neurons. The drug induces characteristic changes in both peak height and duration of the response. Single-channel experiments show that diazepam does not increase the conductance or the open-state durations. High doses of diazepam alter the distribution of closed dwell times in a characteristic but not very dramatic manner (Redmann *et al.*, 1984). Although the data are rather variable, there appears to be a trend towards an increased rate of reopening in the presence of diazepam, as if the drug acted to change the GABA dissociation rate. This is seen in the two-component closed-time distribution, in which the ratio of fast to slow closed times is increased. Figure 4C shows a scheme for a drug acting at an allosteric site to change some property of the transmitter binding. To account for the single-channel observations at high potentiator concentration, the simplified scheme of Fig. 4C was used.

There are many drugs that are themselves agonists. Some, such as muscimol, a mushroom toxin, and (−)-pentobarbital produce channel openings of very long duration in the inhibitory channel. Experiments in which a mixture of two agonists are used are liable to give complicated results. This is not surprising when one looks at the scheme of Fig. 4D. Single-channel experiments with mixed agonists should show multiple open times and might be too complex to analyze quantitatively. The postsynaptic response can still be estimated from the single-agonist, single-channel data because of the saturating-dose constraints. Figure 4D shows the scheme for a mixture of two agonists that is used to illustrate the postsynaptic response when an endogenous transmitter is released in the presence of agonist.

4. RECONSTRUCTING THE POSTSYNAPTIC CURRENT

The macroscopic phenomenon of interest for the postsynaptic membrane is the postsynaptic current. To calculate the shape of this transient, we can assume that transmitter release is instantaneous. In the modeling of the postsynaptic response for the neuromuscular junction, the time course of diffusion across the synaptic cleft and the lateral spread of diffusing transmitter are taken into account (Adams, 1981). This diffusion time affects the leading edge of the postsynaptic current, whereas the decline is dominated by channel activation kinetics. For other synapses,

Postsynaptic Drug Actions

such as the GABA-activated inhibitory synapse, the channel kinetics are slow, and the delays caused by transmitter diffusion can be neglected.

Consider a model such as the biliganded postsynaptic channel of Fig. 2B. If transmitter is released in a saturating amount, all of the receptors are immediately filled, so that at $t = 0$, the system is entirely in state C_1. Now either channels open with rate a or transmitter dissociates with rate d. If we further assume that dissociated transmitter is removed (as, for example, AcCh is hydrolysed by AcCh esterase), then the kinetic scheme for the transient response reduces to the scheme of Fig. 6A. This scheme leads to the second-order differential equation for the probability of the channel being open

$$d^2P_o/dt^2 - (a + b) \, dP_o/dt - bd \, P_o = 0 \quad (16)$$

$$P_o(0) = 0; \, dP_o/dt(o) = -(a + d)$$

A. Postsynaptic Current

$$\xleftarrow{d} C \underset{b}{\overset{a}{\rightleftarrows}} O$$

B. Blocker

$$\xleftarrow{d} C \underset{b}{\overset{a}{\rightleftarrows}} O \underset{d_b}{\overset{k_bB}{\rightleftarrows}} B$$

C. Antagonist

$$\xleftarrow{d} C \underset{b}{\overset{a}{\rightleftarrows}} O$$
$$k_aA \updownarrow d_a$$
$$A$$

D. Potentiator (High Concentration)

$$\xleftarrow{d'} C \underset{b}{\overset{a}{\rightleftarrows}} O$$

E. Two Agonists

$$C_3 \xleftarrow{d_1} C_1 \underset{b_1}{\overset{a_1}{\rightleftarrows}} O_1$$
$$k_2X_2 \searrow d_2$$
$$C_2 \underset{b_2}{\overset{a_2}{\rightleftarrows}} O_2$$

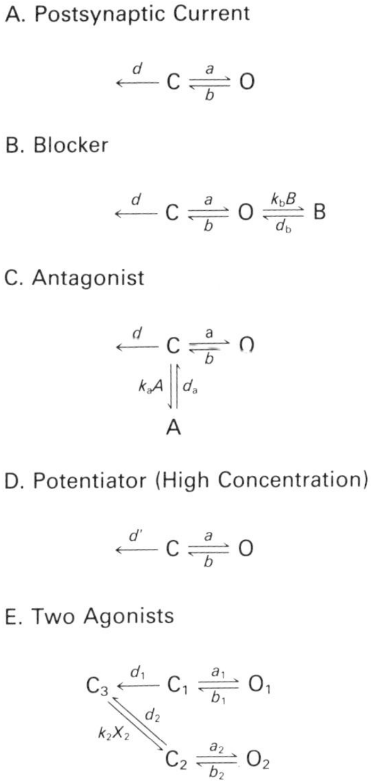

Figure 6. Hypothetical postsynaptic currents in the presence of the various drugs considered in Fig. 4.

Solving equation 16, we obtain

$$P_o(t) = [b/(r_1 - r_2)] [\exp(-r_2 t) - \exp(-r_1 t)] \qquad (17)$$

where the rate constants r_1 and r_2 are given by

$$r_1, r_2 = \tfrac{1}{2}(a + b + d) \pm \tfrac{1}{2}[(a + b + d)^2 - 4bd]^{\frac{1}{2}} \qquad (18)$$

The postsynaptic current, I_s, is just $P_o(t)$ times the number of channels reached by the released transmitter, N_s, times the current flowing through a single open channel, i,

$$I_s = N_s i P_o(t) \qquad (19)$$

The postsynaptic current in the presence of the drugs is calculated by an analogous procedure. The kinetic scheme in the presence of the ambient drug may, however, be considerably more complicated. The calculations are made tractable by the constraint that the endogenous transmitter saturates the available receptors at $t = 0$.

In the presence of a blocker, the scheme for the postsynaptic response becomes that of Fig. 6B, and the response is given by the third-order system

$$dP_o/dt = bP_c - (a + k_b B) P_o + d_b P_B$$

$$dP_c/dt = -(b + k)P_c + aP_o \qquad (20a)$$

$$dP_B/dt = k_b B\, P_o - d_b P_B$$

with initial conditions

$$P_c(0) = 1,\ P_o(0) = P_B(0) = 0 \qquad (20b)$$

The response is now a three-exponential expression, but if we make $k \gg b$, so that the leading edge of the waveform is very fast, then

$$P_o(t) = [(r_4 + a + k_b B)/(r_4 - r_3)] \exp(r_4 t)$$
$$+ [(r_3 + a + k_b B)/(r_3 - r_4)] \exp(r_3 t) \qquad (21)$$

where

$$r_3, r_4 = -\tfrac{1}{2}(a + d_b + k_b B) \pm [(a + d_b + k_b B)^2 - 4ad_b]^{\frac{1}{2}}$$

A potentiator also prolongs the postsynaptic current, but, unlike a blocker, it does not have a falling phase with two components. In order to test whether this explanation of diazepam action is correct, the rates derived from single-channel measurements can be used to model the macroscopic postsynaptic response (Redmann et al., 1985). For the assumption that the potentiator acts to lower the dissociation rate of the endogenous drug, the postsynaptic response can be calculated by the scheme shown in Fig. 6C. This is similar to the basic scheme of equation 13 but with k now taking on the drug-altered value k'. Figure 6 shows some values of p_2 for different choices of k' inferred from single-channel experiments at high drug concentration.

The postsynaptic kinetic scheme in the presence of an applied agonist is shown in Fig. 6. For this version of the interaction between two agonists, the postsynaptic response obeys

$$P_o(t) = \{a_1[1 - (d_2/k_2X_2)K]/(r_5 - r_6)\} [\exp(r_5 t) - \exp(r_6 t)] + (b_2/a_2) K \quad (22a)$$

with

$$K = [1 + (d_2/k_2X_2) + (b_2/a_2)]^{-1} \quad (22b)$$

and

$$r_5, r_6 = -\tfrac{1}{2}(a + b + d) \pm \tfrac{1}{2}[(a + b + d)^2 - 4bd]^{\tfrac{1}{2}} \quad (22c)$$

Equation 22 shows how the amplitude of the postsynaptic current can now be predicted if we know some of the single-channel properties. In this case, the change in the postsynaptic current appears simple, just a superposition of the endogenous transmitter response and the steady-state agonist response. For a more realistic case wherein two transmitters are needed to open a channel, the analysis is more complicated.

5. MACROSCOPIC AND MOLECULAR CONSEQUENCES

In these examples, we have seen how various drugs might change the shape of the postsynaptic response in ways that are predictable from the single-channel data. These shape changes might be characterized by some parameters of the postsynaptic potential so that one might inquire further how the altered postsynaptic potential affects the firing frequency of a nerve with many synaptic inputs. For example, it might be of interest to see whether changes in amplitude and changes in duration have equivalent effects on information transmission. To answer such questions, we need a stochastic model of synaptic integration in the presence of a random stream of excitatory and inhibitory inputs of specified shape. Such models (Gluss, 1967; Johanessma, 1968) assume that the rate of nerve firing depends on the frequency with which a stepping distribution of synaptic inputs exceeds a certain threshold voltage. The membrane potential is then a random variable that can be predicted via a Fokker–Planck equation. Parameters characterizing the random process depend on amplitude and duration in explicitly different ways.

If our goal is to understand the molecular mechanism of gating, we would do better to ask how the finite-state Markov models might emerge from a description of the dynamics of a protein molecule with a vast number of degrees of freedom (Frauenfelder, 1984). Perhaps a distribution such as the closed-time distribution of equation 4 represents the distribution of sojourns of the channel protein in a continuum of substates. Then the observed distribution could depend on the distribution of eigenvalues for some enormous matrix of dynamic transition probabilities. In this case, the distribution in terms of a small finite number of exponentials is the Laplace transform of $F_c(t)$. Thus, an inverse transform of equation 4 would give an explicit formula for the distribution of eigenvalues. Matrices with random distributions of eigenvalues have been considered for other dynamic systems with many degrees of freedom (Dyson, 1962), and the form obtained may be suggestive of some features of the protein motions.

REFERENCES

Adams, P. R., 1981, Acetylcholine receptor kinetics, *J. Membr. Biol.* **58**:161–174.

Armstrong, C. M., and Bezanilla, F., 1974, Charge movement associated with the opening and closing of the activation gates of the Na channels, *J. Gen. Physiol.* **63**:533–552.

Colquhoun, D., and Hawkes, A. G., 1977, Relaxation and fluctuations that flow through drug-operated channels, *Proc. R. Soc. Lond. [Biol.]* **199**:231–262.

Colquhoun, D., and Hawkes, A. G., 1981, On the stochastic properties of single ion channels, *Proc. R. Soc. Lond. [Biol.]* **211**:205–235.

Colquhoun, D., and Hawkes, A. G., 1982, Of bursts of single ion channel openings and clusters of bursts, *Phil. Trans. R. Soc. Lond. [Biol.]* **300**:4–59.

Colquhoun, D., and Hawkes, A. G., 1983, The principles of the stochastic interpretation of ion-channel mechanisms, in *Single-Channel Recording* (B. Sakmann and E. Neher, eds.), Plenum Press, New York, pp. 135–175.

Conti-Tronconi, B. M., and Raftery, M. A., 1982, The nicotinic acetylcholine receptor: Correlation of molecular structure with functional properties, *Annu. Rev. Biochem.* **51**:491–530.

del Castillo, J., and Katz, B., 1957, Interaction at end-plate receptors between different choline derivatives, *Proc. R. Soc. Lond. [Biol.]* **146**:369–381.

Dyson, F. J., 1962, A Brownian-motion model for the eigenvalues of a random matrix, *J. Math. Phys.* **3**:1191–1198.

Ehrenstein, G., Lecar, H., and Nossal, R., 1970, The nature of the negative resistance in bimolecular lipid membranes containing excitability-inducing material, *J. Gen. Physiol.* **55**:119–133.

Frauenfelder, H., 1984, Ligand binding and protein dynamics, in *Structure and Dynamics of Nucleic Acids, Proteins and Membranes* (E. Clementi and R. H. Sarma, eds.), Adenine Press, Guilderland, New York, pp. 369–379.

Fredkin, D. R., Montal, M., and Rice, J., 1983, Identification of aggregated Markovian models: Application to the nicotinic acetylcholine receptor, in *Proceedings of the Berkeley Conference in Honor of Jerzy Neyman and Jack Kiefer.*

Gluss, B., 1967, A model for neuron firing with exponential decay of potential resulting in diffusion equations for probability density, *Bull. Math. Biophys.* **29**:233–243.

Horn, R., 1984, Gating of channels in nerve and muscle: A stochastic approach, in *Ion Channels: Molecular and Physiological Aspects* (W. D. Stein, ed.), Academic Press, New York, pp. 53–97.

Horn, R., and Lange, K., 1983, Estimating kinetic constants from single channel data, *Biophys. J.* **43**:207–233.

Horn, R., and Vandenberg, C. A., 1984, Statistical properties of single sodium channels, *J. Gen. Physiol.* **84**:505–534.

Jackson, M. B., 1985, Stochastic behavior of a many-channel membrane system, *Biophys. J.* **47**:129–137.

Jackson, M. B., Lecar, H., Mathers, D. A., and Barker, J. L., 1982, Single-channel currents activated by γ-aminobutyric acid, muscimol, and ($-$)-pentobarbital in cultured mouse spinal neurons, *J. Neurosci.* **2**:889–894.

Jackson, M. B., Wong, B. S., Morris, C. E., Lecar, H., and Christian, C. N., 1983, Successive openings of the same acetylcholine receptor channel are correlated in open time, *Biophys. J.* **42**:109–114.

Johannesma, P. 1968, Stochastic neural activity, Dissertation, Katholieke Universiteit te Nijmegen, Holland.

Labarca, P., Rice, J., Fredkin, D. R., and Montal, M., 1985, Kinetic analysis of channel gating. Application to the cholinergic receptor channel and the chloride channel, *Biophys. J.* **47**:469–478.

Lecar, H., Morris, C. E., and Wong, B. S., 1982, Single-channel recording of weak cholinergic agonists, *Biophys. J.* **37**:313a.

Lecar, H., Morris, C. E., and Wong, B. S., 1983, Single-channel currents and the kinetics of agonist-induced gating, in *Structure and Function in Excitable Cells* (D. Chang, I. Tasaki, W. Adelman, and R. Leuchtag, eds.), Plenum Press, New York, pp. 159–173.

Morris, C. E., Wong, B. S., Jackson, M. B., and Lecar, H., 1983, Single-channel currents activated by curare in cultured embryonic rat muscle, *J. Neurosci.* **3**:2525–2531.

Neher, E., 1983, The charge carried by single-channel currents of rat cultured muscle cells in the presence of local anesthetics, *J. Physiol. (Lond.)* **339**:663–678.

Neher, E., and Steinbach, J. H., 1978, Local anesthetics transiently block currents through single acetylcholine receptor channels, *J. Physiol. (Lond.)* **277**:153–176.

Noda, M., Shimizu, S., Tanabe, T., Kayano, T., Ikeda, T., Takahashi, H., Nakayama, H., Kanaoka, Y., Minamino, N., Kangawa, K., Matsuo, H., Raftery, M. A., Hirose, T., Inayama, S., Hayashida, H., Miyata, T., and Numa, S., 1984, Primary structure of *Electrophorus electricus* sodium channel deduced from cDNA sequence *Nature* **312**:121–127.

Redmann, G. A., Lecar, H., and Barker, J. L., 1984, Diazepam increases GABA-activated single-channel burst duration in cultured mouse spinal cord neurons, *Biophys. J.* **45**:386a.

Steinbach, J. H., 1980, Activation of nicotinic acetylcholine receptors, in *The Cell Surface and Neuronal Function* (C. W. Cotman, G. Poste and G. L. Nicholson, eds.) Elsevier/North-Holland Biomedical Press, pp. 119–156.

Trautmann, A., 1982, Curare can open and block ionic channels associated with cholinergic receptors, *Nature* **298**:272–275.

van Kampen, N. G., 1981, *Stochastic Processes in Physics and Chemistry*, North-Holland, New York.

Chapter 3

Voltage-Dependent Gating
Gating Current Measurement and Interpretation

Francisco Bezanilla

1. INTRODUCTION

The voltage dependence of ionic conductances such as those of sodium and potassium is the result of the voltage influence on the open and closed times of individual channels. This means that an understanding of the voltage dependence of the fraction of open channels will help to correlate the microscopic (single-channel) and macroscopic properties (total ionic current). The question arises as to how the channel goes from the open to the closed configuration and, in particular, how the voltage across the membrane regulates the state of the channel, a process known as voltage gating.

2. VOLTAGE GATING

The channel may be considered a specialized membrane molecule that has a conductive pathway that allows ions to go through it when it is in the open conformation and is impermeant to ion passage when it is in the closed conformation. The channel may exist in a large number of physical states, some of them closed and others open, with spontaneous

---

FRANCISCO BEZANILLA • Department of Physiology, Ahmanson Laboratory of Neurobiology; and Jerry Lewis Neuromuscular Research Center, University of California at Los Angeles, Los Angeles, California 90024.

transitions occurring among the states. A simple view is to assume that the physical states represent wells of energy of the channel molecule and that the transitions occur when the energy barrier separating the wells is overcome by thermal energy.

How does the voltage regulate the state of the channel? In the first place, there must be a way the channel can sense the presence of the voltage across the membrane. Voltage can modify the position of magnetic fields or electric charges; thus, it is natural to think that the membrane field may be affecting the relative position of charged groups of the channel molecule with a consequent change in the physical state of the channel. In the view of energy wells and barriers, the effect of the voltage is to change the energy profile seen by the charged part of the molecule, also called the voltage sensor.

This view has several important consequences. (1) The probability of finding a channel in a particular physical state will be modified by the strength of the electric field; the time spent in any of the states will also be regulated by the membrane field. (2) When a transition occurs, the movement of the sensor will generate an electric current, which will appear as a capacitive current because the voltage sensor is trapped in the membrane.

These consequences can be tested experimentally. The recording of macroscopic currents as a function of membrane voltage will give an indication of the voltage dependence of the probability of the channels being in the open state. These currents show the fraction of channels in the conducting states, but they do not provide direct information about the transitions between closed states. The kinetics of opening and closing will provide information about the times spent in the open or closed configuration, and a more detailed description of these times is obtained from measurements of closed and open dwell times of single-channel recordings. The movement of charge produced when a channel changes conformation should produce a large and extremely fast current, which will be very difficult to observe but becomes detectable when the charge movement of a large number of channels is recorded. In this case, the sum of a great number of δ functions in current will produce a current decaying in time, which has been called gating current. It is clear that the study of this gating current should give us a good deal of information about the transitions and physical states of the channel even when those transitions occur between closed states and no ionic current is detectable.

3. GATING CURRENT IS A CAPACITIVE CURRENT

The exact nature of the voltage sensor is unknown, but likely candidates are charged groups in the channel molecule. The sodium channel is a large protein with molecular weight of 240,000 containing several α helices (Agnew et al., 1978; Noda et al., 1984). Some of these helices contain charges that are presumably oriented by the membrane field, and their exact position is regulated by the strength of the field; a change in the position of these groups may in turn change the channel from a nonconducting to a conducting state. From the electrical point of view, the ensemble of charged groups in the macromolecule constitutes an electric dipole that can rotate or polarize with the field, producing a current that is recorded as a capacitive current in the external circuit.

In this chapter we concentrate on two aspects of gating currents: (1) measurement, and (2) a summary of gating currents recorded from the sodium channel with interpretations in terms of a sequential model.

4. MEASUREMENT OF GATING CURRENTS

When the membrane voltage is changed suddenly (a step of voltage), two main types of currents are recorded: ionic current and capacitive current. The ionic current is contributed by ions crossing the membrane, for example, through ionic channels. The capacitive current is the current needed to charge the membrane capacitance, and it is carried in the solution by ions that do not cross the membrane. The capacitive current has two main components: (1) the charging of the capacitor constituted of the ionic solutions separated by the insulating membrane, and (2) the rearrangement of charges and dipoles of the dielectric of the membrane. The gating current is part, if not all, of the second component, because the polarization of the macromolecular groups moving under the influence of the membrane field will contribute a dielectric current. The measurement of gating currents is then reduced to a separation of this component from all the other currents crossing the membrane.

The ionic current can be largely eliminated by ion substitution and by the use of toxins. The currents that must be suppressed are the nonlinear ionic currents such as sodium and potassium currents. The first is eliminated by replacing sodium with impermeant ions such as Cs^+ or tetramethylammonium (TMA) and blocked by tetrodotoxin (TTX). The potassium current is suppressed by replacing potassium with TMA inside and tris outside. The remaining current is called ''leakage'' and is carried by the substituting cations and in a minor proportion by the anions (Armstrong and Bezanilla, 1974). After these substitutions, the current recorded is mainly capacitive with a small ionic (leakage) component.

The capacitive current must be separated into its components to isolate the polarization current produced by channel gating. The clue to the separation is the fact that gating is voltage dependent and shows saturation at extreme voltages; therefore, the dielectric polarization produced

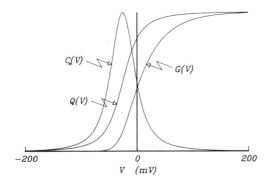

Figure 1. Theoretical representation of the transported charge, $Q(V)$, zero-frequency membrane capacitance, $C_o(V)$, and conductance, $G(V)$, for a sequential model with seven steps. Notice the saturation of $Q(V)$ and $G(V)$ and the shift to the right of G with respect to Q. The subtraction of the linear component of the membrane capacitance can be done at extreme voltages where $Q(V)$ is saturated and $C_o(V)$ approaches zero.

by the gating groups should also show saturation. The sigmoid curve plotted in Fig. 1 shows the expected relationship between the displaced charge and membrane potential (Q–V curve) for this saturating system and corresponds to the area under the gating current transients recorded at a potential V. In the same graph, the bell-shaped curve corresponds to the zero-frequency capacitance as a function of voltage (C_o–V curve) expected from this system. The relationship between Q and C_o is:

$$C_o = dQ/dV$$

There are at least two experimental approaches to studying the voltage-dependent polarization: (1) time-domain measurements and (2) frequency-domain measurements.

4.1. Frequency-Domain Measurement of Gating Currents

Measurements of capacitance can be easily accomplished in the frequency domain, as we show below. The nonlinear charge movement expected from the gating charge reorientation is also part of the capacitance, and the task is to separate this fraction from the total capacitance.

4.1.1. General Transfer-Function Measurement

The technique of transfer-function measurement is a general method to characterize a linear system and, in particular, can be used in an electrical circuit to measure capacitance. The membrane electrical characteristics are nonlinear, but the technique can still be used when the signal applied is small enough, effectively linearizing the characteristics around an operating point.

Consider the system pictured by the box with an input and an output (Otnes and Enochson, 1978)

Voltage-Dependent Gating

$$x(t) \to \boxed{h(t)} \to y(t)$$

where $x(t)$ is an arbitrary signal applied to the input and $y(t)$ is the output produced by the box; $h(t)$ represents the operation performed by the box on $x(t)$ to produce $y(t)$. For example, if the box is a second-order system, we can write

$$D^{(1)}[x(t)] = D^{(2)}[y(t)]$$

with

$$D^{(2)}(y) = (a\, d^2/dt^2 + b\, d/dt + c)y$$

and

$$D^{(1)}(x) = 1$$

Applying the Fourier transform to both sides of the first equation, we get

$$F\{D^{(1)}[x(t)]\} = F\{D^{(2)}[y(t)]\}$$

$$X(\omega) = Y(\omega)\,[\omega^2 a + c + j\omega b]$$

where the following equation defines the Fourier transform of $g(t)$:

$$F[g(t)] = \int_{-\infty}^{\infty} g(t)\exp(-j\omega t)dt = G(\omega)$$

with $\omega = 2\pi f$, f = frequency (in Hertz), and $j = \sqrt{-1}$.
The transfer function $H(\omega)$ is then defined as

$$H(\omega) = Y(\omega)/X(\omega).$$

Taking the circuit pictured in Fig. 2a, one can write the differential equation relating the current to the voltage, and taking the Fourier transform one gets

$$Y(\omega) = I(\omega)/V(\omega) = G + jB = 1/R + j\omega C$$

in which the transfer function Y is called admittance, G the conductance, and B the susceptance. Figure 2a also shows a plot of the real $[G(f)]$ and imaginary $[B(f)]$ parts of Y as a function of the frequency. Notice that the imaginary part of Y divided by ω gives the value of the capacitance $[C(f)]$ of the circuit, which is independent of frequency. This case represents an ideal capacitor with a leak pathway in parallel. If there are

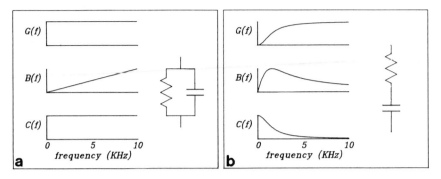

Figure 2. Circuit diagrams and their admittance. a: The parallel C,R combination gives a plot of constant G, a linear increment of B, and a constant C ($C = B/\omega$). b: The series combination, which is a good representation of a charged dipole, gives a more complex dependence of G, B, and C as a function of frequency. In the membrane, one expects the parallel combination of multiple series R,C with the parallel R,C.

polarizable charges or dipoles that are not instantaneously reoriented by the field, the minimum representation of these dielectric properties is a resistor in series with a capacitor (Fig. 2b). The capacitor represents the storage element, and the resistor the dissipative element. In the case of the gating system, it is expected that the charges or dipoles will all be oriented in extreme fields and that this will produce saturation of the polarization. In terms of the equivalent circuit, this implies that the capacitor and the resistor in series are both functions of the voltage; that is, the system is nonlinear.

Taylor and Bezanilla (1979) described the differential equations for the case of a gating particle with two states and a simple barrier separating the two wells of energy. They linearized these expressions in the frequency domain and expressed the transfer function as an admittance and separated its real and imaginary parts. The imaginary part divided by the frequency is the capacitance, but this time the capacitance is frequency dependent, which is a reflection of the kinetic process of transition between the two states of the gating charge. Figure 2b is the diagram of the equivalent circuit for this kinetic model at a given voltage, and the plots of Fig. 2b show the frequency dependence of the conductance [$G(f)$], the susceptance [$B(f)$], and the capacitance ($C = B/\omega$). It is important to note that if there are several dipole systems, each one will contribute to the admittance with a new branch constituted by a series combination of a capacitor and a resistor.

4.1.2. Experimental Procedure

We have used the technique of applying a small voltage perturbation to the axonal membrane under voltage clamp and in ionic conditions that minimize ionic currents. The small perturbation is pseudorandom noise, which contains an equal proportion of a large number of frequencies extending from DC to the maximum bandwidth of the determination (flat power spectrum) and each frequency with random phase. The signal is called "pseudorandom" because each time it is generated it is an exact replica. The methods of generation are varied: the pseudorandom binary sequence (Poussart and Ganguly, 1977) was implemented by Clausen and Fernandez (1981) and used by Fernandez et al. (1982) to measure gating current in the frequency domain; the sum of sinusoids with random phase was implemented by Stimers et al. (1985a) for the same purpose, but with a much extended bandwidth. The advantage of these pseudorandom sequences is that (1) one can use signal-averaging techniques in the time domain to improve signal-to-noise ratio before the Fourier transform is performed and (2) the driving signal (the membrane voltage) may need to be transformed only once because in each trial it is exactly the same, saving one channel in the acquisition system. The procedure is to deliver the pseudorandom sequence to the voltage-clamped axon and take the Fourier transform of the delivered signal (voltage) and of the axon current. The actual transform is done using the fast Fourier transform algorithm (Cooley and Tukey, 1965) in the same computer that acquires the data. The transfer function is calculated as the ratio of the transformed current to the transformed voltage (admittance). The real part of this result is the conductance (G), and the imaginary part (susceptance, B) divided by the frequency is the capacitance (C). This experiment must be done at a series of membrane potentials to extract the voltage dependence of G and C (Fernandez et al., 1982). The capacitance obtained will be the sum of both the voltage-dependent and the voltage-independent capacitance. After measurements are taken at extreme potentials, it is possible to separate these two components by subtracting the admittance obtained at a voltage V from the admittance obtained at a voltage V_1, where V_1 is a potential at which there is no voltage dependence of the capacitance (saturation region of the gating charge, see Fig. 1).

4.2. Time-Domain Measurements of Gating Currents

In the time domain, the measurement usually performed is the current response to a step of voltage. A sudden change in the membrane potential elicits a capacitive current that is the sum of the linear and nonlinear components. The technique to separate the two components is to take advantage of the linear properties of the linear component: if the current for a positive pulse is added to the current for a negative pulse of the same amplitude, the linear components of the currents cancel out, and the resultant current will correspond to the sum of the nonlinear components for both potentials.

In the Q–V curve in Fig. 1, it is clear that the nonlinear charge measured will depend not only on pulse amplitude but also on the potential from which the pulses are started. To get the charge for the potential V, the pulses must be started from a very negative (or very positive) voltage so that the measured charge will not have a contribution of the negative pulse. The original method to measure gating currents (Armstrong and Beezanilla, 1973) used equal-amplitude and opposite-polarity pulses starting at a holding potential of -70 mV, but it was soon realized (Armstrong and Bezanilla, 1974) that the Q–V curve was not flat at -70 mV but saturated at around -150 mV, and a different method was devised to subtract the linear component. The test pulse (P) is started from the usual holding potential (i.e., -70 mV), and the subtraction is done starting at a very negative potential (*ca.* -150 mV). The subtracting pulse cannot be of the same amplitude as the test pulse because the membrane potential would go to extreme negative potentials and the membrane could break down. For this reason, the subtraction is done using four pulses of one-quarter the amplitude of the test pulse ($P/4$), and the added currents from these four pulses are added to or subtracted from the current produced by the test pulse depending on the direction of the voltage pulses (the $P/4$ procedure, Bezanilla and Armstrong, 1977).

For this procedure to be valid, the membrane potential must be accurately controlled to the value imposed, which means that the series resistance present must be compensated for. This is illustrated in Fig. 3. The equivalent circuit in Fig. 3a represents the membrane linear capacitance (C_l) and the gating admittance Y_g in parallel. Figure 3b shows the expected current resulting from a positive pulse and a subtracting pulse

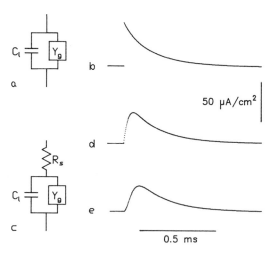

Figure 3. The effects of series resistance on gating current time course. a: Equivalent circuit of the membrane capacitance (C) and the gating admittance (Y_g). b: Gating current transient recorded from the circuit in a. c: Equivalent circuit of the membrane capacitance, gating admittance, and series resistance. d: The calculated gating current for the positive pulse for the circuit of c. e: The subtracted current. This last current is the transient that would be experimentally recorded for this situation with series resistance. Notice that the current in e differs from the actual gating current (d) and is quite different from the expected gating current from a perfect step (b).

in the negative region of the $Q-V$ curve in Fig. 1. This transient represents the current through the gating admittance. When the membrane has a resistance in series, as shown in Fig. 3c, the situation is different. In this case, both C_1 and Y_g are in series with the series resistance R_s, which, in the squid axon, is produced mainly by the Schwann cell layer. The expected current transient diagrammed in Fig. 3e shows that the subtracted current is not the current expected from the gating element (Y_g). The reason is that the imposed step on the total membrane plus series resistance (V_c) is not a step at the membrane (V_m) but rises slowly, and its value is

$$V_m = V_c - I_m R_s$$

The value of I_m, the total membrane current, is different for the test pulse than for the subtracting pulse because the test pulse will elicit gating current and the subtracting pulse will not. The result is that the subtracted current (Fig. 3e) will be different from the current through Y_g (Fig. 3d), which is also different from the expected current without series resistance (Fig. 3b). Most of the error occurs at short times after the step, and it becomes negligible at longer times because the current is small; therefore, early events of the gating process will be masked by the error produced by the series resistance. The solution is to use compensation for series resistance during the measurement (Hodgkin et al., 1952). In the frequency domain, the correction can be done after the total admittance has been calculated (Fernandez et al., 1982) by subtracting the series resistance from the impedance Z, which is the reciprocal of the admittance ($Z = 1/Y$).

5. GATING OF THE SODIUM CHANNEL

An understanding of the gating process requires a full description of all the physical states of the channel and the transitions between the states. A first step in this description is the formulation of a kinetic model of the sodium current throughout the useful range of membrane potentials (Hodgkin and Huxley, 1952). As is shown below, the sodium channel has a large number of states, and there are many different kinetic models that can account for the ionic currents; this means that a model formulation will not be too enlightening. The situation is much improved when the information from macroscopic currents is combined with gating currents and single channel recordings because these measurements restrict the number of possible models that can fit the data. All the measurements should be made from the same preparation to avoid complications produced by differences among different types of cells. The squid axon provides the best signal-to-noise ratio for gating and macroscopic currents, but no single sodium channel records are as yet available. For the potassium channel, the three types of measurements are now possible in the squid axon.

Figure 4a shows the currents recorded with the $P/4$ procedure for a series of depolarizations in a voltage-clamped squid axon with reduced sodium concentration in the external medium and in the absence of potassium in both external and internal media. The currents have a first outward (positive) component, which is a gating current, followed by an inward (negative) sodium current, which increases (activates) and then decreases (inactivates) as the voltage is maintained. Figure 4b shows the gating currents recorded in the absence of sodium and with tetrodotoxin (TTX) present. With these types of measurements we would like to build a model of the gating process of the sodium channel. In what follows, we briefly review our present knowledge of sodium channel gating of the squid axon derived from the available data with emphasis on the results of gating currents.

5.1. The Sodium Channel Has Several Closed States

The sodium current develops with a lag. This rules out the simple two-state model for the sodium channel. The simplest interpretation of the lag is that on depolarization the channel spends time in several closed states before it reaches the open state.

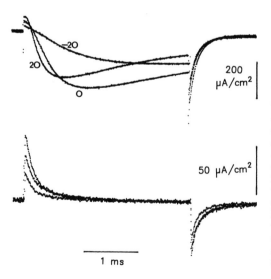

Figure 4. Sodium and gating currents recorded from the squid axon. The upper set are currents recorded in artificial sea water with only one-fifth of the normal sodium concentration. The outward current at the beginning of the pulse is gating current, which is shown in more detail in the lower set after the external solution was made sodium-free and TTX was added. The holding potential was -70 mV, and the pulses were to -20, 0, and 20 mV. Notice the similarity in time course of the *off* gating and sodium currents.

The fraction of open channels versus voltage is shifted to the right of the $Q-V$ curve; this also is a strong indication that the channel has several closed states.

An interesting effect is observed when the initial voltage before the depolarizing pulse is made more negative. In this case the lag increases, indicating that negative membrane potentials favor closed states. Taylor and Bezanilla (1983) found that the gating current is shifted as much as the sodium current as a result of a prepulse; in addition, they found that the charge that moves during the pulse increases as the prepulse is made more negative. These results indicate that there are several closed states that are voltage dependent, and they contribute with charge movement in the negative region of the $Q-V$ curve.

Once it is known that there are several closed states, can we find the number of states and the rate constants of transitions between the states? Unfortunately, there is not a simple answer. In the best case we can eliminate some families of models and propose other types that are not unique. Hodgkin and Huxley (1952) described the activation of the sodium conductance with the m^3 formulation, which can be interpreted as the requirement of the channel to have three identical gating particles in the same position for the channel to be open. The experiments on gating currents are inconsistent with this interpretation. In the m^3 formulation, during deactivation (pulse off) the movement of only one particle is enough to close the channel, which means that the ionic current will subside before the gating current does (time constant of gating should be three times slower than sodium current). The experimental result (Armstrong and Bezanilla, 1977) shows that the gating current is just slightly slower than the sodium current at pulse off. The m^3 formulation is just a special case of a general sequential model for activation in which the rate constants go in the sequence from closed to open $3\alpha, 2\alpha, \alpha$ and $3\beta, 2\beta, \beta$ from open to closed. The above results from gating current experiments rule out this type of sequence but still leave a large number of possibilities available. The results with negative prepulses require a sequence with at least six closed states if the rate constants of the transitions are assumed to be all the same. Although there is very little information about the rate constants in the intermediate transitions, gating current experiments have provided data to determine the relative magnitudes of the first and following transitions and the exclusion of a rate-limiting step.

5.2. The Transition from the Most Closed State Is Fast

The time course of the gating current depends on the transition rates of the gating reaction. In particular, the early part of the gating current transient is very sensitive to the relative rates of the first and following transitions. If the first transition is slower than the following one, the gating current shows a rising phase; if it is faster, no rising phase is observed; and if it is of the same speed, the gating current starts with zero slope before decaying (Armstrong, 1981; Bezanilla and Taylor, 1982). Early experimental results (Armstrong and Bezanilla, 1974; Armstrong and Gilly, 1979) showed a rising phase on the *on* gating current, indicating a slower first transition in the activating sequence. Recent experiments with good membrane potential control have shown that there is no rising phase for the *on* gating current or that it is at least faster than 5 to 10 μsec (Stimers *et al.*, 1984), which is the time it takes to charge the membrane capacity with the voltage-clamp step in squid axon. The presence of a rising phase in previous gating current recordings can be explained as a failure in the longitudinal control of the membrane potential.

Stimers *et al.* (1984) found that the rising phase was associated with the presence of a slow component in the capacitive transient, which in turn suggested a region of membrane clamped with a series resistance. Compensation for series resistance did not eliminate the slow phase of the capacitive transient or the rising phase of the gating current, but hypertonic medium outside or hypotonic inside the axon decreased the slow component of the capacitive transient and the rising phase of the gating current. These results may be explained by an effect of the osmotic gradient that enlarges the periaxonal space, improving the longitudinal membrane potential control between Schwann cell clefts and making the membrane potential homogeneous. Under these conditions, the rising phase is not visible or at least is faster than about 10 μsec. The conclusion is that the first transition is faster than the ones following in the activation sequence.

5.3. Last Step Is not Rate Limiting

One simple way to explain the activation kinetics of the sodium channel is to assume that it is a linear sequence of fast steps leading to a final very slow step that in effect rate-limits the activation sequence. A pulse protocol with two pulses separated by a variable interval (Oxford, 1981) can be used to test the speed of the last transition. The first pulse is used to activate the conductance, and the second pulse is used to test the speed of the activation after a variable recovery interval. If the interval is very short, most of the channels do not have time to return to closed states far from the open state, and in the limit when the interval is very short, the channels can only be in the last closed state, and the second pulse will test the time course of the transition between the last closed state and the open state. If the last transition is rate limiting, the time constant of the activation during the second pulse will be independent of the time interval between pulses. Stimers *et al.* (1983) performed this experiment in pronase-treated axons (to eliminate inactivation), and the result indicated that the time constant of the activation becomes faster as the interval between pulses is made shorter, showing that the last step is not rate limiting. It is not possible to determine exactly the rate constant of the last transition from this type of experiment, but the indication is that it is fast. The conclusion is that the overall speed of the activation kinetics is the expression of the eigenvalues of the system of equations resulting from the multiple transitions, and it is not the reflection of any of the individual rate constants of the elementary transitions of the sequence.

5.4. Inactivation Is Intrinsically Voltage Dependent

The inactivation of the sodium conductance is voltage dependent. Consequently, a gating current associated with its voltage sensor is expected to be observed. Armstrong and Bezanilla (1977) showed that the same gating particles responsible for the activation of the conductance are also involved in the inactivation by going into a different set of kinetic states (charge immobilization). This means that the voltage dependence of the inactivation is a consequence of the voltage dependence of the activation process, and there is no need for an independent voltage sensor for the inactivation. Stimers *et al.* (1985b) showed that a complete absence of voltage dependence of the inactivation step cannot explain the results of macroscopic and gating currents at large depolarizations, and a slight voltage dependence in the inactivation transition had to be included. When there is no voltage dependence in the inactivation step, the peak sodium conductance-versus-voltage curve does not saturate until unrealistically high depolarizations (*ca.* 1 V). Figure 5 shows schematically the reason for the lack of saturation. At high depolarizations, the inactivation step becomes rate limiting, and the sodium current grows with depolarization as the activation phase gets faster. When the inactivation step is made voltage dependent, the peak current does not get larger with depolarization because both the activation and inactivation get faster.

The conclusion is that although the inactivation of the sodium conductance is intrinsically voltage dependent with 0.6 electronic charges moving the entire field of the membrane (Stimers *et al.*, 1985b), most of its voltage dependence is obtained from the activation process. A physical model based on the interaction of the activating particle and the inactivating particle (Bezanilla *et al.*, 1982) has been expanded to include the observation of the voltage dependence of the inactivation step by Stimers *et al.* (1985b).

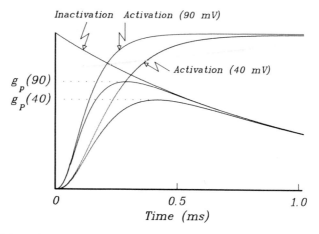

Figure 5. Schematic representation of the effect of voltage-independent inactivation. The curves were calculated with the Hodgkin and Huxley equations (1952). In their equations, the voltage dependence of β_h is negligible at large depolarizations and has the same time course for 40 and 90 mV (exponentially decaying trace). The activation is saturated at 40 mV, and the steady values at 40 mV and 90 mV are the same, although the kinetics are quite different. The result is that the peak sodium conductance does not show saturation in the 40 to 90 mV potential range as the actual activation does.

5.5. A Model for Sodium Channel Gating

Figure 6 shows a state model that takes into account all the possible states required by the two sets of gating particles. From the results of initial condition experiments described above, seven states for the activating gate are required. The inactivating gate has two stable positions,

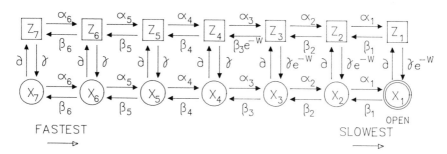

Figure 6. A state diagram of the sodium channel gating. The lower row (Xs) are activating states, and the upper row (Zs) are the inactivated states. All the states except X_1 are nonconducting. State Z_1 is open but inactivated. The progression of rate constant magnitude discussed in the text is schematically shown in the figure. The interaction between the activation and inactivation gates occurs in states 1, 2, and 3, and the interaction energy is W. This accounts for the immobilization of the charge in the inactivated states.

which makes a total of 14 states to represent all the possible physical conformations of the channel.

The effect of prolonged depolarization has not been included in this state diagram. Bezanilla *et al.* (1982) and Fernandez *et al.* (1982) have shown, using time-domain and frequency-domain techniques, that the slow inactivation may be modeled as another interacting particle, which increases the number of states to 28. It is important to realize that these states are the ones required by our present knowledge of the gating properties, and they are most likely the principal states of a much larger set that we cannot distinguish with present techniques. In fact, the rate constant for the elementary transitions between the states of Fig. 6 cannot be determined uniquely from our present experimental information. At most, we can give general rules about the relationships among the different rate constants that are consistent with results of both gating and macroscopic currents. The number of rate constants involved is much too large to hope to obtain a unique fit to the experimental results. The intermediate steps of the activation sequence could also be treated as a continuum, but this is even harder to handle analytically. The solution to this problem may reside in experimental manipulations that simplify the system: elimination of the inactivation and chemical or pharmacological modifications. In addition, new information is now becoming available from the biochemistry and primary structure of the macromolecule that should help in making more realistic physical interpretations of the transitions involved in the activation sequence. Site-directed mutagenesis and reconstitution will be invaluable tools in gaining a molecular picture of channel gating.

ACKNOWLEDGMENTS. Many thanks to Dr. R. E. Taylor for reading and commenting on this manuscript. The experimental data presented in Fig. 4 were obtained in collaboration with Drs. J. R. Stimers and R. E. Taylor. Supported by USPHS grant GM 30376.

REFERENCES

Agnew, W. S., Levinson, S. R., Brabson, J. S., and Raftery, M. A., 1978, Purification of the tetrodotoxin-binding component associated with the voltage-sensitive sodium channel from *Electrophorus electricus* electroplax membranes, *Proc. Natl. Acad. Sci. U.S.A.* **75**:2606–2610.

Armstrong, C. M., 1981, Sodium channels and gating currents, *Physiol. Rev.* **61**:644–683.

Armstrong, C. M., and Bezanilla, F., 1973, Currents related to movement of the gating particles of the sodium channel, *Nature* **242**:459–461.

Armstrong, C. M., and Bezanilla, F., 1974, Charge movement associated with the opening and closing of the activation gates of the Na channels, *J. Gen. Physiol.* **63**:533–552.

Armstrong, C. M., and Bezanilla, F., 1977, Inactivation of the sodium channel. II. Gating current experiments, *J. Gen. Physiol.* **70**:567–590.

Armstrong, C. M., and Gilly, W. F., 1979, Fast and slow steps in the activation of sodium channels, *J. Gen. Physiol.* **74**:691–711.

Bezanilla, F., and Armstrong, C. M., 1977, Inactivation of the sodium channel. I. Sodium current experiments, *J. Gen. Physiol.* **70**:549–566.

Bezanilla, F., and Taylor, R. E., 1982, Voltage dependent gating of sodium channels, in *Abnormal Nerves and Muscles as Impulse Generators* (W. J. Culp and J. Ochoa, eds.), Oxford University Press, New York, pp. 62–79.

Bezanilla, F., Taylor, R. E., and Fernandez, J. M., 1982, Distribution and kinetics of membrane dielectric polarization. I. Long term inactivation of gating currents, *J. Gen. Physiol.* **79**:21–40.

Clausen, C., and Fernandez, J. M., 1981, A low cost method for rapid transfer function measurements with direct application to biological impedance analysis, *Pflugers Arch.* **390**:290–295.

Cooley, J. W., and Tukey, J. W., 1965, An algorithm for the machine calculation of complex Fourier series, *Math. Comput.* **19**:297.

Fernandez, J. M., Bezanilla, F., and Taylor, R. E., 1982, Distribution and kinetics of membrane dielectric polarization. II. Frequency domain studies of gating currents, *J. Gen. Physiol.* **79**:41–67.

Hodgkin, A. L., and Huxley, A. F., 1952, A quantitative description of membrane current and its application to conduction and excitation in nerve, *J. Physiol. (Lond.)* **117**:500–544.

Hodgkin, A. L., Huxley, A. F., and Katz, B., 1952, Measurement of current–voltage relations in the membrane of the giant axon of *Loligo*, *J. Physiol. (Lond.)* **116**:424–448.

Noda, M., Shimizu, S., Tanabe, T., Takai, T., Kayano, T., Ikeda, T., Takahashi, H., Nakayama, H., Kanaoka, Y., Minamino, M., Kangawa, K., Matsuo, H., Raftery, M. A., Hirose, T., Inayama, S., Hayashida, H., Miyata, T., and Numa, S., 1984, Primary structure of *Electrophorus electricus* sodium channel deduced from cDNA sequence, *Nature* **312**:121–127.

Otnes, R. K., and Enochson, L., 1978, *Applied Time Series Analysis, Volume 1. Basic Techniques*, John Wiley & Sons, New York.

Oxford, G. S., 1981, Some kinetic and steady-state properties of the sodium channels after removal of inactivation, *J. Gen. Physiol.* **77**:1–22.

Poussart, D., and Ganguly, U. S., 1977, Rapid measurement of system kinetics an instrument for real-time transfer function analysis, *Proc. IEEE* **65**:741–747.

Stimers, J. R., Bezanilla, F., and Taylor, R. E., 1983, Sodium channel activation in pronase treated axons, *Biophys. J.* **41**:144a.

Stimers, J. R., Bezanilla, F., and Taylor, R. E., 1984, Squid axon sodium channel: Gating current without rising phase, *Biophys. J.* **45**:12a.

Stimers, J. R., Bezanilla, F., and Taylor, R. E., 1985a, Frequency domain measurements of membrane capacitance in squid axons with and without rising phase on the gating current, *Biophys. J.* **47**:31a.

Stimers, J. R., Bezanilla, F., and Taylor, R. E., 1985b, Sodium channel activation in the squid giant axon. Steady state properties, *J. Gen. Physiol.* **85**:65–82.

Taylor, R. E., and Bezanilla, F., 1979, Comments on the measurement of gating currents in the frequency domain, *Biophys. J.* **26**:338–340.

Taylor, R. E., and Bezanilla, F., 1983, Sodium and gating current time shifts resulting from changes in initial conditions, *J. Gen. Physiol.* **81**:773–784.

Chapter 4

Characterizing the Electrical Behavior of an Open Channel via the Energy Profile for Ion Permeation
A Prototype Using a Fluctuating Barrier Model for the Acetylcholine Receptor Channel

George Eisenman and John A. Dani

1. INTRODUCTION

The fundamental importance of the energy profile in determining the permeability parameters (e.g., selectivity, conductance, and binding) of a channel with classical, static structure has been discussed elsewhere (Eisenman and Horn, 1983). Here we examine what happens to the current–voltage and conductance–concentration behaviors when a more dynamic view is taken, namely, one in which significant fluctuations in the energy profile can occur on the same time scale as ion translocation. The surfaces of proteins in solution execute thermally excited motion with a root-mean-square amplitude of 1 to 2 Å (Karplus and McCammon, 1983; McCammon, 1984; McCammon and Karplus, 1983). If such motion extended into the pore, it would occur over a significant fraction of the estimated diameter of biological ionic channels and would be expected to perturb significantly the energy profile for ion permeation. Lauger *et al.* (1980; Lauger, 1984) were the first to extend barrier models of ionic permeation in order to allow for the possibility that the energy profile could

---

GEORGE EISENMAN AND JOHN A. DANI • Department of Physiology, University of California Medical School, Los Angeles, California 90024. *Present address of J.A.D.*: Section of Molecular Neurobiology, Yale University, New Haven, Connecticut 06510.

fluctuate on the same time scale as ionic migration. They showed that when such fluctuations are coupled to ion translocation, even for a singly occupied channel, they lead to behavior usually associated with multiple occupancy, a conclusion supported by Ciani (1984). Such coupling can produce complex dependence of conductance on concentration, with an anomalously low zero-voltage conductance (possibly even a maximum) at high concentration concomitantly with an anomalously high conductance at low concentrations for large negative or positive voltages (Eisenman and Dani, 1985). The exquisite sensitivity of this behavior to the fluctuation frequency is scrutinized further in this chapter.

This chapter is oriented toward the student and the experimentalist and has a twofold purpose. One is to show how a fluctuating barrier model can be used to interpret the current–voltage (I–V) behavior of an open channel over a wide range of concentrations. Acetylcholine receptor (AChR) channel results are used to exemplify how electrical behavior can be modeled in terms of energy profiles, whether static or fluctuating. A second purpose is to explore several particular cases of a general fluctuating two-barrier, one-site model to uncover the novel features of such a model, especially those that might be experimentally testable. Current–voltage (I–V) relations and chord conductances (G) from the AChR channel (Dani and Eisenman, 1984, 1985, 1986) constrain the theory in order to obtain a set of rate constants that adequately describe the data. Various theoretical fits are then analyzed, especially to assess how the frequency of the energy profile fluctuations influence the theoretically predicted behavior.

2. THEORY

The way in which barrier models are used to represent the electrical behavior of a channel has been well described for conventional (static-profile) models in the literature (Hagglund *et al.*, 1984; Hille and Schwarz, 1978; Sandblom *et al.*, 1983; Urban *et al.*, 1980). The convenient matrix method (Hagglund *et al.*, 1984) is extended here to analyze a channel with a fluctuating energy profile, applied within the framework provided by Lauger *et al.* (1980) for a system in which coupling between ion flow and conformational transitions of an ion channel is permitted to occur.

We examine the fluctuating profile model against AChR channel data using a time-varying version of a reaction rate model that has been used with first-order success in the past (Adams, 1979; Dani and Eisenman, 1984; Eisenman *et al.*, 1985; Horn and Brodwick, 1980; Lewis and Stevens, 1979; Marchais and Marty, 1979). The classical model defines the rate constants for ion translocation in terms of an energy profile consisting of a static profile having two barriers and one discrete site (called the

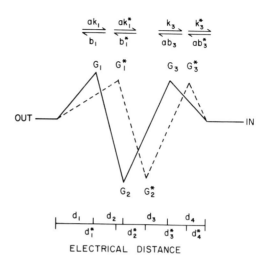

Figure 1. Energy profile for ion translocation for a channel with two conformational conformations, o and *. k_1, b_3, k_1^*, b_3^* are the rate constants for jumping from the solutions into an empty site, and k_3, b_1, k_3^*, b_1^* are the rate constants for jumping from the occupied site into the solutions. The energies underlying these rate constants are defined by the Gs. The voltage dependences are indicated by the distances d and d^*. This figure is the prototype for the energy diagrams to be presented in Figs. 4 and following.

static 2B1S model). We extend this representation to a "fluctuating 2B1S model" by allowing the energy profile of the channel to fluctuate between two such profiles representing two different conformational states, as shown in Fig. 1. The conformation corresponding to the dashed profile will be referred to as conformation*, and that corresponding to the solid profile will be referred to as conformation o. States 0 and X represent the empty and occupied channel in conformation o. States 0* and X* represent the empty and occupied channel in conformation*. (In general, unsuperscripted parameters will refer to the o conformations, whereas the * conformations will be denoted by an asterisk.) The transitions between these two conformational states are schematized by the energy profiles in Fig. 2, and the state diagram for the transitions among the four substates, 0, X, 0*, and X* of the channel is given in Fig. 3.

The rate for an ion to jump into the site is proportional to the aqueous ionic activity (a) times the entry rate constants k_1, b_3, k^*_1, b^*_3, and the rate for an ion to leave the site is proportional to the exit rate constants b_1, k_3, b^*_1, k^*_3 (see Fig. 1). For simplicity, the solutions are assumed to be symmetrical in the present treatment. These eight rate constants are defined in terms of the six Gibbs free energies G_1, G_2, G_3, G^*_1, G^*_2, G^*_3 for the peaks and wells through equations 8–15. The voltage dependences of the rate constants are given through the fraction of the potential traversed during a jump. The relevant electrical distances are defined in Fig. 1 under the constraint that the total potential drop occurs across the length of the channel (equations 17a and 17b).

The rates of fluctuation between the conformational states are defined through the activation energies GM and GN governing the forward and backward rates of transition between conformations o and * for the empty

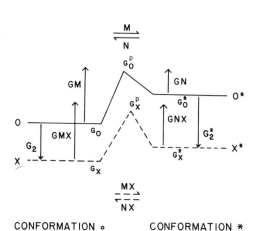

Figure 2. Energy profile for the fluctuations between conformation o and conformation *. M and MX are the forward rate constants for the transition between state o (left) and state * (right) for the empty and occupied channel, respectively, and N and NX are the backward rate constants for this transition. The states are labeled O, X, O*, X*. To avoid clutter, only M and N are indicated. The rate constants are defined in terms of the barrier energies GM, GN, GMX, GNX. The energy levels of the various states of the empty and occupied channel are labeled G_O, G_X, G_O^* and G_X^*. The dashed and solid profiles are separated by the energies G_2 and G_2^*, as defined in equation 25. This figure is the prototype for the right-hand portion of the energy diagrams to be presented in Figs. 7 and following.

channel and GMX and GNX governing the forward and backward rates of transition between conformations o and * for the occupied channel (Fig. 2 and equations 18–23). In Section 2.6 the energy levels in Fig. 2 are defined, and microscopic reversibility is imposed.

2.1. Steady-State Current

The net steady-state current is the same across any barrier; here it is calculated across the left-hand barrier of Fig. 1. The outward ionic flux (to the left) is given by the sum of outward fluxes for both the * and o conformations

$$\text{outward flux} = b_1 P(X) + b^*_1 P(X^*)$$

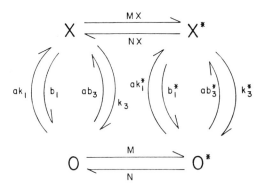

Figure 3. State diagram for the transitions among the four possible states of the channel: X, occupied channel in conformation o; O, empty channel in conformation o; X*, occupied channel in conformation *; O*, empty channel in conformation *.

which for each conformation is proportional to the sum of the probability that the channel is occupied in conformation o [$P(X)$] or conformation * [$P(X^*)$] multiplied by the exit rate constants. The inward flux across this barrier is given by

$$\text{inward flux} = a \cdot [k_1 P(0) + k^*_1 P(0^*)]$$

which is the sum for both conformations of the probability that the channel is empty multiplied by the entrance rate (which is given by the ionic activity, a, times the entrance rate constant for each state).

The net current (I) is the difference of these contributions:

$$I/Q = [b_1 P(X) + b^*_1 P(X^*)] - a \cdot [k_1 P(0) + k^*_1 P(0^*)] \quad (1)$$

where Q is the value of the elementary charge (1.602×10^{-19} coulombs).

The probability of occupancy in each state is defined by the vector **P**,

$$\mathbf{P} = |P(0), P(X), P(0^*), P(X^*)| \quad (2)$$

which is conveniently calculated using the matrix method (see Hagglund et al., 1984) as

$$\mathbf{P} = \mathbf{M}^{-1} \cdot \mathbf{B} \quad (3)$$

where **M** is a matrix and **B** is a vector

$$\mathbf{B} = |0, 0, 0, 0, 1| \quad (4)$$

2.2. Matrix

Equation 3 can be solved numerically by constructing the following matrix:

States	$P(0)$	$P(X)$	$P(0^*)$	$P(X^*)$
0	$-a(k_1+b_3)-M$	$b_1 + k_3$	N	0
X	$a(k_1+b_3)$	$-b_1-k_3-MX$	0	NX
0*	M	0	$-a(k^*_1+b^*_3)-N$	$b^*_1 + k^*_3$
X*	0	MX	$a(k^*_1+b^*_3)$	$-b^*_1-k^*_3-NX$
	1	1	1	1

$$\mathbf{M} = \text{(above matrix)} \quad (5)$$

Each row of the matrix was written down with the help of the state diagram (for details see Hagglund et al., 1984) by expressing the fluxes to the states labeled at the left from the states in the headers of the columns, corresponding to the **P** vector. For example, the first row of equation 5 stands for the transition fluxes from and toward state 0, the total sum being zero (the corresponding element in the **B** vector):

$$-[a(k_1 + b_3) + M]P(0) + (b_1 + k_3)P(X) + NP(0^*) = 0$$

Here, the first term represents all transitions leaving state 0 in Fig. 3 (two ionic fluxes going to state X and one fluctuation going to state 0*). The second term represents the transition to state 0 from state X, which occurs by way of two ionic fluxes leaving state X toward state 0. The third term represents the transition to state 0 from state 0* which is proportional to the backward rate constant N. Notice that all the diagonal elements represent the rate constants for leaving the header state and have negative signs.

The first four rows of this 5×4 matrix was linearly related to each other, as seen by adding all these rows columnwise, thus producing a zero vector. The last row is the normalization equation, which expresses the requirement that the sum of all probabilities of possible states be equal to 1. This row is used, and any of the first four rows is eliminated to produce a 4×4 matrix. Eliminating the third row gives:

States	$P(0)$	$P(X)$	$P(0^*)$	$P(X^*)$
0	$-a(k_1+b_3)-M$	$b_1 + k_3$	N	0
X	$a(k_1+b_3)$	$-b_1-k_3-MX$	0	NX
X*	0	MX	$a(k^*_1+b^*_3)$	$-b^*_1-k^*_3-NX$
	1	1	1	1

$\mathbf{M} = $ (above) (6)

Correspondingly, **B** is redefined as a four-element vector

$$\mathbf{B} = |0, 0, 0, 1| \tag{7}$$

Once the probabilities of equation 2 have been calculated from equations 3, 6, and 7, the I–V behavior is obtained through equation 1 using the following definitions of the rate constants.

2.3. Rate Constants for Ionic Translocation

The rate constants for ionic translocations and their voltage dependences are given in terms of the energy levels and distances of Fig. 1 by equations 8–17:

$$k_1 = A \exp(-G_1 - d_1 U) \tag{8}$$

$$b_1 = A \exp(G_2 - G_1 + d_2 U) \tag{9}$$

$$k_3 = A \exp(G_2 - G_3 - d_3 U) \tag{10}$$

$$b_3 = A \exp(-G_3 + d_4 U) \tag{11}$$

$$k^*_1 = A \exp(-G^*_1 - d^*_1 U) \tag{12}$$

$$b^*_1 = A \exp(G^*_2 - G^*_1 + d^*_2 U) \tag{13}$$

$$k^*_3 = A \exp(G^*_2 - G^*_3 - d^*_3 U) \tag{14}$$

$$b^*_3 = A \exp(-G^*_3 + d^*_4 U) \tag{15}$$

where A is the Eyring frequency factor (Eyring et al., 1949), assumed to be kT/h, and where the voltage U is expressed in reduced ($kT/Q = RT/F$) units, defined in terms of the potential difference (in volts) between the inner and outer solutions as:

$$U = (V_{in} - V_{out})/(kT/Q) \tag{16}$$

Recall that $kT/Q = RT/F = 0.02549$ V at 23°C.

The voltage dependences are expressed in terms of the fractional distances in the potential field, where the relations

$$d_1 + d_2 + d_3 + d_4 = 1 \tag{17a}$$

and

$$d^*_1 + d^*_2 + d^*_3 + d^*_4 = 1 \tag{17b}$$

represent the assumption that the total potential drop occurs across the distances indicated in Fig. 1.

2.4. Fluctuation Rate Constants

The rate constants for the fluctuations (M, N, MX, NX) are assumed, for simplicity, to have no voltage dependence intrinsic to the rearrangement of the channel when it fluctuates between conformation o and conformation *. They are defined in terms of the reduced energies (GM, GN, GMX, GNX) as follows:

$$M = A \exp(-GM) \quad (18)$$

$$N = A \exp(-GN) \quad (19)$$

$$OMX = A \exp(-GMX) \quad (20)$$

$$ONX = A \exp(-GNX) \quad (21)$$

where OMX and ONX are the voltage-independent portions of the fluctuation rate constants MX and NX for the occupied channel. (MX and NX can depend on voltage through equations 23a and 23b even if we assume that there is no voltage dependence intrinsic to the rearrangement of the channel when it fluctuates between conformation o and conformation *.) This is because, even with this simplifying assumption, a voltage dependence associated with the fluctuations occurs if the site occupied by an ion changes its location between the o and * conformations. In this situation, a voltage dependence in the fluctuation rate constants MX and NX for the occupied channel

$$MX = OMX \exp(-0.5AA\ U) \quad (22a)$$

$$NX = ONX \exp(0.5AA\ U) \quad (22b)$$

is introduced as a result of the displacement of a site carrying an ion through the distance AA in the potential field, assuming symmetry in the reaction path:

$$AA = (d^*_1 + d^*_2) - (d_1 + d_2) \quad (23)$$

When the site does not change its location (e.g., when $d^*_1 + d^*_2 = d_1 + d_2$), then AA, of course, equals zero, and there is no voltage dependence of MX and NX, as can be seen in equations 22a and 22b.

2.5. Microscopic Reversibility

From Figs. 1–3 one might think that there are 20 parameters defining the behavior of this system (12 rate constants and eight distances). However, through the constraints imposed in equation 17, one of the distances in each of the two conformational states is a dependent variable, thus reducing the number of free parameters by two. Microscopic reversibility reduces the number of free parameters still further. This constraint has already been imposed on the translocation rate constants through equations 8–15, in which the eight translocation rate constants are defined by six energies, reducing the number of free parameters by two more.

It is worth pausing to point out that microscopic reversibility was introduced implicitly when we used Fig. 1 to define the peak and well energies used in these equations. Microscopic reversibility was imposed when we set the energy levels to be the same on both sides of the membrane for the equilibrium condition of symmetrical solutions and zero applied potential. (The equilibrium energies are defined more generally for any potential difference across the channel in terms of their locations in the potential field given by the ds.)

Lastly, an additional relation among the fluctuation and translocation rate constants may be obtained by applying microscopic reversibility to the state diagram of Fig. 3 (Appendix A of Lauger *et al.*, 1980). This is imposed when we construct Fig. 2, as discussed in Section 2.6, and can be written

$$\exp\left[(1 + AA)\, U\right] = b_3\, b^*_1\, OMX\, N/k_3\, k^*_1\, ONX\, M \qquad (24)$$

Equation 24 reduces by one more the number of independently variable energies. This means that only 15 of the parameters of the model are actually independent (six distances and nine energies).

2.6. Relationship between the Energy Diagrams of Figs. 1 and 2

It is helpful at this point to define the energy profiles in Fig. 2 in terms of peak and well energies to obtain a description consistent with Fig. 1. This has a dual advantage: it provides a graphic and compact way in which to summarize the parameters that describe the experimental data, as seen in the inserts of later figures, and it automatically imposes the microscopic reversibility constraint, as discussed below. The free energy of the system when an ion occupies the channel (G_2 or G^*_2 of Fig. 1) relates to the energies of Fig. 2 in two ways. First, the energy levels for the empty and occupied o conformations (states 0 and X in Fig. 2) are separated by the energy G_2 defined in Fig. 1. Second, the energy levels for the empty and occupied * conformations (states 0* and X* in Fig. 2) are separated by the energy G^*_2 of Fig. 1, i.e.,

$$G_X - G_0 = G_2 \qquad (25a)$$

and

$$G^*_X - G^*_0 = G^*_2 \qquad (25b)$$

where G_0, G_X, G^*_0, and G^*_X are the energy levels for states 0, X, 0*, and X*, respectively.

G_2 fixes the relative levels of the empty and occupied channels in conformation o through equation 25a and enables all the parameters of the model to be inferred from the energy diagrams presented as inserts to later figures. G^*_2 imposes a similar constraint on the levels in conformation * through equation 25b. These two constraints link the fluctuation rate constants and those for translocation and thereby make one of these a dependent variable through the requirement that

$$-G_2 + GM - GN + G^*_2 + GNX - GMX = 0 \qquad (26)$$

which is obtained by summing the energies (indicated by the arrows) around a cycle in Fig. 2. This is an alternate way of imposing the microscopic reversibility requirement of equation 24. It rests on the thermodynamic assumption that at equilibrium there can be no energy gain (or loss) in proceeding around such a cycle.

For explicitness and future convenience, the peak energy for the empty-channel transition between states 0 and 0* in Fig. 2 is defined as G^P_0, and similarly G^P_X represents the peak energy for the occupied-channel transition between states X and X*. Then the fluctuation rate constants can be defined in terms of the following peak and well energies of Fig. 2:

$$GM = G^P_0 - G_0 \qquad (27)$$

$$GN = G^P_0 - G^*_0 \qquad (28)$$

$$GMX = G^P_X - G_X \qquad (29)$$

$$GNX = G^P_X - G^*_X \qquad (30)$$

Recall that the free energy of the system is defined as zero when the channel is unoccupied and in state 0 ($G_0 = 0$) and that the frequencies of translocations (eqs. 8–15) or fluctuations (eqs. 18–21) are inversely related to the exponential of the barrier heights:

$$\text{rate constant} = 6.21 \times 10^{12}/\exp(\text{barrier height}) \qquad (31)$$

where the rate constant is in reciprocal seconds and the barrier height is in kT units.

3. CONFRONTATION WITH EXPERIMENTAL DATA FOR THE AChR CHANNEL

The preceding theoretical equations suffice to confront experimental data. Although the rate constants can be used as adjustable parameters, here the energies and locations of the peaks and well are used. This automatically includes microscopic reversibility, and the results are easily expressed in the compact and visually accessible form of energy profiles. However, it is necessary to make two assumptions in order to express the results in the form of energy profiles. First, a particular value will be assumed for the Eyring frequency factor, A. Second, the partial molar volume of water will be assumed to be the same in the channel as in bulk solution. Neither of these assumptions affects in any way the comparison of data with theory or the values of rate constants we infer. However, for different assumptions, the energy levels in the figures would be changed by a constant amount.

The current (eq. 1) is expressed in picoamperes for comparison with experimental data (activities in moles/liter) by recasting it as

$$I = Q \cdot A \cdot 10^{12} \cdot \{[P(X)b_1/A + P(X^*)b^*_1/A] \\ - a \cdot [P(0)k_1/A + P(0^*)k^*_1/A]\} \quad (1a)$$

where Q is the elementary charge (1.602×10^{-19} coulombs) and A is assumed to equal the Eyring frequency factor $kT/h = 6.21 \times 10^{12}$ sec^{-1} at 23°C. The factor 10^{12} is the number of picoamperes per ampere. Notice that the rate constants in equation 1a have all been divided by A, which means that they correspond to the exponential portions of equations 8–15.

The current expressed in picoamperes at 23°C is therefore

$$I = 994{,}800 \cdot \{[P(X)b_1/A + P(X^*)b^*_1/A] \\ - a \cdot [P(0)k_1/A + P(0^*)k^*_1/A]\} \quad (1b)$$

Note that the rate constants are defined for dimensionless activities. Experimental activities can be converted to dimensionless mole fraction units by dividing the activity in moles/liter by 55.5 moles/liter of water. This assumes the partial molar volume of water in the channel to be the same as in bulk solution. Alternatively, to use molar concentrations, those rate constants that are multiplied by activity must be expressed in units of sec^{-1} per mole/liter.

3.1. Preliminary Exploration Using the Static 2B1S Model

Figure 4A characterizes the I–V behavior of the AChR channel over a wide range of voltages and concentrations using the experimental data of Dani and Eisenman (1984, 1985, 1986) for Cs^+, which are plotted as the jagged curves at the five indicated symmetrical concentrations of Cs^+. Voltage is defined with reference to the inner solution, and consequently outward current is defined positive and plotted upward. The smooth theoretical curves are produced by the static 2B1S model with the indicated energy profile. This profile was obtained by finding the best fit using the Levenberg–Marquardt algorithm (Appendix) to the five Cs^+ I–V relations with all six parameters (three energies and three locations) of the static 2B1S model freely variable (i.e., allowing the barriers and the well to vary their energies and locations in the potential field).

It is surprising how well such a simple model fits the data visually (Hamilton R-factor = 0.08, corresponding to a cumulative relative misfit of 8% as defined in Appendix). However, on closer scrutiny, the following shortcomings become apparent. First, the theoretical behavior shows more curvature than the experimental, and the theory has difficulty fitting the inward-going currents, especially for large negative voltages. Second, the experimentally observed conductance around zero voltage at low concentrations is much larger than that calculated theoretically, as seen from the slopes at the zero crossing for the 7 mM and 20 mM curves. This type of discrepancy could result if negative charges or dipoles increased the concentration of cations near the channel entrance (Dani and Eisenman, 1984, 1985, 1986; Eisenman et al., 1985) or if there is multiple occupancy or if multiionlike behavior is produced by an appropriately fluctuating energy profile (Ciani, 1984; Lauger, 1984), as discussed in Section 3.3.

The above deficiencies do not result from an inherent inability of the static 2B1S model to produce the appropriate shape at a given concentration. For example, Fig. 4B demonstrates that the model can provide an excellent fit at any single concentration, provided that the energy profile is allowed to change with concentration. In this particular case, the locations of site and barriers are fixed at the positions found in Fig. 4A. The energies (G_1, G_2, and G_3) are then found, again using the Levenberg–Marquardt algorithm, that give the best fit to the experimental data at each concentration. This figure demonstrates that the I–V data can be well fitted by the concentration-specific energy profiles indicated above, and Fig. 5 shows the adequacy of an alternative fit obtained previously

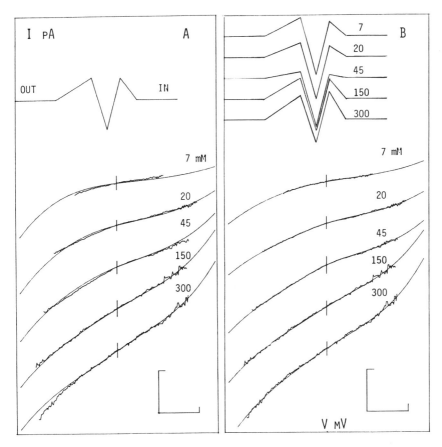

Figure 4. Comparison of experimental and theoretical $I-V$ relations for the static 2B1S model for Cs^+ in symmetrical solutions having the indicated concentrations. The jagged curves are the experimental data for the five indicated symmetrical Cs^+ concentrations (Dani and Eisenman, 1984, 1985, 1986). The smooth curves represent the theoretical expectations. Current is plotted on the Y-axis, and voltage on the X-axis, with zero indicated by the vertical bars. The horizontal scale indicates 100 mV, and the vertical scale 10 pA. The vertical scale also indicates 10 kT for the energy diagram. Part A compares theoretical curves for the static 2B1S model having the indicated energy profile with the experimentally measured $I-V$ behavior. Part B presents theoretical curves calculated for each separate concentration according to the static 2B1S model for the five concentration-specific energy profiles with fixed locations indicated above. Each energy profile corresponds to a theoretical curve, with the bottom profile corresponding to the lowest concentration and the uppermost profile to the highest concentration.

(Eisenman *et al.*, 1985) allowing the locations and energies both to vary with concentration.

To identify parameters that have little effect on the fitting, and to find acceptable ranges for the parameters to be used in the fluctuating 2B1S model, locations of site and barrier were systematically varied.

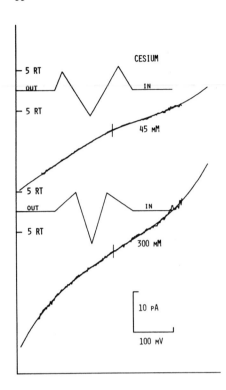

Figure 5. A demonstration that the static 2B1S model can produce the proper shapes to fit the $I-V$ data at 45 mM and 300 mM Cs^+ when the site and barriers are allowed to vary their locations and energy levels. (Reproduced, with permission, after Fig. 5 of Eisenman et al., 1985.)

Figure 6A illustrates excellent fits at each concentration when the locations of the site and barriers are fixed at the positions obtained for 45 mM Cs^+ in Fig. 5 and only the three energy levels are varied. Figures 6B and 6C extend this finding to a centrally located site for two different barrier locations. They demonstrate the adequate fits for two very different asymmetric barrier shapes (having, respectively 45% and 5% voltage dependences of the entry and exit steps). In addition, the site location can be displaced as much as 15% from the center with little deterioration of fit quality (data not shown). Clearly, in exploring the effect of fluctuations, it will be reasonable to locate the site in the middle of the field, and the fits should be quite insensitive to the precise location (and shape) of the barriers.

From the above it is apparent that the present $I-V$ data can be fit by allowing the energy profile of a static 2B1S model to change with concentration in a variety of ways, and it is conceivable (Record et al., 1978; Von Hippel and Schleich, 1969) that ionic interactions with the glycoprotein channel could lead to such concentration-dependent conformational changes. However, rather than interpreting the data of Figs. 4 and 5 as indicative of conformational changes that alter the energy profile as a

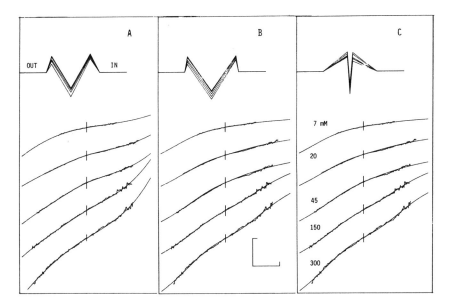

Figure 6. Comparison of theoretical curves for the static 2B1S model having the indicated energy profiles with the experimentally measured $I-V$ behavior. The scales are the same as in Fig. 4. Described in the text.

function of concentration, here we pursue explanations only in terms of fluctuating energy profiles that do not vary with concentration. This is, of course, not the only approach. Indeed, we have shown elsewhere (Eisenman *et al.*, 1985) how the static 2B1S model can be extended to allow for the site and/or the barriers to have a significant width and thereby to provide an adequate fit to the data with a concentration-independent energy profile. Section 3.2 shows how a similar improvement can be accomplished by using a fluctuating 2B1S model.

3.2. An Examination of the Fluctuating 2B1S Model

The presently available $I-V$ data for the AChR channel are not sufficiently detailed to allow a unique fit to the fluctuating 2B1S model with all 15 independent parameters allowed to be variable. To reduce the number of free parameters, some or all of the energies or locations can be fixed to gain a sense of the general behavior of the model. Fixing the six peak and well energies (G_1, G_2, G_3, G^*_1, G^*_2, G^*_3) or the six locations governing ion translocation reduces the number of free parameters from 15 to nine. Fixing the locations is a particularly attractive strategy in that the zero-voltage properties are totally independent of these. This means that such interesting zero-voltage behavior as the concentration dependence of conductance at 0 mV (which is explored in detail below) is determined solely by the energies. In this way, we circumvent the anticipated inability to determine locations uniquely for the fluctuating 2B1S model. Instead, by fixing them somewhat arbitrarily and concentrating on the energies, we at least obtain a description of the behavior of the fluctuating 2B1S model that must hold with increasing precision as zero voltage is approached.

Accordingly, the behavior of the fluctuating profile model is explored by allowing only the energy levels, but not the locations, of the barriers and site to vary. The location of the site is fixed in the middle for both the o and * conformations and the barriers at positions intermediate between those in Figs. 6B and C.† This immediately reduces the number of free parameters of the model to nine. Recall that fixing the site at the same locus in conformations * and o (which makes $AA = 0$ in eq. 22) also removes any voltage dependence arising from fluctuations in the occupied conformation.

Figure 7A shows how this constrained fluctuating 2B1S model describes the I–V behavior, and Fig. 7B represents the conductance–concentration (G–C) behavior. We confine our discussion here to the I–V data and examine the G–C behavior in Section 3.3. From the good visual agreement between the theoretical curves and experimental I–V data, it is clear that the fluctuating 2B1S model with fixed barrier locations and a centrally located site and with only the energy levels variable can adequately represent the behavior over a wide range of voltages and concentrations. The energy profiles at the top of Fig. 7A represent the values for the best fit (R − factor = 0.048) using Ken Lange's implementation of a variable-metric error minimization algorithm (Powell, 1977).

Consistent with the findings for the static 2B1S model in Section 3.1, Fig. 8A shows good fits when the barriers are symmetrical (all ds = 0.25), and Fig. 8B shows good fits when the site is moved away from a central location. Figures 9A and B show the G–C behavior corresponding to Figs. 8A and B and can be compared with Fig. 7B. These figures are discussed in the next section.

A feature shared by the fluctuation profiles found for all the particular locations of site and barriers is the asymmetry of the energy profile in each of the fluctuation states. In all cases this is found to have a larger

---

† We have explored a range of locations for the sites and barriers, and although slightly better fits were obtained (Eisenman, 1986) for a site 0.35 of the electrical distance from the outer surface, for simplicity a centrally located site is used here. For our present task of exploring the most general features of the model, the data do not seem to demand an asymmetrically located site.

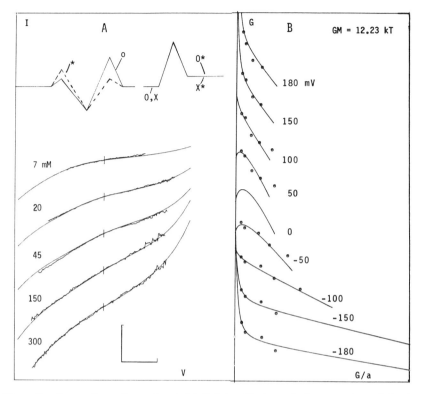

Figure 7. Comparison of experimental I–V data with theoretical expectations of the fluctuating 2B1S model with the indicated energy profiles, which correspond to the definitions of Fig. 1. The solid profile for ion translocation is for conformation o, the dashed profile for conformation *. Note that the two profiles for the fluctuations are almost indistinguishable. For reasons of compactness, in this and subsequent figures the separation between the energy levels for states O and X has not been indicated in the right-hand diagram. This separation can be read directly from the depth of the well of the translocation profile for conformation o (solid profile) in A, since it is given directly by G_2, as noted in Section 2.6. The X–X* profile is lower than the O–O* profile by this amount. The manner of plotting the I–V data is the same as in Fig. 4. The fit (Hamilton R-factor = 0.048) was obtained using the variable metric method (Appendix) with the distances d_1 to d_4 fixed at 0.141, 0.359, 0.316, 0.184 for both the o and * conformations. ONX was chosen to be the dependent variable in equation 24, and the remaining nine energies were allowed to be independently variable. The rate constants (in Hz) corresponding to this fit have the following values at zero voltage: k_1, 7.2×10^{11}; b_1, 9.7×10^8; k_3, 2.8×10^6; b_3, 2.1×10^9; k_1^*, 5.8×10^{10}; b_1^*, 8.5×10^7; k_3^*, 1×10^9; b_3^*, 6.8×10^{11}; M, 3×10^7; N, 5×10^8; MX, 4.3×10^7; NX, 7.8×10^8. Part B plots the theoretical expectations (curves) for chord conductance (G) versus specific conductance (G/a) for the indicated voltages (the 0 mV curve is the average of the virtually identical curves computed for $+0.2$ mV and -0.2 mV). For comparison, the points represent experimental values computed from the I–V data as described in Section 3.4. For A, the horizontal scale denotes 100 mV and the vertical scale 10 pA. For B the horizontal scale indicates 2000 pS/M and the vertical scale 50 pS.

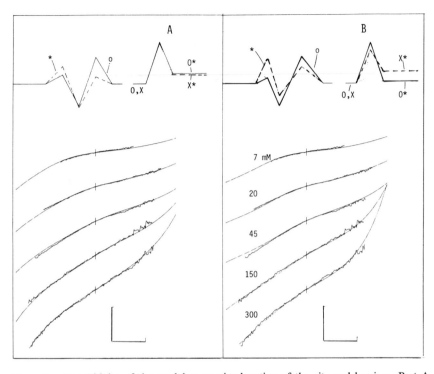

Figure 8. Insensitivity of the model to precise location of the site and barriers. Part A demonstrates that a model with a centrally located site and symmetrical barriers can represent the experimental data nearly as well (Hamilton R-factor = 0.063) as the model with a centrally located site and asymmetric barriers of Fig. 7. The ds for both conformations o and * were $d_1 = d_2 = d_3 = d_4 = 0.25$. Part B demonstrates the adequacy of fit (Hamilton R-factor = 0.051) for a model with an asymmetrically located site. The ds for both conformations o and * were $d_1 = d_2 = 0.175$; $d_3 = d_4 = 0.325$. Here the site lies 35% of the way from the outside, a position that was previously found to correspond to a sharp minimum of the standard deviation in a more general fit (12 free parameters) in which the site was allowed to vary its location under the constraint that $d_1 + d_2 = d_1^* + d_2^*$ and allowing the locations of the barriers to vary independently, and differently, for the two conformations.

barrier facing one side and a smaller barrier facing the other in one conformation. In the other conformation, the situation reverses. This is true whether the site is in the middle of the channel or located nearer one end (Fig. 8A vs. 8B) or whether the barriers are symmetrical or asymmetric (Fig. 8A vs. Fig. 7A). Such an asymmetric energy profile, which is different between conformations o and *, appears to be a particular requirement for a maximum to occur in conductance, as shown by Lauger *et al.* (1980), who found it necessary for at least one of their conformations to be strongly asymmetric, and by Ciani (1984), who showed that for coupling to occur between the fluxes of two ions, the barrier heights must be

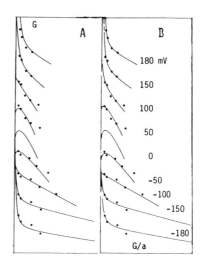

Figure 9. Conductance versus conductance/activity plots corresponding to Fig. 8.

asymmetric (and relatively different) in the two conformations. We speculate in Section 4.2 on a further implication of these particular fluctuations in profile and their frequency of occurrence.

3.3. A Complex, Multiionlike, Nonmonotonicity of Conductance is a Function of Concentration Is Expected Theoretically

One of the most interesting aspects of the behavior of fluctuating profile models is their ability to produce a multiionlike behavior in a one-ion channel. This manifests itself as flux coupling (Ciani, 1984) or as a nonmonotonic dependence of conductance on concentration (Lauger, 1984; Lauger et al., 1980). In marked contrast to this is the behavior expected for any one-ion model with a static profile, which must possess a monotonic Michaelis–Menten type of behavior for the conductance measured at a given (fixed) voltage. This can be easily proven to be true regardless of the number of barriers or the complexity of profile (see Lauger, 1973). On an Eadie–Hofstee (E–H) type plot, where chord conductance (G) at a given voltage is plotted against specific conductance (G/a), a monotonic behavior appears as a simple straight line (Eisenman et al., 1978),† and nonmonotonic behavior, on the other hand, appears as a curve, which may be concave or convex.

The markedly curvilinear behavior for the theoretical expectations in Figs. 7B, 9A, and 9B indicates multiionlike behavior for the single-

---

† This is strictly true only in the absence of surface potential effects which can distort the concentrations at the mouth of the channel from those in the bulk solution.

occupancy fluctuating 2B1S model. This behavior is seen when the fluctuation frequencies are in a range near the ion translocation rate, as is discussed in Section 3.6. Note the general similarity among the E–H plots of Figs. 7B, 9A, and 9B for the various site and barrier locations.† In all three cases, the E–H plots change shape from voltage to voltage, exhibiting a conductance maximum around zero voltage and a nonlinear "foot" at high voltages.‡ This kind of behavior, with its unprecedented voltage sensitivity, is quite different from what would be expected from a surface potential effect, which should produce a foot but no maximum at any voltage. Such behavior, therefore, provides an experimentally testable feature of the fluctuating 2B1S model, enabling it to be distinguished from static one-ion models.

The maximum predicted at high concentrations near zero voltage in Figs. 7 and 9 has not yet been reported for the AChR channel, although a foot at low concentrations like that predicted at sufficiently high positive and negative voltages has been observed (Dani and Eisenman, 1984, 1985). Such a maximum should certainly be searched for experimentally, but its absence would not exclude the fluctuating 2B1S model since a good fit to the conductance data presented in Fig. 15 has almost no maximum. The really crucial test of this model is the exquisite sensitivity to voltage near $V = 0$ of the theoretically expected changes in G versus G/a shape.

3.4. Experimentally Observed Chord Conductances at Various Voltages as a Function of Concentration

Although our I–V experiments were not designed for measuring conductances with the precision that may be required for testing the above theoretical expectations, particularly around zero voltage, it is still in-

---

† Note that in the case of the centrally located site (Fig. 7 and Fig. 8A), the depth of the site is approximately the same in both conformations regardless of the location of the barriers. Also, note that the energy profiles for the fluctuations are very similar for both the empty and occupied states. The case for the asymmetrically located site of Fig. 8B differs from the case of the central site in the following details. First, in contrast to the preceding cases in which the depth of the well was the same in both o and * conformations, there is a decrease in binding affinity between conformation o and conformation *, indicated by the higher energy well for the dashed energy profile for ion translocation. From the findings for the centrally located site it is not possible to identify the o and * conformations with any particular physical states (e.g., "polarized" and "unpolarized"). However, from the stronger affinity of the site in conformation o than in conformation * in Fig. 8B, one could speculate that conformation o is the polarized conformation for the asymmetrically located site.

‡ Apparently from the findings presented here, a one-site model can exhibit changes from concave to convex behavior with changing voltage but apparently cannot exhibit an inflection at any given voltage.

structive to compute chord conductances (G) at various voltages from our present $I-V$ data to see to what extent the concentration dependence of the conductances is in accord with theoretical expectations. Figure 10 presents experimental $I-V$ data as jagged curves, and Fig. 11 presents the jagged chord conductances corresponding to these data. The experimental chord conductances are clearly inadequate for potentials less than 50 mV on either side of 0 since experimental scatter causes them to "blow up" around zero voltage. We tried a number of filtering procedures to circumvent this problem. The most satisfactory is fitting a sixth-degree polynomial with a first coefficient of 0 to the data. Such a representation is shown by the smooth curves in Figs. 10 and 11 for the currents and conductances. The polynomial is seen to fit the $I-V$ data quite well, neither losing any important detail nor apparently producing any spurious inflections. Thus, a polynomial representation is quite satisfactory for representing existing current data, although it should not be trusted for extrapolation beyond the range of measurements. The corresponding polynomially generated "experimental" chord conductances at the selected voltages, indicated by dots in Fig. 11, have been used for the experimental data points in all plots of G versus G/a (Figs. 7, 9, etc.).

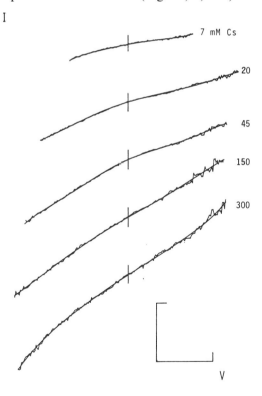

Figure 10. Polynomial fit to the $I-V$ data. A sixth-degree polynomial (shown by the smooth curves) has been fitted to the experimental $I-V$ data (jagged curves). The values of the polynomial coefficients are as follows. For 7 mM Cs^+: 0, 0.018185, -5.1559×10^{-5}, -3.6301×10^{-7}, -5.5769×10^{-10}, 5.0109×10^{-11}, 8.4567×10^{-15}. For 20 mM Cs^+: 0, 0.029512, -1.0297×10^{-4}, 3.1398×10^{-7}, 2.4845×10^{-9}, -5.2836×10^{-12}, -2.0558×10^{-14}. For 45 mM Cs^+: 0, 0.042429, -1.7338×10^{-4}, 2.7955×10^{-7}, 7.0617×10^{-9}, -3.926×10^{-12}, -1.1786×10^{-13}. For 150 mM Cs^+: 0, 0.055623, -2.8101×10^{-5}, 4.6889×10^{-7}, 7.1876×10^{-10}, -8.3102×10^{-12}, -1.9079×10^{-14}. For 300 mM Cs^+: 0, 0.063437, -3.4177×10^{-5}, 1.3494×10^{-7}, 5.5423×10^{-10}, 8.4384×10^{-12}, -5.4653×10^{-16}.

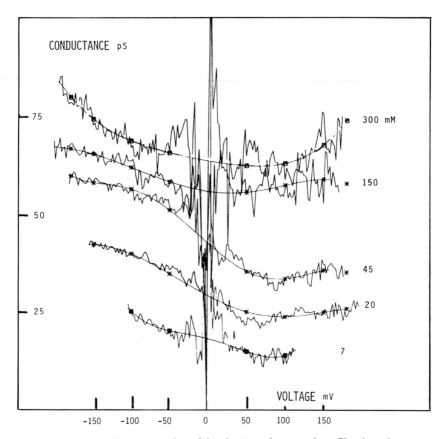

Figure 11. Polynomial representation of the chord conductance data. Chord conductances have been calculated using the polynomials of Fig. 10 and are plotted as smooth curves for comparison with the experimental data (jagged curves) calculated from the *I–V* data. The values at the voltages indicated by points are used as experimental data for comparison with theory in Figs. 7, 9, etc.

We consider that such chord conductances are the most reliable that we can extract from our present data. Notice that we will not plot any experimental chord conductances for voltages smaller in amplitude than 50 mV because we do not presently trust the polynomial filtering procedure for smaller voltages.

3.5. Comparison of Theoretical and Experimental Chord Conductances

The experimental conductances (data points) are compared with the theoretical expectations (smooth curves) for various locations of site and barriers in Figs. 7 and 9. They are also compared for different ranges of fluctuation rate constants in Figs. 14 and 15, presented in Section 3.6. The best visual fit to the conductance data is found for a fluctuation rate corresponding to $GM = 12.9\,kT$ in Fig. 14, which is replotted for emphasis at the right of Fig. 15. Clearly, the observed conductance data are compatible with the model.

For present purposes it should be pointed out that we reach the same general conclusions about the effect of fluctuations regardless of whether we consider Fig. 7 or Fig. 15 to be more correct. However, notice how extremely sensitive the G–C behavior is to the fluctuation rate constant M, which is related to the barrier height GM through equation 31. The two figures differ chiefly in that the value of GM of 12.23 kT in Fig. 7 implies that $M = 3 \times 10^7$ Hz, whereas the value of GM of 12.9 kT in Fig. 15 implies that $M = 1.6 \times 10^7$ Hz.

3.6. The Frequency Range over Which Coupling Occurs between Channel Fluctuations and Ion Translocations

Only in a particular range of frequencies will fluctuations couple to ion transport to produce multiionlike behavior. For example, if the fluctuations are too rapid relative to ion migration, the system will behave like a classical one-ion channel with an energy profile that is the average of the two rapidly fluctuating conformations (Lauger et al., 1980). Also, since an essential condition for coupling is that transitions between the two conformations occur in both the empty and occupied states (Lauger, 1984), very slow fluctuations do not result in coupling. In general, it is to be expected that fluctuations of particular interest will be those occurring with a rate comparable to that of ion translocation. For example, it can be seen in the matrix (eq. 5) that both fluctuation and translocation rates contribute significantly to the individual elements on the diagonal when the fluctuation rate constant MX is comparable to the sum of the exit rate constants for the ions ($b_1 + k_3$) or when the fluctuation rate constant M is comparable to the ionic activity times the sum of the entrance rate constants for the ions [$A(k_1 + b_3)$].

Accordingly, we depart from the previous strategy of letting a sophisticated algorithm find the best fit to a given model and here examine the range of behavior encountered when we deliberately force the fluctuation rates into ranges where the best fit is not obtained. By so doing, we define a range of frequencies in which the channel protein must fluctuate if, with the present model, it is to produce the experimentally observed behavior.

Figures 12–16 illustrate the effect of varying the fluctuation rate constants by setting the activation energy GM for the empty channel for the

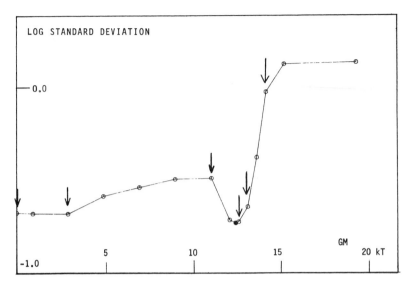

Figure 12. Adequacy of fit as a function of the energy barrier *GM*. This figure plots the logarithm of the standard deviation computed for the best fits using the variable-metric algorithm as a function of the energy barrier *GM* for the forward rate constant of the fluctuation between conformations o and * of the empty channel. A series of fits is shown, which were carried out for each of the indicated values of *GM* with the distances fixed at the values of Fig. 7 and with the eight remaining energies variable. Values of the other activation energies for the fluctuations (*GN*, *GMX*, *GNX*) as well as of the translocation rate constants for the various values of *GM* are given in Fig. 16. Notice that an optimum fit occurs in the narrow region around 12.3 *kT* [the best fit at 12.23 *kT* (filled circle) corresponds to that of Fig. 7]. Values of *GM* only slightly larger than this lead to very poor fits; and for values between 5 and 11 *kT*, the fit is also poor. However, for values smaller than 3.5 *kT*, the fit becomes better, although not as good as at the optimum. The arrows indicate values of *GM* for which the *I–V* and *G–C* behaviors will be examined in Figs. 13 and 14.

fluctuation from the o to the * conformation at various fixed values and then finding the values for the eight remaining energies that give the best fit to the experimental data for a given value of *GM*. We first summarize what is contained in these figures and then examine the individual figures in detail. Figure 12 plots the adequacy of fit (indicated by the log of the standard deviation for the *I–V* data) as a function of *GM* ranging from 0.0 to 18.9 *kT*. Figures 13 and 14 present the corresponding *I–V* and *G–C* behavior, respectively, for representative values of *GM*. Figure 15 presents the *I–V* and *G* versus *G/a* behavior for what appears to be the best set of rate constants for the present model. This figure is discussed further in Section 4.1. Figure 16 summarizes graphically the variations in all the other rate constants that occur as *GM* is varied.

It can be seen from the size of the standard deviation in Fig. 12 that the best fit to the *I–V* data (filled circle) occurs for *GM* = 12.23 *kT*

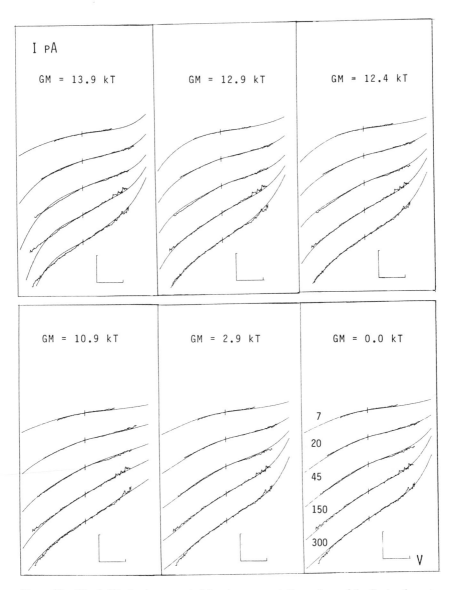

Figure 13. The *I–V* behavior expected for six representative values of the fluctuation rate *M* from state O to O*. The fits correspond to the values of *GM* indicated by arrows in Fig. 12. The best fit to the *I–V* data in this figure is seen at *GM* = 12.4 kT.

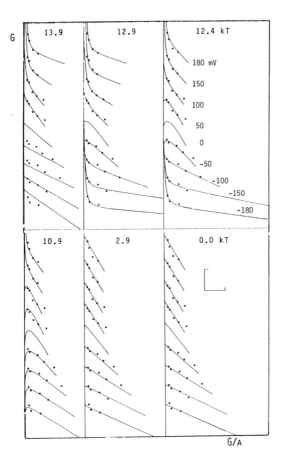

Figure 14. Eadie–Hofstee (or Scatchard)-like plot of chord conductance (G) versus specific conductance (G/a) to illustrate the complex nonmonotonic behavior characteristic of the fluctuating barrier model. The fits correspond to the values of GM indicated by arrows in Fig. 12. The scale indicates $Y = 50$ pS and $X = 2000$ pS/M. The best fit to the present conductance data appears to occur at $GM = 12.9$ kT (corresponding to $M = 1.55 \times 10^7$ Hz).

(corresponding to 3×10^7 Hz), which is the fit that was presented in Fig. 7.† The fit deteriorates rapidly for values only 1.5 kT different from this. This is seen more clearly in Fig. 13, where I–V data and theoretical curves are compared for the six representative values of GM indicated by the arrows in Fig. 12. The upper subfigures are for values larger than the optimum, and the lower subfigures are for values smaller than the optimum. Note in Fig. 12 that the fit becomes much worse for slowing (in-

---

† The best fit to the I–V data is not necessarily the best fit to the G–C behavior. This is because the G–C fit assigns more weight to low-voltage data, as discussed in the Appendix. Thus, the best fit to the I–V data, which occurs for the parameters of Fig. 7, is a poorer representation of the G–C data, which are best described by Fig. 15 despite the poorer statistical fit of the parameters of this figure, as seen by the trends in Fig. 12 (the standard deviation for the parameters of Fig. 7 correspond to the filled circle in Fig. 12, whereas those of Fig. 15 correspond to the second open circle to the right of the filled circle).

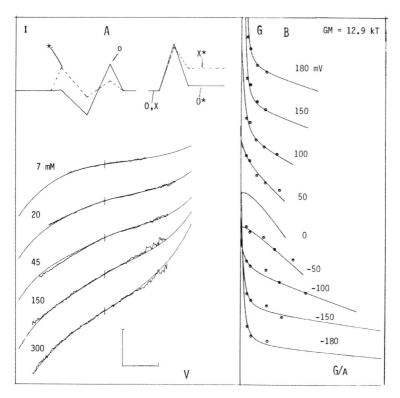

Figure 15. The $I-V$ and Eadie–Hofstee plots for $GM = 12.9\ kT$, plotted in the same manner as Fig. 7. The fluctuation rate constants are $M = 1.55 \times 10^7$, $N = 6.2 \times 10^7$, $MX = 3.75 \times 10^7$, $NX = 1.83 \times 10^{10}$ Hz.

creases in GM) than for speeding up (decreases in GM) the fluctuation rate from its optimum. Indeed, when the fluctuation rate is at its maximum ($GM = 0.0\ kT$), the fits on the $I-V$ plots (see Fig. 13) are only slightly worse than the optimal. In fact, at high fluctuation frequencies, the differences in theoretically expected behavior between $GM = 12.4\ kT$ and $GM = 0.0\ kT$ only show up clearly when the concentration dependence of the conductance is plotted, as is done in Fig. 14.

This figure illustrates the sensitivity of the nonmonotonic behavior in the theoretical plots to small variations in the fluctuation rate. Examining the effect of slowing the fluctuation rate from its near optimum value of $GM = 12.4\ kT$ in the upper part, we note the following. First, at $GM = 12.4\ kT$ a maximum is clearly shown in the theoretical curves for $+50$ mV, 0 mV, and -50 mV. However, on increasing GM by only 0.5 kT to 12.9 kT, the maximum disappears. When GM is increased 1 kT more (to 13.9 kT), there is no longer any sign even of the saturation, and for

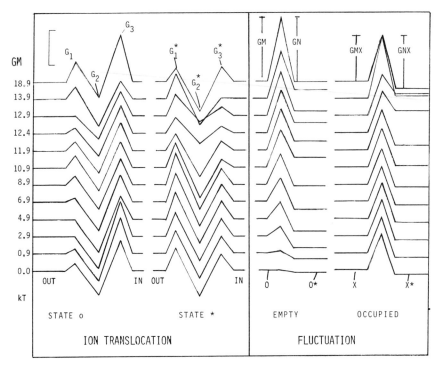

Figure 16. Energy diagrams summarizing the rate constants for the best fits to the data for the *GM* values indicated at the left. See Figs. 1 and 2 for definitions. For simplicity of presentation, the displacement (G_2) between levels X and O for the occupied and empty channels is not shown.

negative voltages a monotonic (straight line) behavior is expected. The fit is so poor at 13.9 kT and higher, particularly at negative voltages, that we can confidently exclude fluctuation rates slower than 5.7×10^6 Hz for the present model.

It can be seen from the lower portion of Fig. 14 that as the fluctuation rate of the empty channel becomes faster than optimal, the maximum becomes more prominent, and the foot is lost. For $GM = 10.9$ kT a maximum is expected for all voltages more negative than +100 mV, and the foot is seen only above +150 mV. It should be noted that for sufficiently fast fluctuations, the fit to straight-line Eadie–Hofstee plots in Fig. 14 becomes quite acceptable (see curves for *GM* at 0 and 2.9 kT).† How-

---

† It is of interest that, even in this situation, the rate constants for fluctuations of the occupied state are still comparable to those of ion translocation, as can be seen from the comparable heights of the barriers in Fig. 16 (recall eq. 31). Note that a rate constant in the range 10^8–10^{10} Hz (corresponding to the shallow maximum in Fig. 12) can probably be ruled out by present conductance data.

ever, the curvilinear E–H plots for GM ranging between 12.4 kT and 12.9 kT are a better representation of the data than the linear E–H plots for $GM = 0.0\ kT$.†

Our primary purpose here has been to examine what consequences fluctuations in the energy profile might have on the general behavior of a channel rather than to try to interpret such behavior as if it were really occurring. If channel fluctuations do have a major influence on transport, and the present model is sufficient, then the present data may in fact provide an initial hint of the actuality of fluctuations in the 10^7-Hz frequency range.‡

4. DISCUSSION

4.1. Physical Implications of the Best-Fit Parameters

Figure 12 indicates that the best fits to the model with the I–V data weighted as we have (see Appendix) occur around 12.23 kT. However, the best visual fit to the conductance data in Fig. 14 occurs for a value of GM of 12.9 kT, which is 0.67 kT larger. This fit is presented in Fig. 15 for both the I versus V (left) and G versus G/a (right) behaviors. Comparing Figs. 7 and 15, it is seen that the I–Vs are better fit in Fig. 7, whereas the conductances are better fit in Fig. 15. This difference reflects the fact that the parameters of Fig. 15 represent the lower-voltage data better than those of the fit of Fig. 7, and vice versa (see Appendix).

The best fits of the present model occur with GM in the range 12.23–12.9

---

† Since the behavior around zero voltage is completely independent of the locations of the site and barriers, as noted in Section 3.2, the fluctuation frequencies that can produce the multiionlike saturation (or maximum) in the zero-voltage E–H plots of Figs. 7 and 15 are of particular interest. Note that the usual saturation of conductance in a plot of G versus a appears as straight-line behavior on an E–H plot. A saturation on an E–H plot corresponds to a nonideality of behavior that is formally equivalent to an additional apparent binding constant (Eisenman, 1986).

‡ So far we have discussed the effect of varying the fluctuation frequency solely in terms of the rate constant M for the fluctuation of the empty channel from the o to the * conformation. It is also necessary to scrutinize what values were found for the other rate constants for the variations of M examined above. This can be done with the aid of Fig. 16, which summarizes the values of all the rate constants computed for the indicated values of GM in the compact form of energy profile plots. We note that associated with the optimum in the fluctuation frequency M around 1.6×10^7 to 3×10^7 Hz are rate constants for N in the range 6.2×10^7 Hz to 5×10^8 Hz, for MX from 3.8×10^7 Hz to 4.3×10^7 Hz, and for NX from 7.8×10^8 Hz to 1.8×10^{10} Hz. A parallel exploration to that of Figs. 12–16 was carried out by varying the fluctuation rate MX of the occupied, instead of the empty, channel (Eisenman, 1986) and showed that acceptable fits to the I–V data could be obtained only over a relatively narrow range of GMX between 10 and 12.2 kT, consistent with the energy profiles in Fig. 16.

kT (corresponding to a frequency of fluctuations between 1.6×10^7 Hz and 3×10^7 Hz). To decide between these alternatives will require more extensive measurements at low concentrations and high voltages as well as more precise measurements at high concentrations and low voltages.† Such measurements will test whether or not fluctuations of the energy profile of the AChR channel are really suggested by its steady-state electrical behavior. If they are, the proposed experiments will help define more precisely in what frequency range such fluctuations must occur.

It must be noted that an alternative to the one-ion fluctuating 2B1S model might be a fixed-barrier model occupiable by two (or more) ions or a model that introduces charge effects because of the vestibules (J. Dani, unpublished data). The discrimination between these possibilities may not be possible from steady-state electrical measurements of the type presented here but might be accomplished by other methods. For example, definition of locations of bound ions using NMR or X-ray techniques would be expected to show two occupiable loci for a fixed-profile two-ion channel but only one locus for the fluctuating-profile one-ion case.

The fluctuation rate constants M, N, MX, NX corresponding to the energy profiles in Figs. 7 and 15 all lie in the range 10^7 to 10^9 Hz. Fluctuations at such frequencies occur commonly in proteins: helix–coil transitions typically have rate constants around 10^9 Hz (Karplus and McCammon, 1983), and "hinge-bending" relative motion of different globular regions occurs in the frequency range 10^7–10^{11} Hz (McCammon, 1984). A particularly tempting speculation would be to identify the two conformations of the model with two different tilts of the five amphiphilic putative channel-forming subunits suggested by Guy (1984) and by Young *et al.* (1985). The units lining the channel are postulated to be in an arrangement like that proposed by Unwin and Zamphighi (1980) for the gap junction channel. A rocking between two different tilts is attractive for the following two reasons: because it might occur at the appropriate frequency since the concerted motions of relatively large regions of the protein that would be involved seem analogous to "hinge-bending," and because, for an appropriate pivot point, such a rocking would be compatible with the observed shifts in translocation energy profiles common to all the different situations examined here (see Figs. 7, 8, 15, and 16) where in one conformation the larger barrier faces the inside solution whereas in the other conformation the larger barrier faces the outside solution. This is discussed further in Section 4.2.

Although there is as yet no direct evidence that AChR channels

---

† Such measurements might be carried out (E. Neher, personal communication) using a lock-in amplifier and applying small sinusoidal signals at kiloherz frequencies (e.g., 10 mV at 1000 Hz) around the desired voltage.

fluctuate at 10^7 Hz, and the arguments presented here can only be taken as being consistent with this possibility, it is worth noting that a high-frequency component of excess channel noise has already been reported in the megahertz (or higher) region (Sigworth, 1982, 1985), although it is clear that the better-characterized low-frequency (1000 Hz) fluctuations observed in the same studies are, on present evidence, too slow to correspond to any of the processes inferred here from the steady-state electrical behavior. In principle, fluctuations between two conducting states at 10^7 Hz might also be revealed in the skewing of the amplitude histograms for appropriately filtered channels (Yellen, 1984).

4.2. Other Implications of the Asymmetric Energy Profiles

The asymmetric profiles inferred for ionic translocation in Section 3.2 are of interest because for all three cases of Figs. 7 and 8 they are asymmetric and different in the o and * conformations, with the inner barrier being larger than the outer in the conformation o and vice versa in conformation *. They suggest a channel whose structure fluctuates so that it has a wider mouth facing the inside in one conformation and a wider mouth facing the outside in the other conformation.

4.3. General Conclusions

Although we cannot make any claims to uniqueness with so many free parameters in the model, we believe that the particular cases studied have uncovered a general class of behavior that should be common to all two-state fluctuating profile models and that appears not to depend on the particular simplifying assumptions used here. We therefore hope the present results will help encourage further experimental and theoretical studies.

APPENDIX

A.1. Error Minimization Algorithms

We used several nonlinear least-squares algorithms on an IBM 3033 or a VAX 11/730 computer to minimize the sum square error between theory and experiment. The most powerful of these algorithms was a variable metric method (Powell, 1977) implemented on the VAX by Ken Lange, which converged better than Chris Clausen's APL implementation of the Levenberg–Marquardt algorithm that we have previously used with the IBM 3033 (Eisenman and Sandblom, 1983; Eisenman *et al.*, 1985).

Lange's algorithm also enabled us to bound the parameters conveniently and was used for all the fits to the fluctuating 2B1S model presented here. A minor limitation with this algorithm was its restriction to 12 as the maximum number of independent variables. A valuable feature of all the algorithms used is their ability to display the correlation matrix for the parameters to indicate the degree to which they are correlated (and hence unseparable if the correlation is too high).

So as not to have an overwhelming number of data points in the computations, we reduced the raw data (originally gathered at 2-mV intervals) by roughly a factor of 10. This was done by first removing high-frequency jitter by five iterations of a Hamming filter. The filtered data were then shortened by taking the values only at 20-mV intervals except for the 7 mM data (where the interval was 4 mV). All experimental data presented in the figures for comparison with theory are the full raw data at 2-mV intervals.

A.2. Weighting the Data

We tried various procedures to give different weights to the data, ultimately settling, at the suggestion of Avi Ring, for weighting the data inversely as the absolute value of the current for all voltages whose absolute value exceeded 50 mV (and at the weighting value for the current at 50 mV for all smaller Vs). This procedure counteracts the undue weight that would otherwise be given to the measurements at large voltages and high concentrations. An alternative procedure would be to compare the theoretically expected and experimentally observed conductances, as has been done in the case of gramicidin (Eisenman and Sandblom, 1983). Comparison of Figs. 7 and 15 indicates that the parameters that yield the best fit visually for the $I-V$ representation are not identical to those that yield the most satisfactory representation for the conductance behavior as a function of concentration. This is because curve fits to the $I-V$ data give greater weight to the higher current data at high voltages, whereas curve fits for G versus G/a give equal weight to the data at all voltages. We did not do this because we did not consider the conductances to be our primary data. Such a procedure would, of course, lead to a better agreement for $G-C$ than for $I-V$ behavior.

A.3. The Hamilton R-Factor

The Hamilton R-factor (see pages 157–160 of Hamilton, 1964) measures the cumulative square difference between theoretical and experimental values, expressing this in fractional form as the relative misfit between the data and the theoretical expectations (an R-factor of 0.05

signifies a 5% misfit). The shortened data were used for computing R-factors. When R-factors are given in the text for curve fits based on the variable metric algorithm, they are based on the value calculated for the best-fit parameters with this algorithm after a single iteration of the Clausen algorithm with only one parameter variable. It should be noted that the R-factor also depends on the weights assigned to the data, and all R-factors cited in the text were calculated for currents weighted equally at all voltages but with the lower-concentration data weighted more heavily (relative weights being 5:2:2:1:1 for 7 mM, 20 mM, 45 mM, 150 mM, and 300 mM Cs, respectively). This procedure gives approximately equal weight to the shape at each concentration.

ACKNOWLEDGMENTS. This work has been supported by the NSF (PCM 81-09702 and BNS 84-11033), the USPHS (GM 24749), and a USPHS (NIRA) award to John Dani. We acknowledge valuable scientific discussions with Peter Lauger, Sergio Ciani, and Dick Horn and thank Ken Lange and Chris Clausen for providing their algorithms for nonlinear regression analysis. We thank Chris Clausen and Avi Ring for their time and help in the FORTRAN reprogramming for the VAX 11/730 of functions originally written in APL for the IBM 3033 and acknowledge the critical and helpful comments of Sergio Ciani and Avi Ring on the manuscript.

REFERENCES

Adams, P. R., 1979, A completely symmetrical barrier model for endplate channels, *Biophys. J.* **25**:70a.
Ciani, S., 1984, Coupling between fluxes in one-particle pores with fluctuating energy profiles, *Biophys. J.* **46**:249–252.
Dani, J. A., and Eisenman, G., 1986, Electrical behavior of the acetylcholine receptor channel for mono- and divalent cations, *J. Gen. Physiol.* (submitted).
Dani, J. A., and Eisenman, G., 1985, Acetylcholine receptor channel permeability properties for mono- and divalent cations, *Biophys. J.* **47**:43a.
Dani, J. A., and Eisenman, G., 1986, Electrical behavior of the acetylcholine receptor chanel for mono- and divalent cations, *J. Gen. Physiol.* (submitted).
Eisenman, G., 1986, Electrical signs of rapid fluctuations in the energy profile of an open channel, in *Proceedings of International Symposium on Ion Transport through Membranes. Nagoya, Japan, 1985,* Academic Press, Tokyo.
Eisenman, G., and Dani, J. A., 1985, A fluctuating 2 barrier 1 site model compared with electrical data from the acetylcholine receptor channel, in *Water and Ions in Biological Systems,* Proceedings of the third International Conference. (A. Pullman, V. Vasilescu, and L. Packer, eds.), Union of Societies for Medical Sciences, pp 437–450. *Bucharest, Romania.*
Eisenman, G., and Horn, R., 1983, Ionic selectivity revisited: The role of kinetic and equilibrium processes in ion permeation through channels, *J. Membr. Biol.* **76**:197–225.
Eisenman, G., and Sandblom, J., 1983, Energy barriers in ionic channels: Data for gramicidin A interpreted using a single-file (3B4S'') model having 3 barriers separating 4 sites, in

Physical Chemistry of Transmembrane Ion Motions (G. Spach, ed.), Elsevier, Amsterdam, pp. 329–348.

Eisenman, G., Sandblom, J., and Neher, E., 1978, Interactions in cation permeation through the gramicidin channel Cs, Rb, K, Na, Li, Tl, H, and effects of anion binding, *Biophys. J.* **22**:307–340.

Eisenman, G., Dani, J. A., and Sandblom, J., 1985, Recent studies on the energy profiles underlying permeation and ion selectivity of the gramicidin and acetylcholine receptor channels, in: *Ion Measurement in Physiology and Medicine* (M. Kessler, D. K. Harrison, and J. Hoper, eds.), pp. 54–66, Springer-Verlag, New York.

Eyring, H., Lumry, R., and Woodbury, J. W., 1949, *Rec. Chem. Prog.* **10**:100–114.

Guy, H. R., 1984, A structural model of the acetylcholine receptor channel based on partition energy and helix packing calculations, *Biophys. J.* **45**:249–261.

Hagglund, J. V., Eisenman, G., and Sandblom, J. P., 1984, Single-salt behavior of a symmetrical 4-site channel with barriers at its middle and ends, *Bull. Math. Biol.* **46**:41–80.

Hamilton, W. C., 1964, *Statistics in Physical Science*, Ronald Press, New York.

Hille, B., and Schwarz, W., 1978, Potassium channels as multi-ion single-file pores, *J. Gen. Physiol.* **72**:409–442.

Horn, R., and Brodwick, M., 1980, Acetylcholine-induced current in perfused rat myoballs, *J. Gen. Physiol.* **75**:297–321.

Karplus, M., and McCammon, J. A., 1983, Dynamics of proteins: Elements and function, *Annu. Rev. Biochem.* **53**:263–300.

Lauger, P., 1973, Ion transport through pores: A rate-theory analysis, *Biochim. Biophys. Acta* **311**:423–441.

Lauger, P., 1984, Channels with multiple conformational states: Interrelations with carriers and pumps, *Curr. Top. Membr. Transport* **21**:309–326.

Lauger, P., Stephan, W., and Frehland, E., 1980, Fluctuations of barrier structure in ionic channels, *Biochim. Biophys. Acta* **602**:167–180.

Lewis, C. A., and Stevens, C. F., 1979, Mechanism of ion permeation through channels in a postsynaptic membrane, in *Membrane Transport Processes*, Vol. 3 (C. F. Stevens and R. W. Tsien, eds.), Raven Press, New York, pp. 133–151.

Marchais, D., and Marty, A., 1979, Interaction of permeant ions with channels activated by acetylcholine in *Aplysia* neurones, *J. Physiol. (Lond.)* **297**:9–45.

McCammon, J. A., 1984, Protein dynamics, *Rep. Prog. Phys.* **47**:1–46.

McCammon, J. A., and Karplus, M., 1983, The dynamic picture of protein structure, *Acc. Chem. Res.* **16**:187–193.

Powell, M. J. D., 1977, A fast algorithm for nonlinearity constrained optimization calculations, in: *Numerical Analysis, Lecture Notes in Mathematics No. 630, Dundee 1977* (G. A. Watson, ed.), Springer-Verlag, Berlin.

Record, M. T., Jr., Anderson, C. F., and Lohman, T. M., 1978, Thermodynamic analysis of ion effects on the binding and conformational equilibria of proteins and nucleic acids: The roles of ion association or release, screening, and ion effects on water activity, *Q. Rev. Biophys.* **11**:103–178.

Sandblom, J., Eisenman, G., and Hagglund, J., 1983, Multioccupancy models for single filing ionic channels: Theoretical behavior of a four-site channel with three barriers separating the sites, *J. Membr. Biol.* **71**:61–78.

Sigworth, F. J., 1982, Fluctuations in the current through open ACh-receptor channels, *Biophys. J.* **37**:309a.

Sigworth, F. J., 1985, Open channel noise. I. Noise in acetylcholine receptor currents suggests conformational fluctuations, *Biophys. J.* **47**:709–720.

Unwin, P. N. T., and Zamphighi, G., 1980, Structure of the junction between communicating cells, *Nature* **283**:545–549.

Urban, B. W., Hladky, S. B., and Haydon, D. A., 1980, Ion movements in gramicidin pores. An example of single-file transport, *Biochim. Biophys. Acta* **602**:331–354.

Von Hippel, P. H., and Schleich, T., 1969, The effects of neutral salts on the structure and conformational stability of macromolecules in solution, in *Biological Macromolecules*. Vol. 2: *Structure and Stability of Biological Macromolecules* (S. N. Timasheff and G. Fasman, eds.), Marcel Dekker, New York, pp. 417–574.

Yellen, G., 1984, Ionic permeation and blockade in Ca^{2+}-activated K^+ channels of bovine chromaffin cells, *J. Gen. Physiol.* **84**:157–186.

Young, E. F., Ralston, E., Blake, J., Ramachandran, J., Hall, Z. W., and Stroud, R. M., 1985, Topological mapping of acetylcholine receptor: Evidence for a model with five transmembrane segments and a cytoplasmic COOH-terminal peptide, *Proc. Natl. Acad. Sci. U.S.A.* **82**:626–630.

Chapter 5

The Use of Specific Ligands to Study Sodium Channels in Muscle

Enrique Jaimovich

1. INTRODUCTION

In the last 15 years the study of the ion channels as receptors for toxins, drugs, and hormones has yielded most of the information we now have on the biochemistry of these particular membrane protein molecules. Chemically gated channels are natural receptors and have been studied as such in several ways (see Barrantes, Chapter 24). In order to use this approach with voltage-gated channels, we need to have ligands that will bind to the channel with a fairly high specificity. Fortunately, nature has developed such molecules, at least for some voltage-gated channels, and binding of the ligand to the channel can be expressed according to the following reaction scheme:

$$\text{Ch} + \text{L} \underset{k_{-1}}{\overset{k_1}{\rightleftharpoons}} \text{Ch-L} \qquad (1)$$

where Ch is the channel molecule, L is the ligand, Ch-L represents the channel–ligand complex, and k_1 and k_{-1} are the rate constants for the

---

ENRIQUE JAIMOVICH • Departamento de Fisiología y Biofísica, Facultad de Medicina, Universidad de Chile, Santiago, Chile.

association and dissociation reactions of the ligand with the channel, respectively.

In this chapter, we are concerned with reaction scheme 1 in equilibrium conditions, where we can measure bound and free ligand concentrations.

From reaction scheme 1 we have

$$K_d = [Ch][L]/[Ch-L] = k_{-1}/k_1 \qquad (2)$$

where K_d is the dissociation equilibrium constant. If we define $B =$ [Ch-L], bound ligand concentration, $B_{max} = $ [Ch-L] + [Ch], maximal binding capacity, and $F = $ [L], free ligand concentration, then equation 2 can be written as:

$$B = B_{max} F/(K_d + F) \qquad (3)$$

which is a rectangular hyperbola that can be linearized in an Eadie–Scatchard form.

$$B = -K_d B/F + B_{max} \qquad (4)$$

In the case of two independent receptors for the same ligand, binding is described by a sum of two different hyperbolas and equation 3 becomes

$$B = B_{max_1} F/(K_{d_1} + F) + B_{max_2} F/(K_{d_2} + F) \qquad (5)$$

Figure 1 shows the Eadie–Scatchard plot for a single receptor (left) and the theoretical curve obtained when there are two independent receptors that differ by tenfold in their K_ds and by threefold in their B_{max} values (right). Using this kind of analysis, we can study ion channels as receptors that have high affinity for toxins or drugs, and we can work with the receptor in isolated membrane vesicles or in solubilized protein fractions in the same way as, and sometimes even better than, when it is forming part of the intact cell surface. Using the ligand–receptor system, one can ask questions that deal with the density, distribution, homogeneity, and ontogenesis of ion channels and also with the molecular characteristics of the channel.

2. MOLECULAR PHARMACOLOGY OF THE SODIUM CHANNEL IN MUSCLE

The voltage-gated sodium channel present in nerve and muscle membranes has a particularly rich pharmacology, mostly based on toxins that

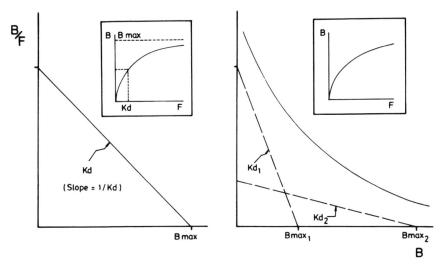

Figure 1. Equilibrium binding curves. Left: Single receptor interaction with a ligand displaying a rectangular hyperbola (inset) and a straight line in the Eadie–Scatchard plot (main panel). Right: Same analysis for the interaction of two independent receptors and the ligand. $K_{d_2} = 10\ K_{d_1}$; $B_{max_2} = 3\ B_{max_1}$.

evolution has selected as specific weapons to reach a target as vital as nerve conduction. The most widely studied toxins are tetrodotoxin (TTX) and saxitoxin (STX) (Fig. 2). For binding studies, tritiated STX has been routinely employed (Rogart, 1981), although its use presents some problems, namely, a low specific radioactivity and uncertain radiochemical purity. The first active, stable, and radiochemically pure derivatives of tetrodotoxin were synthesized by Chicheportiche et al. (1980). These derivatives (Fig. 2) were used in some of the experiments described below.

Besides TTX and STX, several other toxins interact specifically with the sodium channel in skeletal muscle fibers. As observed with nerve membranes, sea anemone toxin (ATX) and some scorpion toxins (referred to as α toxins; Catterall, 1980; Hille, 1984) slow the inactivation of the sodium current. In combination with the family of lipid-soluble toxins that include veratridine, batrachotoxin (BTX), and grayanotoxin, these toxins have been used to induce sodium fluxes that are sensitive to TTX in cultured muscle cells (Catterall, 1980).

A different family of polypeptide scorpion toxins were shown to block sodium currents in skeletal muscle (see below). This group includes toxins obtained from the venom of two American scorpions: *Centruroides sufussus sufussus* toxin II (Css II) and *Tityus serrulatus* toxin γ (TyTx γ). They have been named β scorpion toxins (Hille, 1984).

The binding of these different toxin families to their receptors is

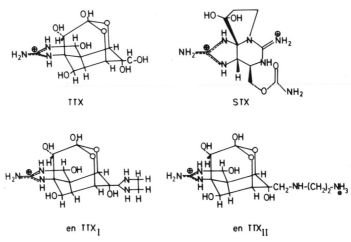

Figure 2. Molecular structure of guanidinium toxins. Abbreviations used: TTX, tetrodotoxin; STX, saxitoxin; en TTX, ethylene diamino tetrodotoxin, synthesized and purified as described (Chicheportiche *et al.*, 1980; Jaimovich *et al.*, 1983).

independent of the presence of high concentrations of toxins from a different family (Jaimovich *et al.*, 1982; Barhanin *et al.*, 1984). Thus, a given toxin can be displaced only by a toxin molecule of the same family. On this basis, a model for the toxin-specific receptors of the sodium channel has been postulated (Fig. 3). It shows three independent receptor sites accessible from the external medium and at least one site accessible only to lipid-soluble toxins. The model depicted in Fig. 3 attempts to show the different mechanisms of action of these toxins, but the limited evidence available only allows us to suggest that toxins of group 1 (STX, TTX, and derivatives) plug the channel either directly or allosterically (Hille, 1984) and that toxins from both group 2 (scorpion toxins and ATX) or

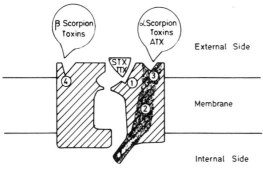

Figure 3. A model for the toxin receptors of the muscle sodium channel. For details, see text.

group 3 (BTX, veratridine) modify channel inactivation but interact with different receptor sites. Finally, the evidence obtained in nerve indicates that toxins from group 4 (β scorpion toxins) interact with channel activation (Meves et al., 1982).

It is interesting to note that the effect of these toxins on node of Ranvier (Meves et al., 1982; Cahalan, 1975) is quite different from the blockade of sodium currents seen in muscle. In isolated muscle fibers, the effect of β scorpion toxins on the surface sodium current is indistinguishable from that of TTX, at least in a given range of membrane potentials (Jaimovich et al., 1982; Barhanin et al., 1984). In node of Ranvier, a late current component, presumably resulting from persistent activation of the channels, is observed in the presence of β scorpion toxins (Meves et al., 1982; Cahalan, 1975). There are, in addition, other less specific receptor sites in muscle sodium channels, such as those for local anesthetics (Catteral, 1980), which are not pictured in this model.

3. SODIUM CHANNEL IN CARDIAC MUSCLE: ARE ALL SODIUM CHANNELS ALIKE?

A low sensitivity to TTX has been described for the sodium component of the cardiac action potential (Dudel et al., 1967; Cohen et al., 1981), although binding studies of TTX derivatives to membranes isolated from heart muscle, to muscle homogenates, and to heart cells in culture show high-affinity binding sites (Lombet et al., 1981; Renaud et al., 1983). However, Renaud et al. (1983) were able to show that low-affinity TTX binding sites are also present in these preparations and that this low-affinity receptor seems to be related to the physiologically relevant channels of cardiac muscle cells. Renaud et al. (1983) showed that there are two independent TTX receptors in heart muscle (K_{d_1} = 10^{-9} M and K_{d_2} = 10^{-7} M; Table I). They measured the dose–response curve for TTX inhibition of veratridine-stimulated sodium fluxes in heart cells and

Table I. Tetrodotoxin Binding to Heart Muscle

	High affinity		Low affinity	
Preparation used	B_{max} (fmol/mg)	K_d (nM)	B_{max} (fmol/mg)	K_d (nM)
Adult rat ventricle homogenate	60	1.5	600	175
Newborn rat ventricle homogenate	None		495	125
Heart cells in culture	31	1.6	408	135

found that half-maximal inhibition was reached at TTX concentrations higher than 10^{-7} M. Thus, it can be concluded that functional sodium channels in the heart are low-affinity TTX receptors; they differ in this respect from the sodium channels of nerve and of normal skeletal muscle. The function of the high-affinity TTX receptors present in heart muscle remains to be established. At some point it was suggested that they originate from nerve endings (Catteral and Coppersmith, 1981), but their presence in cultured heart cells (Renaud et al., 1983) rules out this possibility. One possible explanation is that these high-affinity TTX receptors represent sodium channel precursor structures that cannot be activated by voltage or by toxins from groups 2 or 3 but that will be converted into functional channels by some unknown physiological trigger.

Recently, Tanaka et al. (1984) estimated the affinity of saxitoxin from binding studies in subcellular fractions of cardiac muscle from frog, chicken, and turtle. They find low-affinity receptors in both chicken and turtle muscles and high-affinity receptors in frog heart membranes; they also describe high-affinity cytosolic receptors in the latter tissue.

Sodium channels with low affinity for TTX are also present in denervated mammalian skeletal muscle and in cultured rat muscle fibers (Redfern and Thesleff, 1971; Pappone, 1980), where the concentration of TTX for half-maximal inhibition is of the order of 10^{-6} M, quite similar to the effective dose in heart muscle. As in the heart, receptors with high (10^{-9} M) and low affinity for TTX coexist in cultured rat skeletal muscle (Frelin et al., 1983), although the amount of low-affinity receptors has not been measured quantitatively. it seems likely that there are at least two types of sodium channels in skeletal muscle, characterized by their different affinities for TTX. Unlike cardiac muscle sodium channels, the functional state of skeletal muscle channels can be regulated by innervation.

4. SURFACE AND TUBULAR SODIUM CHANNELS IN SKELETAL MUSCLE

Sodium channels have a role in the spreading of excitation along two different structures in the muscle fiber, the surface membrane, which propagates the action potential along the fiber axis, and the transverse tubule (T-tubule) membrane, which is responsible for the inward spread of excitation in the radial direction before coupling to contraction takes place. A new distinction between two types of sodium channels, both with high affinity for TTX, can be made in this case.

Years ago, Jaimovich et al. (1976) published measurements of TTX receptor densities in frog skeletal muscle using a bioassay method to measure free TTX concentrations. They found that TTX binding was reduced to approximately one-half of the control value when muscle fibers were "detubulated" by a glycerol-induced osmotic shock, a finding that suggests that an important fraction of the receptors is actually located in the T-tubular system. Considering that the area for the T-tubule is larger than that of surface membranes, the actual density of channels (receptor sites) present in the tubules can be estimated as being between one-half and one-fifth the surface density. An interesting conclusion of these studies is that the density of sodium channels varies in different regions of a continuous membrane system (see below).

The use of TTX derivatives in skeletal muscle has allowed us to make another distinction between surface and T-tubular sodium channels. In isolated frog muscle fibers kept under voltage clamp, it is possible to distinguish two separate inward sodium current components by using the appropriate experimental conditions. A fast signal that has been attributed to the surface current and a slow current component that may reflect an ill-controlled potential at the T-tubular level can be observed (Caillé *et al.*, 1978). Tension developed by the fiber can be recorded simultaneously and is related mainly to the tubular sodium-current (Caillé *et al.*, 1978, 1981).

Derivative I of ethylene diamino TTX (en TTX) (Jaimovich *et al.*, 1983) produces a blockade of the surface sodium current at low concentrations of the toxin with no change in the late sodium current or in contractile tension. On the other hand, derivative II of en TTX has the opposite effect; it inhibits tubular currents and contractions before affecting the early surface current. An effect similar to that of en TTX_1 was obtained with β scorpion toxins (*Centruroides sufussus sufussus* toxin II and *Tityus* toxin γ) (Jaimovich *et al.*, 1982; Barhanin *et al.*, 1984), which also have a blocking effect on muscle sodium currents but selectively affect surface sodium channels and have no effect on either tubular currents or contraction (Table II).

We can isolate membrane fractions containing either purified T-tubules (Rosemblatt *et al.*, 1981) (see Hidalgo, Chapter 6) or containing

Table II. Sodium Channel Blockers in Frog Skeletal Muscle

	Blockade (%)			
Toxin used	Fast Na^+ current	Tail Na^+ current	Contraction	Reference
TTX, 5 min, 6 nM	100	90	85	Jaimovich *et al.* (1982)
en TTX_1, 4.5 min, 5 nM	55	5	0	Jaimovich *et al.* (1983)
en TTX_{11}, 2 min, 0.4 nM	20	50	65	Jaimovich *et al.* (1983)
Css II, 6.5 min, 20 nM	95	0	0	Jaimovich *et al.* (1982)
TyTx, 10 min, 1 nM	90	0	0	Barhanin *et al.* (1984)

both surface and T-tubular membranes (Jaimovich et al., 1983; Jaimovich et al., 1985). A single component of en TTX$_{II}$ binding was seen for T-tubular membranes, whereas two distinct receptor populations were seen for the mixed membrane fraction (Fig. 4). The "low"-affinity binding site was ascribed to surface membrane, and it was selectively modified by treating the membranes with low concentrations of amino-group modifier reagents (Jaimovich et al.,1985).

From these results we could also confirm that the density of TTX receptors in T-tubular membranes is lower than that in surface membranes, as was previously postulated from whole-fiber experiments (Jaimovich et al., 1976).

Differences in the properties of the sodium channels are not the only differences between surface and T-tubular membranes; several biochemical markers seem to be different in these preparations. Among these are receptors for dihydropyridines, calcium channel blockers in different systems (Fleckenstein, 1983), which seem to be preferentially located in the T-tubule membrane (Fosset et al., 1983).

5. MODELS FOR SODIUM CHANNELS IN MUSCLE MEMBRANES

With the evidence discussed here, we can build a model for sodium channels as toxin receptors in muscle. We have to postulate that not all

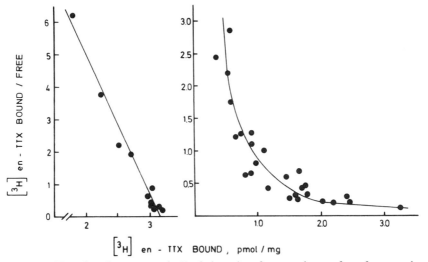

Figure 4. Tetrodotoxin receptors in T-tubule and surface membranes from frog muscle. Data taken from Jaimovich et al., 1985. The Eadie–Scatchard analysis shows (left) a single class of receptors for T-tubule membranes and (right) two types of receptors for a mixed membrane fraction that contains both T-tubules and surface membranes.

sodium channels are alike; they can be divided in those with high affinity and those with low affinity for TTX. The former group in skeletal muscle can be subdivided into surface and tubular channels with regard to their affinity for TTX derivatives. Sodium channels are not homogeneously distributed among the muscle membranes. Almers *et al.* (see Almers and Stirling, 1984) and recently Beam *et al.* (1985) found that sodium currents measured with a loose-patch technique are not homogeneously distributed in the surface of the muscle fiber but concentrate at the end-plate region (Beam *et al.*, 1985). A possible model postulates that sodium channels are synthesized in particular spots in the surface membrane and then migrate within the membrane plane a certain distance, which is a function of the mean life of the channel (Almers and Stirling, 1984). A gradient of channel density will appear so that the mean density of the T-tubules will be lower than that of the surface membranes, as the former are far from the site of synthesis.*

Such a model does not explain the differences in affinity for various toxins seen in surface and tubular receptors. A further complexity should be added, namely, that changes in affinity parallel channel aging or that differences in affinity reflect differences in the lipid content and composition of the membranes (see Hidalgo, Chapter 6). Some very elegant experiments were performed by Gundersen *et al.* (1983) in which voltage-dependent sodium currents were recorded in frog oocytes injected with cat muscle messenger RNA. Unlike brain mRNA, muscle mRNA seems to induce two types of sodium currents in the same membrane. The authors' interpretation (Gundersen *et al.*, 1983) is that the two types of current may correspond to "surface" and "tubular" sodium channels.

The overall evidence favors a model for skeletal muscle in which surface and tubular sodium channels not only have restricted mobility from one membrane system to the other but also have distinct molecular characteristics, reflected in (1) a different affinity for TTX derivatives, (2) a high affinity for scorpion toxins limited to surface membranes, and (3) a different density, being higher in the surface than in the T-tubules, which may reflect the restricted mobility.

It would be interesting to know if in cardiac muscle or in denervated skeletal muscle there is also some topological distribution of sodium channels with high and low affinity for TTX. Another important question to be solved concerns the possible functional differences of the various species of sodium channels we have described here. A system that may prove

---

* An attempt to measure sodium channel mobility in the plane of the membrane was made by Stuhmer and Almers (1982) using photobleaching through loose-patch micropipettes on frog muscle fibers. Their conclusion was that such mobility, if it exists, is undetectable with this kind of technique, and the diffusion coefficient should be less than 10^{-12} cm^2/sec.

to be useful in answering these questions in the near future is the fusion of membrane vesicles of known origin to artificial bilayers to study directly the behavior of the reconstituted channels. Progress in this area is beginning to take place (Moczydlowski et al., 1984).

It is interesting to point out that purified sodium channel proteins (reviewed by Agnew, 1984) isolated either from skeletal muscle membrane fractions that probably contain both surface and tubular elements (Barchi et al., 1979) or from cardiac muscle membranes (Lombet and Lazdunski, 1984) have similar apparent molecular weights and do not differ at the present level of analysis from those isolated from eel electroplax or mammalian brain (Agnew, 1984). It seems reasonable to think that major differences in the properties of the Na^+ channel as a receptor will imply only minor changes in the protein molecules that compose the channel or in their environment.

REFERENCES

Agnew, W. S., 1984, Voltage-regulated sodium channel molecules, *Annu. Rev. Physiol.* **46**:517–530.

Almers, W., and Stirling, C., 1984, Distribution of transport proteins over animal cell membranes, *J. Membr. Biol.* **77**:169–186.

Barchi, R. L., Weigele, J. B., Chalikian, D. M., and Murphy, L. E., 1979, Muscle surface membranes. Preparative methods affect apparent chemical properties and neurotoxin binding, *Biochim. Biophys. Acta* **550**:59–76.

Barhanin, J., Ildefonse, M., Rougier, O., Videla Sampaio, S., Giglio, J. R., and Lazdunski, M., 1984, Tityus γ toxin, a high affinity effector of the Na^+ channel in muscle, with a selectivity for channels in the surface membrane, *Pflugers Arch.* **400**:22–27.

Beam, K. G., Caldwell, J. H., and Campbell, D. T., 1985, Na channels in skeletal muscle concentrated near the neuromuscular junction, *Nature* **313**:588–590.

Cahalan, M. D., 1975, Modification of sodium channel gating in frog myelenated nerve fibres by *Centruroides sculpturatus* scorpion venom, *J. Physiol. (Lond.)* **244**:511–534.

Caillé, J., Ildefonse, M., and Rougier, O., 1978, Existence of a sodium current in the tubular membrane of frog twitch muscle fibre; implication in excitation contraction, *Pflugers Arch.* **374**:167–177.

Caillé, J., Ildefonse, M., Rougier, O., and Roy, G., 1981, Surface and tubular membrane sodium currents in frog twitch muscle fibre; implication in excitation contraction coupling, *Adv. Physiol. Sci.* **5**:389–409.

Catterall, W. A., 1980, Neurotoxins that act on voltage-sensitive sodium channels in excitable membranes, *Annu. Rev. Pharmacol. Toxicol.* **20**:15–44.

Catterall, W. A., and Coppersmith, J. 1981, High-affinity saxitoxin receptor sites in vertebrate heart: Evidence for sites associated with autonomic nerve endings, *Mol. Pharmacol.* **20**:526–532.

Chicheportiche, R., Balerna, M., Lombet, A., Romey, G., and Lazdunski, M., 1980, Synthesis of new, highly radioactive tetrodotoxin derivatives and their binding properties to the sodium channel, *Eur. J. Biochem.* **104**:617–625.

Chen, C. J., Bean, B. P., Colatsky, T. J., and Tsien, R. W., 1981, Concentration dependence of tetrodotoxin action on sodium currents and maximum upstroke velocity in rabbit cardiac Purkinje fibers, *J. Gen. Physiol.* **78**:383–411.

Dudel, J., Pepek, K., Dudel, R., and Trantwein, E., 1967, The effect of tetrodotoxin in the membrane current in cardiac muscle (Purkinje fibers), *Pflugers Arch.* **295**:213–226.

Fleckenstein, A., 1983, History of calcium antagonists, *Circ. Res.* **52**:3–16.

Fosset, M., Jaimovich, E., Delpont, E., and Lazdunski, M., 1983, (^{3}H)Nitrendipine receptors in skeletal muscle. Properties and preferential localization in transverse tubules, *J. Biol. Chem.* **258**:6086–6092.

Frelin, C., Vigne, P., and Lazdunski, M., 1983, Na$^+$ channels with high and low affinity tetrodotoxin binding sites in the mammalian skeletal muscle cell, *J. Biol. Chem.* **258**:7256–7259.

Gundersen, C. B., Miledi, R., and Parker, I., 1983, Voltage-operated channels induced by foreign messenger RNA in *Xenopus* oocytes, *Proc. R. Soc. Lond. [Biol.]* **220**:131–140.

Hille, B., 1984, *Ionic Channels of Excitable Membranes*, Sinauer Associates, Sunderland, MA.

Jaimovich, E., Venosa, R. A., Shrager, P., and Horowicz, P., 1976, Density and distribution of tetrodotoxin receptors in normal and detubulated frog sartorius muscle, *J. Gen. Physiol.* **67**:399–416.

Jaimovich, E., Ildefonse, M., Barhanin, J., Rougier, O., and Lazdunski, M., 1982, *Centruroides* toxin, a selective blocker of surface Na$^+$ channels in skeletal muscle: Voltage clamp analysis and biochemical characterization of the receptor, *Proc. Natl. Acad. Sci. U.S.A.* **79**:3896–3900.

Jaimovich, E., Chicheportiche, R., Lombet, A., Lazdunski, M., Ildefonse, M., and Rougier, O., 1983, Differences in the properties of Na$^+$ channels in muscle surface and T-tubular membranes revealed by tetrodotoxin derivatives, *Pflugers Arch.* **397**:1–5.

Jaimovich, E., Liberona, J. L., and Hidalgo, C., 1985, Characterization of two types of tetrodotoxin receptors in isolated frog muscle membranes, *Biophys. J.* **47**:192a.

Lombet, A., and Lazdunski, M., 1984, Characterization, solubilization, affinity labeling and purification of the cardiac Na$^+$ channel using *Tityus* toxin, *Eur. J. Biochem.* **141**:651–660.

Lombet, A., Renaud, J. F., Chicheportiche, R., and Lazdunski, M., 1981, A cardiac tetrodotoxin-binding component. Biochemical identification, characterization and properties, *Biochemistry* **20**:1279–1285.

Meves, H., Rubly, M., and Watt, D. D., 1982, Effect of toxins isolated from the venom of the scorpion *Centruroides sculpturates* on the Na currents of the node of Ranvier, *Pflugers Arch.* **393**:56–62.

Moczydlowski, E., Garber, S., and Miller, C., 1984, Batrachotoxin activates Na$^+$ channels in planar lipid bilayers. Competition of tetrodotoxin block with Na$^+$, *J. Gen. Physiol.* **84**:665–686.

Pappone, P. A., 1980, Voltage clamp experiments in normal and denervated mammalian skeletal muscle fibers, *J. Physiol. (Lond.)* **306**:377–410.

Redfern, P., and Thesleff, S., 1971, Action potential generation in denervated rat skeletal muscle II. The action of tetrodotoxin, *Acta Physiol. Scand.* **82**:70–78.

Renaud, J. F., Kazozoglu, T., Lombet, A., Chicheportiche, R. Jaimovich, E., Romey, G., and Lazdunski, M., 1983, The Na$^+$ channel in mammalian cardiac cells. Two kinds of tetrodotoxin receptors in rat heart membranes, *J. Biol. Chem.* **258**:8799–8805.

Rogart, R., 1981, Sodium channels in nerve and muscle membrane, *Annu. Rev. Physiol.* **43**:711–725.

Rosemblatt, R., Hidalgo, C., Vergara, C., and Ikemoto, N., 1981, Immunological and biochemical properties of transverse tubular membranes isolated from rabbit skeletal muscle, *J. Biol. Chem.* **256**:8140–8144.

Stuhmer, W., and Almers, W., 1982, Photobleaching through glass micropipettes: Sodium channels without lateral mobility in the sarcolemma of frog skeletal muscle, *Proc. Natl. Acad. Sci. U.S.A.* **79**:946–950.

Tanaka, J. C., Dogle, D. D., and Barr, L., 1984, Sodium channels in vertebrate hearts. Three types of saxitoxin binding sites in heart, *Biochim. Biophys. Acta* **775**:203–214.

Chapter 6

Isolation of Muscle Membranes Containing Functional Ionic Channels

Cecilia Hidalgo

1. INTRODUCTION

The physiology of muscle cells has been extensively studied, and there is growing evidence obtained from observations on whole muscle fibers, as well as with isolated membrane preparations, that ionic channels have a crucial role in the process of excitation–contraction coupling.

2. EXCITATION–CONTRACTION COUPLING

It is generally accepted that depolarization of the transverse tubule (T-tubule) membrane system of skeletal muscle is the electrical signal that precedes a rapid and transient release of calcium from the terminal cisternae region of the sarcoplasmic reticulum (SR) (Schneider and Chandler, 1973; Costantin, 1970). Muscle fiber contraction follows the binding of calcium to troponin and parvalbumin. Subsequent muscle relaxation takes place by a decrement of intracellular calcium concentration brought about

---

CECILIA HIDALGO • Department of Muscle Research, Boston Biomedical Research Institute, and the Department of Neurology, Harvard Medical School, Boston, Massachusetts 02114; *present address:* Departmento de Fisiología y Biofísica, Facultad de Medicina, Universidad de Chile; and Centro de Estudios Científicos de Santiago, Santiago, Chile.

by the powerful Ca^{2+}-ATPase of the SR membrane, which transports calcium back into the SR lumen (Martonosi, 1984).

The nature of the signal transmitted as the T-tubule–SR junction is a major unsolved problem. Voltage-dependent charge movement within the T-tubule membrane may be an initial voltage-sensitive step to gate calcium release from SR (Schneider and Chandler, 1973). Since charge movement and calcium release occur in different membranes, two basic linking mechanisms have been proposed: a mechanical model that involves the feet or pillars connecting the T-tubule and SR membranes (Chandler *et al.*, 1976) and an "intracellular agonist" model (Bianchi, 1968; Stephenson, 1982; Fabiato and Fabiato, 1977) in which calcium or another intracellular messenger, such as inositol trisphosphate (Vergara *et al.*, 1985), would gate the calcium release process. Because of the complex nature of the muscle membrane systems involved in excitation–contraction (E–C) coupling, electrophysiological studies have limitations. The T-tubule and SR membranes are not accessible to conventional electrophysiological recordings. Therefore, the properties of the ionic channels present in these membranes are best studied in isolated membrane preparations. However, in order to carry out studies of ion channels in isolated membranes and to correlate the information thus obtained with the physiology of the intact muscle cell, it is essential to establish unambiguously the origin and the purity of the preparations.

3. IONIC CHANNELS AND E–C COUPLING

Electrophysiological measurements in muscle fibers as well as studies with isolated muscle membrane vesicles clearly indicate the presence of a variety of ionic channels in the different muscle membranes. Some of these ionic channels have a direct role in E–C coupling, whereas others might participate either indirectly or not at all in the process.

3.1. Ionic Channels of T-Tubule and Surface Membranes

3.1.1. Sodium Channels

The propagation of the action potential from the motor end plate region along the surface of the muscle cell and into the fiber via the T-tubule system involves openings and closings of voltage-dependent sodium channels. Isolated membrane vesicles containing unknown proportions of T-tubule and surface membranes have sodium channels as evidenced by toxin binding (Barchi, 1982; Jaimovich *et al.*, 1983), bilayer incorporation (Moczydlowski *et al.*, 1984), and reconstitution studies (Barchi *et al.*, 1984).

Recent reconstitution studies have shown that sodium channels of a membrane fraction isolated from rat muscle and presumably enriched in T-tubules are blocked by *Conus* toxin GIIIA, whereas sodium channels from brain are not (Moczydlowski, 1985). Furthermore, sodium channels in surface membranes differ in their density and pharmacological characteristics from those located in T-tubule membranes (see Jaimovich, Chapter 5). Our studies indicate that T-tubule membranes free of surface membranes have sodium channels with different binding properties towards TTX derivatives from those present in surface membranes (Jaimovich *et al.*, 1985). The molecular origin of these differences is unknown. Recent studies indicate that sodium channels purified from rabbit T-tubules or from rat plasma membranes or rat brain have the same physical and functional characteristics (Kraner *et al.*, 1985). It is thus tempting to speculate that the differences in lipid composition within surface membranes and T-tubules, especially the high cholesterol content of the latter, might give rise to the differences in behavior of the sodium channels present in T-tubules relative to those of muscle surface membranes or brain membranes. The physiological implications, if any, of the peculiar behavior of T-tubule sodium channels are likewise unknown.

3.1.2. Calcium Channels

The depolarization of the T-tubule membrane causes opening of calcium channels present in T-tubules (Nicola Siri *et al.*, 1980; Almers and Palade, 1981). Two classes of calcium channels with different kinetics of openings have been described in frog muscle, a slow and a fast calcium channel (Cota and Stefani, 1985). The voltage dependence of activation of the fast channel and its time course of opening make it likely that a significant calcium current is activated during an action potential (Cota and Stefani, 1985).

Preliminary studies in our as well as in other laboratories have shown the existence of calcium channels in isolated T-tubule membranes incorporated into lipid bilayers (Affolter and Coronado, 1985) or into the tip of patch-clamp pipettes using the tip-dip method (B. Suarez-Isla, G. Riquelme and C. Hidalgo, unpublished data). Nitrendipine binding studies indicate that T-tubule membranes have a considerably higher density of receptors for this well-known calcium channel blocker (Fosset *et al.*, 1983; see also Table IV) than other membranes studied, including muscle surface membranes. Although considerable advances have been made lately towards the molecular characterization of the nitrendipine receptor (Borsotto *et al.*, 1984; Curtis and Catterall, 1984), it remains to be established whether T-tubule calcium channels have a physiological role in E–C coupling.

Furthermore, recent studies (Schwartz *et al.*, 1985) measuring nitrendipine binding to whole muscle fibers suggest that the density of nitrendipine binding sites is too high to account for the calcium currents measured in muscle. However, these studies are experimentally difficult to do, since whole-muscle binding studies of a lipid-soluble compound such as nitrendipine are subject to significant error as a result of the large contribution of nonspecific binding sites.

3.1.3. Potassium Channels Activated by Calcium

In addition to calcium channels, isolated T-tubule membranes contain a calcium-activated potassium channel (Latorre *et al.*, 1982; Moczydlowski and Latorre, 1983a). This channel has the necessary characteristics to give origin to late afterpotentials in skeletal muscle. At present, we do not know whether this channel is exclusively located in T-tubules or is present in the surface membrane of adult muscle as well. Calcium-activated potassium channels are ubiquitous and play an important role in a number of physiological processes (Latorre *et al.*, 1985), although they do not seem at present to have a direct role in the E–C coupling process of skeletal muscle.

3.2. Ionic Channels of SR

Regarding the role of SR ionic channels in E–C coupling, it has been shown that isolated SR membranes have a potassium channel (Miller, 1983), which probably avoids the build-up of a significant membrane potential difference during calcium release. It is generally believed that release of calcium from the SR takes place through a calcium channel that would open in response to T-tubule depolarization. However, although numerous hypotheses concerning the gating and permeation mechanisms for calcium release have been proposed (Martonosi, 1984; Endo, 1977; Ford and Podolski, 1970), the translocation mechanisms underlying the calcium release process remain unknown. A complete understanding of E–C coupling in muscle will not be achieved without a detailed picture of how calcium is released from the SR lumen into the myoplasmic space during muscle activation.

Several calculations of calcium current densities during the process of calcium release by SR *in vivo* have been done (Martonosi, 1984; Endo, 1977; Baylor *et al.*, 1983). For example, Baylor *et al.* (1983) estimated that if there are calcium channels in SR they should have a peak single-channel current of 0.02–0.03 pA. All these estimates, though consistent, are based on two critical unknown parameters: the number of channels and their distribution. This is a consequence of the macroscopic nature of the methods involved that generate flux values averaged over a large number of channels. Thus, they can only suggest that (1) calcium fluxes during activated release *in vivo* may be comparable to those measured in excitable cells and (2) a calcium release channel should have a small conductance. These observations cannot prove conclusively the existence of a channel for calcium release. However, present experiments (Orozco *et al.*, 1985) have indicated the presence of a cationic channel in SR membranes with a small conductance selective for calcium and barium and activated by caffeine. The estimated conductance of this channel near physiological calcium concentrations lies in the range of 0.4–0.8 pS. Thus, this channel could sustain the estimated flux levels for activated calcium release, and its preliminary characterization suggests that it could make a plausible contribution to calcium release from SR.

However, in order to associate this conductance with the putative release channel, several criteria have to be satisfied. Specifically, its regulation profile by known and postulated agonists and antagonists of calcium release should be investigated. The presence of a calcium channel activated by ATP has recently been described in isolated SR membranes incorporated in lipid bilayers (Smith *et al.*, 1985). Whether these two calcium channels of SR represent the same entity remains to be established.

4. ISOLATION OF MUSCLE MEMBRANES

From the preceding discussion, it follows that channels for a given ion are present in different muscle membranes. Thus, sodium channels are present in both surface membranes and T-tubules, and calcium channels are present in both T-tubules and SR. The emerging evidence indicates that these ionic channels behave differently: sodium channels of T-tubules are not identical to sodium channels of surface membranes (Jaimovich, Chapter 5), and calcium channels of SR are quite different from calcium channels of T-tubules (Orozco *et al.*, 1985; Smith *et al.*, 1985; Affolter and Coronado, 1985). Hence, it becomes apparent that in order to study the behavior of ionic channels using isolated membrane preparations, it is crucial to isolate highly purified, well-characterized membranes.

The isolation of highly purified SR vesicles from rabbit muscle has been accomplished (Meissner, 1975; Fernandez *et al.*, 1980). Likewise, methods are available to isolate T-tubule membranes from rabbit (Lau *et al.*, 1977; Rosemblatt *et al.*, 1981; Hidalgo *et al.*, 1983), chicken (Scales and Sabbadini, 1979), and frog muscle (Hidalgo et al., 1985b), and several criteria indicate that some of the isolated T-tubule preparations are essentially free of SR and surface membrane contamination (see below). The isolation of surface membranes devoid of contamination with T-tubules represents a difficult problem because of the low area of surface membrane relative to the T-tubular area. So far there are no descriptions in the literature of procedures to isolate highly purified and well-preserved surface membranes, as described in Section 4.3.

4.1. Isolation of SR

Several criteria must be satisfied in order to ascertain the purity of the isolated SR vesicles. Two types of SR vesicles can be isolated by separating the isolated SR microsomes according to their density in sucrose gradients (Meissner, 1975). Heavy SR vesicles contain calsequestrin, a low-affinity calcium-binding protein located in the vesicular lumen and presumably attached to the SR membrane and are thus more dense in sucrose gradients than light SR vesicles, which do not contain calsequestrin. Heavy SR vesicles are derived from the terminal cisternae of the SR, whereas light SR vesicles are derived from longitudinal SR (Meissner, 1975).

The main criteria of purity of isolated SR vesicles are as follows:

1. Protein composition in SDS-containing polyacrylamide gels characterized by a predominance of the 100,000-dalton band of the Ca^{2+}-ATPase. Light SR contains 90% or more Ca^{2+}-ATPase, and heavy SR contains 60–70% Ca^{2+}-ATPase and 15–20% calsequestrin (Meissner, 1975).
2. The Ca^{2+}-ATPase as the only ATPase activity (2–3 μmol Pi mg^{-1} min^{-1} measured at 22°C with the ionophore A23187).
3. ATP-dependent Ca^{2+}-transport stimulated 50- to 100-fold by oxalate or phosphate. A well-sealed light SR preparation isolated from rabbit muscle displays, in the presence of 5 mM oxalate, transport rates of 3–6 μmol Ca^{2+}/min per mg protein at 22°C (Meissner, 1975). A stoichiometry of Ca^{2+} transported per ATP hydrolyzed of 2.0 is obtained at 22°C (Berman, 1982).
4. Cholesterol content lower than 0.03 μmol/mg protein (Rosemblatt et al., 1981). It is likely that these very low cholesterol levels represent minor contamination of SR with the cholesterol-rich T-tubule membranes.
5. Phospholipid contents of 0.95–1.15 μmol Pi/mg protein for light SR and of 0.65–0.95 μmol/mg protein for heavy SR (Meissner, 1975). Phospholipid composition showing 70–75% PC, 15–20%

PE, and 8–10% PI. Virtual absence of PS and SM (Rosemblatt et al., 1981).

4.2. Isolation of T-Tubules

Homogenization of muscle tissue under well-controlled conditions causes selective fragmentation of SR and T-tubule membranes without detaching the surface membrane from the other components of the sarcolemma. The resulting SR and T-tubule vesicles can be isolated as a microsomal fraction free of contractile proteins and mitochondria by differential centrifugation. Whether the microsomal fraction contains T-tubules still attached to the SR, forming triads, or free T-tubules seems to depend on whether the microsomes are incubated with 0.6 M KCl to remove contractile proteins, since loss of the particles attaching the T-tubules to SR has been described following incubation of microsomes with 0.6 M KCl (Campbell et al., 1980). T-tubules are subsequently separated from SR by sedimentation in sucrose density gradients (Scales and Sabbadini, 1979; Rosemblatt et al., 1981). If T-tubules are still attached to SR, it is necessary to dissociate them by passage through a French press (Lau et al., 1977) prior to separation in density gradients. The possibility that these two types of isolation procedures yield, in one case, mostly junctional T-tubules (Lau et al., 1977) and, in the other, nonjunctional T-tubules (Scales and Sabbadini, 1979; Rosemblatt et al., 1981) has to be considered as well.

Although there is general agreement regarding what constitutes a highly purified SR preparation, the situation is less clear for T-tubules and even less so for surface membranes. Nevertheless, in the last few years several criteria have emerged that allow establishment of the purity of isolated T-tubules. A list of these criteria follows:

1. Protein composition in SDS-containing polyacrylamide gels (Fig. 1) characterized by a complex pattern (Lau et al., 1977; Hidalgo et al., 1983).
2. A Na^+,K^+-ATPase activity (ouabain sensitive) with values in the presence of detergents of 30–60 μmol mg^{-1} hr^{-1} at 37°C (as shown in Table I, the T-tubule preparations described initially have lower Na^+,K^+-ATPase activity).
3. A Mg^{2+}-ATPase (azide insensitive) with values of 2–4 μmol mg^{-1} min^{-1} at 25°C (T-tubules isolated by disruption of triadic junctions have much lower activity, Table II).
4. Lack of measurable Ca^{2+}-ATPase (less than 0.01 μmol mg^{-1} min^{-1} at 25°C with A23187, Table II). Although it was initially reported that T-tubules had Ca^{2+}-ATPase activity, recent reports describe no measurable Ca^{2+}-ATPase (Table II).

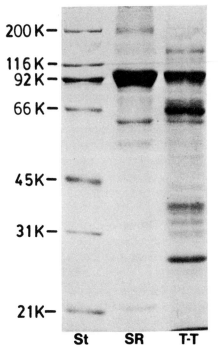

Figure 1. Sodium dodecyl sulfate electrophoretic pattern of isolated SR and T-tubule membranes. The main protein components of SR are the Ca^{2+}-ATPase (subunit molecular weight 100,000) and calsequestrin (55,000). The T-tubules have a more complex protein composition, and it is likely that several bands correspond to cytoplasmic proteins trapped inside the T-tubular lumen during isolation (Sabbadini and Okamoto, 1983; Beeler et al., 1983). The main protein components of T-tubules have subunit molecular weights of 107,000, 68,000, 66,000, 39,000, and 30,000. Minor bands of subunit molecular weights 225,000, 140,000, 55,000, 41,000, 37,000, 36,000 and 33,000 are also present. The T-tubules and SR membranes were isolated from rabbit skeletal muscle as described by Hidalgo et al. (1983) and by Fernandez et al. (1980), respectively.

5. An ATP-dependent Ca^{2+} transport with low rates (10 nmol mg^{-1} min^{-1} at 22°C) not stimulated by oxalate or phosphate, calmodulin dependent, and inhibited by orthovanadate ($K_{0.5} = 0.5$ μM) (Hidalgo et al., 1986a); ATP-dependent Na^+ transport (Lau et al., 1979b).
6. Cholesterol content of 1.0–1.1 μmol/mg protein (Table I). Cholesterol-to-phospholipid molar ratio of 0.5–0.6.
7. Phospholipid composition characterized by 15% SM and 10% PS in addition to PC (45%), PE (25%), and PI (10%) (Lau et al., 1979a; Rosemblatt et al., 1981).
8. ATP-dependent ouabain binding. In the presence of unmasking detergents, up to 100–200 pmol/mg at 37°C. $K_d = 20$–40 nM (Table III).
9. High density of nitrendipine binding sites: 50–150 pmol/mg at 10°C. $K_d = 1$–2 nM (Table IV).
10. A single class of en-TTX_{II} binding sites with $K_d = 0.5$ nM. In the presence of unmasking detergents, $B_{max} = 4$–5 pmol/mg (Table V).

Table I. Na^+,K^+-ATPase[a] Activity and Cholesterol Content of Isolated T-Tubules

Source	Unmasking agent	Na^+,K^+-ATPase (μmol mg^{-1} hr^{-1})	Cholesterol (μmol mg^{-1})	Na^+,K^+-ATPase/ cholesterol	Molar ratio C/PL
Lau et al. (1977, 1979); rabbit (from triads)	2 M NaI	6.2	0.64	9.7	0.40
Narahara et al. (1979); frog	None	8–21 (30°C)			
Scales and Sabbadini (1979); chicken	None	3.2 (37°C)			
Rosemblatt et al. (1981); rabbit	Triton X-100	12.0 (37°C)	0.90	13.3	0.55
Sunnicht and Sabbadini (1982); Sabbadini and Okamoto (1983); chicken	Valinomycin, monensin	26.2 (37°C)	0.94	27.9	0.86
Glossman et al. (1983); guinea pig	None	11.85 (37°C)			
Mitchell et al. (1983); rabbit (from triads)	SDS	39.0 (37°C)			
Kirley and Schwartz (1984); rabbit	None	48	0.94	51.1	
Hidalgo et al. (1986b)					
rabbit	Saponin	50.4 (37°C)	1.01	49.9	0.54
frog	Saponin	38.4 (37°C)	1.13	34.0	0.55

[a] Na^+,K^+-ATPase activity represents ouabain-sensitive activity; the temperature at which this activity was measured is given in parentheses when specified by the authors. Na^+,K^+-ATPase and cholesterol are given per mg protein. C/PL: cholesterol-to-phospholipid ratio.

Table II. Mg^{2+}-ATPase[a] and Ca^{2+}-ATPase Activity of Isolated T-Tubules

Source	Mg^{2+}-ATPase (μmol mg^{-1} min^{-1})	Ca^{2+}-ATPase (μmol mg^{-1} min^{-1})
Lau et al. (1977); rabbit (from triads)	0.431 (37°C)	0.922 (37°C, A23187)
Narahara et al. (1979); frog		<0.13 (30°C)
Scales and Sabbadini (1979); chicken	0.32 (37°C)	1.44 (37°C, A23187)
Malouf and Meissner (1979); chicken[b]	3.9 (25°C)	0.2 (25°C, A23187)
Malouf et al. (1981); chicken[b]	8 (32°C)	0 (32°C, A23187)
Rosemblatt et al. (1981); rabbit	5.35 (32°C)	0.56 (32°C)
Hidalgo et al. (1983); rabbit	1.2–3.4 (25°C)	<0.01 (25°C, A23187)
Sabbadini and Okamoto (1983); chicken	3.9 (25°C)	0–0.15 (37°C, X537A)
Glossman et al. (1983); guinea pig		0.16 (37°C, A23187)
Beeler et al. (1983); rat[b]	8.9[c] (37°C)	0
Kirley and Schwartz (1984); rabbit	12.7	2.33
Baskin and Kawamoto (1984); chicken (from triads)	0.43	0.12
Michalak et al. (1984); rabbit	0.052 (37°C)	0.024 (37°C, A23187)
Hidalgo et al. (1986b);		
rabbit	3.4 (25°C)	<0.01 (25°C, A23187)
frog	3.8[c] (25°C)	<0.01 (25°C, A23187)

[a] The Mg^{2+}-ATPase activity of T-tubules is a Ca^{2+}-or-Mg^{2+}-ATPase; it utilizes both MgATP and CaATP as substrates (Malouf and Meissner, 1979; Hidalgo et al., 1983). Assay temperatures and addition of ionophore to measure Ca^{2+}-ATPase activity are shown when specified by the authors.
[b] These authors used low-density membrane fractions from skeletal muscle. Presumably, these fractions contain mostly T-tubules, although the authors did not specify whether they contained T-tubules and/or surface membranes.
[c] These values represent initial rates; in both cases the reaction rate decreases with time.

Table III. Binding of Ouabain[a] to Isolated T-Tubules

Source	Unmasking agent	Ouabain bound B_{max} (pmol mg^{-1})	K_d (nM)
Lau et al. (1979b); rabbit	DOC	37	52.3
Jaimovich et al. (1986);			
rabbit	Saponin	169	32
frog	Saponin	215	10.4
frog	DOC	53	47

[a] Data for ouabain binding represent ATP-dependent ouabain binding; only binding data measured as a function of ouabain concentration are given. In addition to these measurements, several descriptions of ouabain binding at a single ouabain concentration have been given in the literature. Thus, Kirley and Schwartz (1984), using T-tubules from rabbit, reported maximal ouabain binding of 8.6 pmol mg^{-1}, calculated by extrapolation from binding measurements at 10^{-7} M ouabain without detergents. Mitchell et al. (1983) reported 118 pmol mg^{-1} of ouabain bound to T-tubules from rabbit, using SDS as unmasking agent and a total ouabain concentration of 10^{-6} M. Brandt et al. (1985b), using T-tubules isolated from rabbit muscle, reported values of 11.9 pmol mg^{-1} of ouabain bound, measured using SDS as unmasking agent and a total ouabain concentration of 5.4 × 10^{-8} M.

Table IV. Binding of Dihydropyridines to Isolated T-Tubules

Source	Ligand	T (C°)	B_{max} (pmol mg^{-1})	K_d (nM)
Fosset et al. (1983); rabbit	[^{3}H]Nitrendipine	10	56.0	1.7
Glossman et al. (1983); guinea pig	[^{3}H]Nimodipine	37	11.7	1.30
	[^{3}H]Nimodipine + diltiazem	37	61.6	1.29
Kirley and Schwartz (1984); rabbit	[^{3}H]Nitrendipine	10	12	1.7
	[^{3}H]Nitrendipine + diltiazem	10	17.5	1.44
Curtis and Catterall (1984); rabbit	[^{3}H]Nitrendipine	4	2.1–5.9	
Galizzi et al. (1984a); rabbit	[^{3}H]Nitrendipine	10	55	1.8
Borsotto et al. (1984); rabbit	[^{3}H]PN200-110	10	90	0.2
Galizzi et al. (1984b); rabbit	[^{3}H]Nitrendipine	10	60–80	
Barhanin et al. (1984); rabbit	[^{3}H]Nitrendipine	10	77.0	
Brandt et al. (1985a); rabbit	[^{3}H]Nitrendipine	23	27.2	3.5
Jaimovich et al. (1986);				
rabbit	[^{3}H]Nitrendipine	10	120.3	1.7
frog	[^{3}H]Nitrendipine	10	125.0	1.0
frog	[^{3}H]Nitrendipine	37	44.5	3.9
frog	[^{3}H]Nitrendipine + diltiazem	37	131	3.8
Kraner et al. (1985); rabbit	[^{3}H]Nitrendipine		41	
Chin and Beeler (1985); rat	[^{3}H]Nitrendipine		45	

Table V. Binding of Sodium Channel Blockers to Isolated T-Tubules

Source	Ligand	B_{max} (pmol mg^{-1})	K_d (nM)	Unmasking agent
Jaimovich et al. (1983); frog	en-TTX$_I$	0.6–0.7	20–40	None
	en-TTX$_{II}$	0.6–0.7	0.2–0.8	None
Moczydlowski and Latorre (1983b); rabbit	STX	22.0	1.0	Lubrol PX
Barhanin et al. (1984);				
frog	Tityus γ-toxin	0.04	0.02	None
rabbit	Tityus γ-toxin	0.01	0.5	None
rabbit	en-TTX$_{II}$	0.70	0.6	None
Jaimovich et al. (1986);				
rabbit	en-TTX$_{II}$	3.5	0.5	Saponin
frog	en-TTX$_{II}$	4.8	0.5	Saponin
Kraner et al. (1985); rabbit	STX	3.3		

Of all these T-tubular markers, the three most specific markers are the high content of cholesterol, the presence of nitrendipine binding sites in high density, and the single class of high affinity en-TTX$_{II}$ binding sites.

4.3. Isolation of Surface Membranes

Different methods have been described to isolate plasma membranes from skeletal muscle. Most investigators have used extraction with high salt of a low-speed pellet of muscle homogenates, a procedure first described by Kono and Colowick (1961). The rationale for this isolation is that following homogenization of the muscle tissue, the plasma membrane forming part of the sarcolemma sediments at low speed. To dissociate the plasma membrane from the collagen fibers and other sarcolemmal constituents, this heavy fraction is incubated in high salt (0.4 M LiBr at pH 8.5) for several hours. The dissociated plasma membrane is subsequently obtained by differential centrifugation as a microsomal fraction sedimenting at 50,000–100,000 × g. Following the initial description of the method, several groups, including ours, added another sucrose gradient purification step to the preparation (Table VI). Other methods have been described to isolate surface plasma membranes (Table VI), including a recently described procedure (Seiler and Fleischer, 1982) that replaces the long extraction in high salt by a shorter extraction time (3 hr) in 0.6 M KCl instead of 0.4 M LiBr.

Most of these plasma membrane preparations were isolated before methods were available to isolate T-tubules and thus before the establishment of suitable T-tubule markers. Hence, it is not possible to ascertain whether the isolated plasma membrane preparations described in the literature contain solely surface membranes or are contaminated with T-tubules. Our own observations with frog plasma membrane preparations, isolated by the high-salt extraction procedure followed by sucrose density gradient purification, indicate that the resulting membranes contain 40–50% T-tubules as judged by nitrendipine and en-TTX$_{II}$ binding studies (Liberona and Jaimovich, 1985). It is likely that the heavy fraction obtained following mild homogenization of muscle contains all the surface plasma membrane and a fraction of the T-tubules (20% or less), which presumably were not broken free from their attachment site to the surface membranes. Furthermore, the more recent surface plasma membrane preparation of Seiler and Fleischer (1982) was not characterized regarding nitrendipine binding, a marker that, according to all the reported evidence (Fosset *et al.*, 1983; Brandt *et al.*, 1985a; Chin and Beeler, 1985), is solely present in T-tubules. Seiler and Fleischer (1982) concluded that their plasma membranes were free of T-tubules solely on the basis of morphological observations.

In addition to the problem of T-tubule contamination, the high-salt extraction step, either prolonged with 0.4 M LiBr or shorter with 0.6 M KCl, removes the characteristic square arrays (Schmalbruch, 1979) of intramembrane particles of skeletal muscle surface membranes (Saito *et al.*, 1984) and results in significant inhibition of the Mg^{2+}-ATPase activity (Boegman *et al.*, 1970). Thus, as mentioned above, methods are not available at present to isolate highly purified and well-preserved surface membranes from skeletal muscle.

In spite of all the above limitations, and keeping in mind that the isolated membranes have lost structural characteristics and some enzymatic activities and presumably contain variable amounts of T-tubules, certain features emerge following examination of the markers described in plasma membrane preparations, as discussed below.

4.4. Surface Membrane Markers

4.4.1. Na^+,K^+-ATPase

The ouabain-sensitive Na^+,K^+-ATPase activity of surface membranes seems to be equal to or slightly higher than that of T-tubules. Values as high as 40–60 μmol mg^{-1} hr^{-1} have been described (Table VI), with one report describing a value as high as 155 μmol mg^{-1} hr^{-1} (Table VI). However, as becomes apparent following examination of Table VI, there is a wide variability in the reported values of Na^+,K^+-ATPase activity. Some of this variability can be attributed to the different procedures used to measure the enzyme, since only in a few cases were detergents used to unmask the latent activity of sealed vesicles. Likewise, the cholesterol content of the plasma membranes shows a wide range of values (Table VI), indicating varying degrees of purity of the preparations.

In an attempt to compare the different preparations, we calculated the ratio of Na^+,K^+-ATPase to the cholesterol content (Table VI) with the assumption that both are present only in surface membranes and T-tubules and not in SR. A comparison of these values shows that most ratios fall in the range of 30 to 90 (Table VI). By comparison, most of the ratios of Na^+,K^+-ATPase activity to cholesterol content for the T-tubules fall in the range of 28–55, with no values higher than 55 (Table I). If we assume that plasma membrane preparations contain T-tubules, which themselves have higher cholesterol contents than plasma membranes (compare the values given in Table I with those of Table VI), we may assume that a T-tubule-free surface membrane preparation would have a significantly higher Na^+,K^+-ATPase-to-cholesterol ratio than those given in Table VI. In fact, our frog plasma membrane preparation, which has about 50% surface membranes and 50% T-tubules, has a ratio of about

Table VI. Na$^+$,K$^+$-ATPase Activity and Cholesterol Content of Isolated Plasma Membranes

Source	Method	Na$^+$,K$^+$-ATPase (μmol mg^{-1} hr^{-1})	Cholesterol (μmol mg^{-1})	Na$^+$,K$^+$-ATPase/ Cholesterol	Molar ratio C/PL
Boegman et al. (1970); frog	High salt	41–90	0.816	50.2–110.3	0.12
Peter (1970); Fiehn et al. (1971); rat	EGTA extraction	3.5	0.68	5.15	0.96
McNamara et al. (1971); hamster	High salt	13.0[a]			
Sulakhe et al. (1971); hamster	High salt	24.0			
Severson et al. (1972); rabbit	High salt	3.2; 8.0[a]	0.34	9.4; 23.5	0.57
Kidwai et al. (1973); rat	Filtration, density	15.0[a]	0.18	83	0.11
Andrew and Appel (1973); rat	High salt, density	25.9			
Schapira et al. (1974); rat	High salt, density	11.05	0.34	32.5	0.65
Festoff and Engel (1974); rat	High salt, density	28.6[a]			
Pinkett and Perlman (1974); rat	Density	26.0			
Nagotomo and Peter (1975); rat	EGTA extraction	3.1			
Reddy et al. (1976); rat	High salt, density	40–65[a]			
Agapito and Cabezas (1977); rabbit	High salt, density	5.1	0.19	26.8	0.43
frog	High salt, density	1.9	0.44	4.3	0.57

Reference; species	Method				
Barchi et al. (1977); rat	Density	31.0			
De Kretser & Livett (1977); mouse	EGTA extraction	3.3	0.01		
Walaas et al. (1977); rat	Filtration, density	155			
Raible et al. (1978); chicken	Density	11.0			
Reddy et al. (1978); human	High salt, density	8.91			
Reddy and Engel (1979); rat	High salt, density	44			
Barchi et al. (1979); rat	Density	4.7	0.124	37.9	
	High salt, density	45.2	0.487	92.8	
Seiler and Fleischer (1982); rabbit	KCl extraction, density	$56.5^{a,b}$	0.92	58.2	0.39
Desnuelle et al. (1983); rabbit	KCl extraction, density	2.58			
De Boland et al. (1983); chicken	High salt, density	25–30	0.52	52.9	0.55
Sabbadini and Okamoto (1983); chicken	High salt, density	73.4	0.59^d	124.4	0.38^d
Rock et al. (1984); rabbit	KCl extraction, density	$40.2; 52.2^a$			
Michalak et al. (1984); rabbit	KCl extraction, density	$40^{a,b}$			0.20; 0.30
Hidalgo et al. (1986b); frog	High salt, density	$37^{a,c}$	0.62	59.7	

[a] Only in these cases was the temperature of the assay specified by the authors as 37°C.
[b] Measured in the presence of SDS.
[c] Measured in the presence of saponin.
[d] From Summicht and Sabbadini (1982).

60. If we correct for the T-tubules present, which have a ratio of 34, the frog surface membranes would have a Na^+,K^+-ATPase-to-cholesterol ratio of about 90.

4.4.2. Mg^{2+}-ATPase

As shown by Boegman et al. (1970), treatment with high salt inhibits drastically the Mg^{2+}-ATPase of plasma membranes without having any effect on the Na^+,K^+-ATPase. Thus, the Mg^{2+}-ATPase is not a suitable marker for these plasma membrane preparations.

4.4.3. Cholesterol Content

As discussed above, plasma membrane preparations have lower cholesterol content (Table VI) than T-tubules (Table I).

4.4.4. ATP-Dependent Ouabain Binding

A few experiments have been described reporting ouabain binding to plasma membranes (Table VII). High values of binding, similar to those seen in some T-tubule preparations, have been found by Mitchell et al. (1983) and by us (Jaimovich et al., 1986).

4.4.5. Nitrendipine Binding

As is becoming increasingly clear (Fosset et al., 1983; Brandt et al., 1985a; Chin and Beeler, 1985), nitrendipine binding is restricted to T-tubules. Most reports give values in the range of 40–120 pmol mg^{-1}. Thus, by determining nitrendipine binding to isolated plasma membranes, it is possible to ascertain the extent of contamination with T-tubules.

Table VII. Binding of Ouabain to Isolated Plasma Membranes[a]

Source	Unmasking agent	B_{max} (pmol mg^{-1})	K_d (nM)
Desnuelle et al. (1983); rabbit	None	10.5	800–1100
Jaimovich et al. (1985); frog	Saponin	162.8	9

[a] Ouabain binding represents ATP-dependent ouabain binding. Only binding data measured as a function of ouabain concentration are given in the table. In addition, ATP-dependent ouabain binding values of 16–20 pmol mg^{-1} were described by Rock et al. (1984), using DOC as unmasking agent and a ouabain concentration of 1 μM; Mitchell et al. (1983) reported 118 pmol mg^{-1} of ATP-dependent ouabain binding measured with SDS at 1 μM ouabain.

4.4.6. Binding of Sodium Channel Blockers

Studies carried out with frog muscle fibers indicate that T-tubules have a lower number of TTX binding sites than surface membranes (Jaimovich et al., 1976). Studies with isolated plasma membranes give 7.0 pmol/mg of TTX or STX bound as the highest number (Table VIII). By comparison, isolated T-tubules give values of 3–4 pmol/mg with the exception of one report (Moczydlowski and Latorre, 1983b), which gave values of 22 pmol/mg. Again assuming that our frog plasma membrane preparations have 50% T-tubules and correcting by the T-tubule binding, we can extrapolate to binding values for TTX of more than 10 pmol/mg, about two to three times higher than those of T-tubules and in agreement with studies done in whole frog muscle (Jaimovich et al., 1976).

4.4.7. Other Markers

Several other markers have been measured in plasma membranes. Among the enzymatic markers are acetylcholinesterase (Table IX), 5'-nucleotidase (Table X), and adenylate cyclase (Table XI). For comparison, the values for T-tubules are also included in these tables. In addition, other receptors have been measured (Table XII). Again, T-tubule and plasma membrane values are compared.

Table VIII. Binding of Sodium Channel Blockers to Isolated Plasma Membranes

Source	Ligand	B_{max} (pmol mg^{-1})	K_d (nM)	Unmasking agent
Barchi et al. (1979); rat				
light fraction	STX	0.35–0.50	1.6	None
high-salt fraction	STX	6.7	1.5	None
Barchi et al. (1980); rat	STX	6.1		NP-40
Jaimovich et al. (1982); frog	Centruroides toxin	0.22	0.4	None
Desnuelle et al. (1983); rabbit	en-TTX$_{II}$	1.38	1.3–1.8	None
Jaimovich et al. (1983)[a]; frog	STX	0.4–0.5	3	None
	en-TTX$_I$	0.4–0.5	4	None
	en-TTX$_{II}$	0.4–0.5	8	None
Barhanin et al. (1984);				
frog	en-TTX$_{II}$	0.4	4	None
frog	Tityus γ-toxin	0.375	0.012	None
rabbit	Tityus γ-toxin	0.450	0.3	None
Liberona and Jaimovich	en-TTX$_{II}$	2.5	0.3	Saponin
(1986)[b]; frog	en-TTX$_I$:ZI	6.5	6.0	Saponin

[a] Two classes of en-TTX$_{II}$ binding sites were seen, but only the low-affinity sites were characterized in detail.
[b] The authors resolved the binding of en-TTX$_{II}$ into two components. B_{max} and K_d values for each component are given in the table.

Table IX. Acetylcholinesterase of Membrane Preparations Isolated from Skeletal Muscle

Source	Acetylcholinesterase (nmol mg^{-1} min^{-1})
Plasma membranes	
Severson et al. (1972); rabbit	37
De Kretser and Livett (1977); mouse	0.53
Seiler and Fleischer (1982); rabbit	275 (SDS)[a]
Rock et al. (1984); rabbit	310–315
Transverse tubules	
Hidalgo and Riquelme, unpublished observations	
rabbit	174.3 (saponin)
frog	19.3 (saponin)

[a] The type of detergent used to unmask latent activity is given in parentheses.

It is apparent that the reported values of acetylcholinesterase activity are similar for both T-tubules and plasma membranes isolated from rabbit skeletal muscle, whereas T-tubules isolated from frog muscle give significantly lower values (Table IX). The values of 5′-nucleotidase activity cover a wide range, with the lowest value of activity representing less than 1% of the highest value (Table X). Furthermore, 5′-nucleotidase activity has not been reported in highly purified T-tubules, so it is not possible to ascertain with the present data whether this enzymatic activity

Table X. 5′-Nucleotidase Activity of Membrane Preparations Isolated from Skeletal Muscle

Source	5′-Nucleotidase (nmol mg^{-1} min^{-1})	Preparation
Kidwai et al. (1973); rat	96.7	P.M.[a]
Schapira et al. (1974); rat	27.5	P.M.
Pinkett and Perlman (1974); rat	102–134	P.M.
Agapito and Cabezas (1977);		
rabbit	2.8	P.M.
frog	15.0	P.M.
De Kretser and Livett (1977); mouse	1.50	P.M.
Walaas et al. (1977); rat	26.7	P.M.
Malouf and Meissner (1979); chicken	180	P.M. and/or T-tubules
Malouf et al. (1981); chicken	125	P.M. and/or T-tubules
Beeler et al. (1983); rat	16	P.M. and/or T-tubules
Rock et al. (1984); rabbit	186; 92	P.M.
Sabbadini and Okamoto (1983);	68	T-tubules
chicken	70	P.M.

[a] P.M. stands for plasma membranes.

Table XI. Adenylate Cyclase Activity of Membrane Preparations Isolated from Skeletal Muscle[a]

	Adenylate cyclase (pmol mg^{-1} min^{-1})	Preparation
Severson et al. (1972); rabbit	810 (basal)	P.M.[b]
Walaas et al. (1977); rat	1380 (basal)	P.M.
Reddy et al. (1978); rat[a]	324 (NaF)	P.M.
Caswell et al. (1978); rabbit	2300 (NaF)	T-tubules
Grefrath et al. (1978); rat	148 (basal)	P.M.
Raible et al. (1978); chicken	5455 (basal)	Light membranes
Reddy and Engel (1979); rat	68 (basal)	P.M.
Malouf et al. (1981); chicken	120 (NaF)	P.M. and/or T-tubules
Seiler and Fleischer (1982); rabbit	330 (basal)	P.M.
Mitchell et al. (1983); rabbit	375 (basal)	T-tubules
Rock et al. (1984); rabbit	910;1460 (NaF)	P.M.

[a] Using less purified preparations, Festoff et al. (1977) reported adenylate cyclase activities in the presence of NaF of 143 and 122 pmol mg^{-1} min^{-1} for slow and fast muscle plasma membranes, respectively.
[b] P.M. stands for plasma membranes.

is different in T-tubules and surface membranes. Regarding adenylate cyclase activity, either basal or measured with NaF, the values for T-tubules seem to be similar to those reported for plasma membranes (Table XI).

In addition to these enzymatic activities, binding of bungarotoxin (for

Table XII. Binding of Different Ligands to Membrane Preparations Isolated from Skeletal Muscle

Ligand, source	Binding (pmol mg^{-1})	Preparation
Bungarotoxin		
Barchi et al. (1979); rat	0.35	P.M.[a] and or T-tubules
Andrew et al. (1974); rabbit	1.04	P.M. and or T-tubules
Jaimovich et al. (1986);		
frog	15.2	P.M.
frog	5.0	T-tubules
rabbit	4.0	T-tubules
DHA		
Grefrath et al. (1978); rat	1.23	P.M.
Reddy and Engel (1979); rat	0.62	P.M.
Caswell et al. (1978); rabbit	0.40–0.61	T-tubules
Mitchell et al. (1983); rabbit	1.69	P.M.
Insulin		
Rock et al. (1984); rabbit	2.4–3.7 × 10^{-3}	P.M.

[a] P.M. stands for plasma membranes.

acetylcholine receptor), of DHA (for β-adrenergic receptors), and of insulin has been measured (Table XII). Only for DHA binding are there sufficient data to conclude that β-adrenergic receptors seem to be present in the same density in T-tubules as in plasma membranes.

5. CONCLUDING REMARKS

As discussed in detail in this chapter, it is now possible to isolate highly purified SR and T-tubule membranes from skeletal muscle. It is likely that in the near future methods will be developed to isolate surface membranes with a similar degree of purity. The availability of highly purified and functionally well-preserved isolated muscle membranes, together with the techniques available to study ionic channels in membrane vesicles, will contribute significantly to our understanding of the physiology of E–C coupling in skeletal muscle. In fact, significant advances in the characterization of the ionic channels present in muscle membranes have already been made using isolated membrane vesicles.

ACKNOWLEDGMENTS. This research was supported by grant HL23007 from the National Institutes of Health, U.S.A., and by a grant from The Tinker Foundation, Inc., U.S.A., to the Centro de Estudios Científicos de Santiago, Chile.

REFERENCES

Affolter, H., and Coronado, R., 1985, Planar bilayer recording of single calcium channels from purified muscle transverse tubules, *Biophys. J.* **47**:434a.

Agapito, M. T., and Cabezas, J. A., 1977, Isolation and chemical composition of sarcolemmal membranes from rabbit and frog skeletal muscle, *Int. J. Biochem.* **8**:811–817.

Almers, W., and Palade, P. T., 1981, Slow calcium and potassium currents across frog muscle membrane: Measurements with a vaseline-gap technique, *J. Physiol. (Lond.)* **312**:159–176.

Andrew, C. G., and Appel, S. H., 1973, Macromolecular characterization of muscle membranes. I. Proteins and sialic acid of normal and denervated muscle, *J. Biol. Chem.* **248**:5156–5163.

Andrew, C. G., Almon, R. R., and Appel, S. H., 1974, Macromolecular characterization of muscle membranes. Acetylcholine receptor of normal and denervated muscle. 1974, *J. Biol. Chem.* **249**:6163–6165.

Andrew, C. G., Almon, R. R., and Appel, S. H., 1975, Macromolecular characterization of muscle membranes. Endogenous membrane kinase and phosphorylated protein substrate from normal and denervated muscle, *J. Biol. Chem.* **250**:3972–3980.

Barchi, R. L., 1982, Biochemical studies of the excitable membrane sodium channel, *Int. Rev. Neurobiol.* **23**:69–101.

Barchi, R. L., Bonilla, E., and Wong, M., 1977, Isolation and characterization of muscle membranes using surface-specific labels, *Proc. Natl. Acad. Sci. U.S.A.* **74:**34–38.

Barchi, R. L., Weigele, J. B., Chalikian, D. M., and Murphy, L. E., 1979, Muscle surface membranes. Preparative methods affect apparent chemical properties and neurotoxin binding, *Biochim. Biophys. Acta* **550:**59–76.

Barchi, R. L., Cohen, S. A., and Murphy, L. E., 1980, Purification from rat sarcolemma of the saxitoxin-binding component of the excitable membrane sodium channel, *Proc. Natl. Acad. Sci. U.S.A.* **77:**1306–1310.

Barchi, R. L., Tanaka, J. C., and Furman, R. E., 1985, Molecular characteristics and functional reconstitution of muscle voltage-sensitive sodium channels, *J. Cell. Biochem.* **26:**135–146.

Barhanin, J., Ildefonse, M., Rougier, O., Vilela Sampaio, S., Giglio, J. R., and Lazdunski, M., 1984, Tityus γ-toxin, a high affinity effector of the Na^+ channel in muscle, with a selectivity for channels in the surface membrane, *Pflugers Arch.* **400:**22–27.

Baskin, R. J., and Kawamoto, R., 1984, Stereological analysis of transverse tubules and sarcoplasmic reticulum isolated from normal and dystrophic skeletal muscle, *Biochim. Biophys. Acta* **771:**109–118.

Baylor, S. M., Chandler, W. K., and Marshall, M. W., 1983, Sarcoplasmic reticulum calcium release in frog skeletal muscle fibres estimated from arsenazo III calcium transients, *J. Physiol. (Lond.)* **344:**625–666.

Beller, T. J., Gable, K. S., and Keefer, J. M., 1983, Characterization of the membrane bound Mg^{2+}-ATPase of rat skeletal muscle, *Biochim. Biophys. Acta* **734:**221–234.

Berman, M. C., 1982, Energy coupling and uncoupling of active calcium transport by sarcoplasmic reticulum membranes, *Biochim. Biophys. Acta* **694:**95–121.

Bianchi, C. P., 1968, *Cell Calcium*, New York, Appleton–Century–Crofts.

Boegman, R. J., Manery, J. F., and Pinteric, L., 1970, The separation and partial purification of membrane bound $(Na^+ + K^+)$-dependent Mg^{2+}-ATPase and $(Na^+ + K^+)$-independent Mg^{2+}-ATPase from frog skeletal muscle, *Biochim. Biophys. Acta* **203:**506–530.

Borsotto, M., Barhanin, J., Norman, R. I., and Lazdunski, M., 1984, Purification of the dihydropyridine receptor of the voltage-dependent Ca^{2+} channel from skeletal muscle transverse tubules using (+) [^{3}H] PN 200-110, *Biochem. Biophys. Res. Commun.* **122:**1357–1366.

Brandt, N., Kawamoto, R. M., and Caswell, A. H., 1985a, Dihydropyridine binding sites on transverse tubules isolated from triads of rabbit skeletal muscle, *J. Receptor. Res.* **5:**155–170.

Brandt, N. R., Kawamoto, R. M., and Caswell, A. H., 1985b, Effects of mercaptans upon dihydropyridine binding sites on transverse tubules isolated from triads of rabbit skeletal muscle, *Biochem. Biophys. Res. Commun.* **127:**205–212.

Campbell, K., Franzini-Armstrong, C., and Shamoo, A., 1980, Further characterization of light and heavy sarcoplasmic reticulum vesicles. Identification of the 'sarcoplasmic reticulum feet' associated with heavy sarcoplasmic reticulum vesicles, *Biochim. Biophys. Acta* **602:**97–116.

Caswell, A. H., Baker, S. P., Boyd, H., Potter, L. T., and Garcia, A. M., 1978, β-Adrenergic receptor and adenylate cyclase in transverse tubules of skeletal muscle, *J. Biol. Chem.* **253:**3049–3054.

Chandler, W. K., Rakowski, R. F., and Schneider, M. F., 1976, A non-linear voltage dependent charge movement in frog skeletal muscle, *J. Physiol. (Lond.)* **254:**245–283.

Chin, H., and Beeler, T., 1985, Nitrendipine binding to transverse tubule and plasma membrane vesicles purified from rat skeletal muscle, *Biophys. J.* **47:**265a.

Costantin, L. L., 1970, The role of sodium current in the radial spread of contraction in frog muscle fibers, *J. Gen. Physiol.* **55:**703–715.

Cota, G., and Stefani, E., 1985, Fast and slow Ca channels in twitch muscle fibers of the frog, *Biophys. J.* **47**:65a.

Curtis, B. M., and Catterall, W. A., 1984, Purification of the calcium antagonist receptor of the voltage-sensitive calcium channel from skeletal muscle transverse tubules, *Biochemistry* **23**:2113–2118.

De Boland, A. R., Gallego, S., and Boland, R., 1983, Effects of vitamin D-3 on phosphate and calcium transport across and composition of skeletal muscle plasma cell membranes, *Biochim. Biophys. Acta* **733**:264–273.

De Kretser, T. A., and Livett, B. G., 1977, Skeletal muscle sarcolemma from normal and dystrophic mice. Isolation, characterization and lipid composition, *Biochem. J.* **168**:229–237.

Desnuelle, C., Lombet, A., Liot, D., Maroux, S., and Serratrice, G., 1983, Complete monitoring of the purification of the plasma membrane from rabbit skeletal muscle, *Biochem. Biophys. Res. Commun.* **112**:521–527.

Endo, M., 1977, Calcium release from the sarcoplasmic reticulum, *Physiol. Rev.* **57**:71–108.

Fabiato, A., and Fabiato, F., 1977, Calcium release from the sarcoplasmic reticulum, *Circ. Res.* **40**:119–129.

Fernandez, J. L., Rosemblatt, M., and Hidalgo, C., 1980, Highly purified sarcoplasmic reticulum vesicles are devoid of Cs^{2+}-independent ('basal') ATPase activity, *Biochim. Biophys. Acta* **599**:552–568.

Festoff, B. W., and Engel, W. K., 1974, *In vitro* analysis of the general properties and junctional receptor characteristics of skeletal muscle membranes. Isolation, purification and partial characterization of sarcolemmal fragments, *Proc. Natl. Acad. Sci. U.S.A.* **71**:2435–2439.

Festoff, B. W., Oliver, K. L., and Reddy, N. B., 1977, *In vitro* studies of skeletal muscle membranes. Adenylate cyclase of fast and slow twitch muscle and the effects of denervation, *J. Membr. Biol.* **32**:331–343.

Fiehn, W., Peter, J. B., Mead, J. F., and Gan-Elepano, M., 1971, Lipids and fatty acids of sarcolemma, sarcoplasmic reticulum and mitochondria from rat skeletal muscle, *J. Biol. Chem.* **246**:5617–5620.

Ford, L. E., and Podolsky, R. J., 1970, Regenerative calcium release within muscle cells, *Science* **167**:58–69.

Fosset, M., Jaimovich, E., Delpont, E., and Lazdunski, M., 1983, [^{3}H]Nitrendipine receptors in skeletal muscle. Properties and preferential localization in transverse tubules, *J. Biol. Chem.* **258**:6086–6091.

Galizzi, J.-P., Fosset, M., and Lazdunski, M., 1984a, [^{3}H]Verapamil binding sites in skeletal muscle transverse tubule membranes, *Biochem. Bioophys. Res. Commun.* **118**:239–245.

Galizzi, J.-P., Fosset, M., and Lazdunski, M., 1984b, Properties of receptors for the Ca^{2+}-channel blocker verapamil in transverse-tubule membranes of skeletal muscle. Stereospecificity, effect of Ca^{2+} and other inorganic cations, evidence for two categories of sites and effect of nucleoside triphosphates, *Eur. J. Biochem.* **144**:211–215.

Glossman, H., Ferry, D. R., and Boschek, C. B., 1983, Puurification of the putative calcium channel from skeletal muscle with the aid of [^{3}H]-nimodipine binding, *Arch. Pharmacol.* **323**:1–11.

Grefrath, S. P., Smith, P. B., and Appel, S. H., 1978, Characterization of the β-adrenergic receptor and adenylate cyclase in skeletal muscle plasma membranes, *Arch. Biochem. Biophys.* **188**:328–337.

Hidalgo, C., Gonzalez, M. E., and Lagos, R., 1983, Characterization of the Ca^{2+}- or Mg^{2+}-ATPase of transverse tubule membranes isolated from rabbit skeletal muscle, *J. Biol. Chem.* **258**:13937–13945.

Hidalgo, C., Gonzalez, M. E., and Garcia, A. M., 1986a, Calcium transport by transverse tubules from rabbit skeletal muscle, *Biochim. Biophys. Acta* **854**:279–286.

Hidalgo, C., Parra, C., Riquelme, G., and Jaimovich, E., 1986b, Transverse tubules from frog skeletal muscle. Purification and properties of vesicles sealed with the inside-out orientation, *Biochim. Biophys. Acta* **855**:79–88.

Jaimovich, E., Venosa, R. A., Shrager, P., and Horowitz, P., 1976, Density and distribution of tetrodotoxin receptors in normal and detubulated frog sartorius muscle, *J. Gen. Physiol.* **67**:399–416.

Jaimovich, E., Ildefonse, M., Barhanin, J., Rougier, O., and Lazdunski, M., 1982, Centruroides toxin, a selective blocker of surface Na^+ channels in skeletal muscle: Voltage-clamp analysis and biochemical characterization of the receptor, *Proc. Natl. Acad. Sci. U.S.A.* **79**:3896–3900.

Jaimovich, E., Chicheportiche, R., Lombet, A., Lazdunski, M., Ildefonse, M., and Rougier, O., 1983, Differences in the properties of Na^+ channels in muscle surface and T-tubular membranes revealed by tetrodotoxin derivatives, *Pflugers Arch.* **397**:1–5.

Jaimovich, E., Donoso, P., Liberona, J. L., and Hidalgo, C., 1986, Ion pathways in transverse tubules. Quantification of receptors in membranes isolated from frog and rabbit skeletal muscle, *Biochim. Biophys. Acta* **855**:89–98.

Kidwai, A. M., Radcliffe, M. A., Lee, E. Y., and Daniel, E. E., 1973, Isolation and properties of skeletal muscle plasma membranes, *Biochim. Biophys. Acta* **298**:593–607.

Kirley, T. L., and Schwartz, A., 1984, Solubilization and affinity labeling of a dihydropyridine binding site from skeletal muscle: Effects of temperature and diltiazem on [^{3}H]dihydropyridine binding to transverse tubules, *Biochem. Biophys. Res. Commun.* **123**:41–49.

Kono, T., and Colowick, S. P., 1961, Isolation of skeletal muscle cell membrane and some of its properties, *Arch. Biochem. Biophys.* **93**:520–533.

Kraner, S. D., Tanaka, J. C., Roberts, R. H., and Barchi, R. L., 1985, Purification and functional reconstitution of the voltage-sensitive sodium channel from rabbit T-tubular membranes, *Biophys. J.* **47**:440a.

Latorre, R., Vergara, C., and Hidalgo, C., 1982, Reconstitution in planar lipid bilayers of a Ca dependent K channel from transverse tubule membranes isolated from rabbit skeletal muscle, *Proc. Natl. Acad. Sci. U.S.A.* **79**:605–609.

Latorre, R., Alvarez, O., Cecchi, X., and Vergara, C., 1985, Properties of reconstituted ion channels, *Annu. Rev. Biophys. Biophys. Chem.* **14**:79–111.

Lau, Y. H., Caswell, A. H., and Brunschwig, J.-P., 1977, Isolation of transverse tubules by fractionation of triad junctions of skeletal muscle, *J. Biol. Chem.* **252**:5565–5574.

Lau, Y. H., Caswell, A. H., Brunschwig, J.-P., Baerwald, R. J., and Garcia, M., 1979a, Lipid analysis and freeze-fracture studies on isolated transverse tubules and sarcoplasmic reticulum subfractions of skeletal muscle, *J. Biol. Chem.* **254**:540–546.

Lau, Y. H., Caswell, A. H., Garcia, A. M., and Letellier, L., 1979b, Ouabain binding and coupled sodium, potassium and chloride transport in isolated transverse tubules of skeletal muscle, *J. Gen. Physiol.* **74**:335–349.

Liberona, J. L., and Jaimovich, E., 1986, Tetrodotoxin receptors in membranes isolated from frog skeletal muscle. Differential sensitivity to amino group reagents. *Arch. Biochem. Biophys.* (submitted).

Malouf, N. N., and Meissner, G., 1979, Localization of an Mg^{2+}-or-Ca^{2+}-activated ('basic') ATPase in skeletal muscle, *Exp. Cell Res.* **122**:233–250.

Malouf, N. N., Samsa, D., Allen, P., and Meissner, G., 1981, Biochemical and cytochemical comparison of surface membranes from normal and dystrophic chickens, *Am. J. Pathol.* **105**:223–231.

Martonosi, A. N., 1984, Mechanisms of Ca^{2+}-release from sarcoplasmic reticulum of skeletal muscle, *Physiol. Rev.* **64**:1240–1320.

McNamara, D. B., Sulakhe, P. V., and Dhalla, N. S., 1971, Properties of the sarcolemmal calcium-ion-stimulated adenosine triphosphatase of hamster skeletal muscle, *Biochem. J.* **125**:525–530.

Meissner, G., 1975, Isolation and characterization of two types of sarcoplasmic reticulum vesicles, *Biochim. Biophys. Acta* **389**:51–68.

Michalak, M., Famulski, K., and Carafoli, E., 1984, The Ca^{2+}-pumping ATPase in skeletal muscle sarcolemma. Calmodulin dependence, regulation by cAMP-dependent phosphorylation, and purification, *J. Biol. Chem.* **259**:15540–15547.

Miller, C., 1983, Integral membrane channels: Studies in model membranes, *Physiol. Rev.* **63**:1209–1242.

Mitchell, R. D., Volpe, P., Palade, P., and Fleischer, S., 1983, Biochemical characterization, integrity and sidedness of purified skeletal muscle triads, *J. Biol. Chem.* **258**:9867–9877.

Moczydlowski, E. G., 1985, Na-channel block by μ-conotoxin GIIIa: A peptide toxin specific for skeletal muscle, *Biophys. J.* **47**:190a.

Moczydlowski, E., and Latorre, R., 1983a, Gating kinetics of Ca^{2+}-activated K^+ channels from rat muscle incorporated into planar lipid bilayers: Evidence for two voltage-dependent Ca^{2+} binding reactions, *J. Gen. Physiol.* **82**:511–542.

Moczydlowski, E., and Latorre, R., 1983b, Saxitoxin and ouabain binding activity of isolated skeletal muscle membranes as indicators of surface origin and purity, *Biochim. Biophys. Acta* **732**:412–420.

Moczydlowski, E., Garber, S., and Miller, C., 1984, Batrachotoxin-activated Na^+ channels in planar lipid bilayers. Competition of tetrodotoxin block by Na^+, *J. Gen. Physiol.* **84**:665–686.

Nagatomo, T., and Peter, J. B., 1975, $(Na^+ + K^+)$-ATPase in subcellular fractions of rat skeletal muscle, *Exp. Neurol.* **49**:345–355.

Narahara, H. T., Vogrin, V. G., Green, J. D., Kent, R. A., and Gould, M. K., 1979, Isolation of plasma membrane vesicles, derived from transverse tubules, by selective homogenization of subcellular fractions of frog skeletal muscle in isonotic media, *Biochim. Biophys. Acta* **552**:247–261.

Nicola Siri, L., Sanchez, J. A., and Stefani, E., 1980, Effect of glycerol treatment on the calcium current of frog skeletal muscle, *J. Physiol. (Lond.)* **305**:87–96.

Orozco, C., Suarez-Isla, B., Froehlich, J. P., and Heller, P. F., 1985, Calcium channels in sarcoplasmic reticulum (SR) membranes, *Biophys. J.* **47**:57a.

Peter, J. B., 1970, A $(Na^+ + K^+)$ ATPase of sarcolemma from skeletal muscle, *Biochem. Biophys. Res. Commun.* **40**:1362–1367.

Pinkett, M. O., and Perlman, R. L., 1974, Phosphorylation of muscle plasma membrane protein by a membrane-bound protein kinase, *Biochim. Biophys. Acta* **372**:379–387.

Raible, D. G., Cutler, L. S., and Rodan, G. A., 1978, Localization of adenylate cyclase in skeletal muscle sarcoplasmic reticulum and its relation to calcium accumulation, *FEBS Lett.* **85**:149–152.

Reddy, N. B., and Engel, W. K., 1979, In vitro characterization of skeletal muscle β-adrenergic receptors coupled to adenylate cyclase, *Biochim. Biophys. Acta* **585**:343–359.

Reddy, N. B., Engel, W. K., and Festoff, B. W., 1976, In vitro studies of skeletal muscle membranes. Characterization of a phosphorylated intermediate of sarcolemmal $(Na^+ + K^+)$ ATPase, *Biochim. Biophys. Acta* **433**:365–382.

Reddy, N. B., Oliver, K. L., Festoff, B. W., and Engel, W. K., 1978, Adenylate cyclase system of human skeletal muscle. Subcellular distrubution and general properties, *Biochim. Biophys. Acta* **540**:371–388.

Rock, E., Lefaucheur, L., and Chevallier, J., 1984, Isolation and characterization of plasma membranes from rabbit skeletal muscle, *Biochem. Biophys. Res. Commun.* **123**:216–222.
Rosemblatt, M., Hidalgo, C., Vergara, C., and Ikemoto, N., 1981, Immunological and biochemical properties of transverse tubule membranes isolated from rabbit skeletal muscle, *J. Biol. Chem.* **256**:8140–8148.
Sabbadini, R. A., and Okamoto, V. R., 1983, The distribution of ATPase activities in purified transverse tubule membranes, *Arch. Biochem. Biophys.* **223**:107–119.
Saito, A., Seiler, S., and Fleischer, S., 1984, Alterations in the morphology of rabbit skeletal muscle plasma membranes during membrane isolation, *J. Ultrastruct. Res.* **86**:277–293.
Scales, D., and Sabbadini, R. A., 1979, Microsomal T system. A stereological analysis of purified microsomes derived from normal and dystrophic skeletal muscle, *J. Cell. Biol.* **83**:33–46.
Schapira, G., Dobocz, I., Piau, J. P., and Delain, E., 1974, An improved technique for preparations of skeletal muscle cell plasma membranes, *Biochim. Biophys. Acta* **345**:348–358.
Schmalbruck, H., 1979, 'Square arrays' in the sarcolemma of human skeletal muscle, *Nature* **281**:145–146.
Schneider, M. F., and Chandler, W. K., 1973, Voltage-dependent charge movement in skeletal muscle: A possible step in excitation–contraction coupling, *Nature* **242**:244–246.
Schwartz, L. M., McCleskey, E. W., and Almers, W., 1985, Dihydropyridine receptors in muscle are voltage-dependent, but most are not functional calcium channels, *Nature* **314**:747–751.
Seiler, S., and Fleischer, S., 1982, Isolation of plasma membrane vesicles from rabbit skeletal muscle and their use in ion transport studies, *J. Biol. Chem.* **257**:13862–13871.
Severson, D. L., Drummond, G. I., and Sulakhe, P. V., 1972, Adenylate cyclase in skeletal muscle. Kinetic properties and hormonal stimulation, *J. Biol. Chem.* **247**:2949–2958.
Smith, J. S., Coronado, R., and Meissner, G., 1985, A nucleotide stimulated calcium conducting channel from sarcoplasmic reticulum incorporated into planar lipid bilayers, *Biophys. J.* **47**:451a.
Stephenson, E. W., 1982, The role of free calcium ion in calcium release in skinned muscle fibers, *Can. J. Physiol. Pharmacol.* **60**:417–426.
Sulakhe, P. V., Fedelesova, M., McNamara, D. B., and Dhalla, N. S., 1971, Isolation of skeletal muscle membrane fragments containing active Na^+–K^+ stimulated ATPase: Comparison of normal and dystrophic muscle sarcolemma, *Biochem. Biophys. Res. Commun.* **42**:793–800.
Sumnicht, G. E., and Sabbadini, R. A., 1982, Lipid composition of transverse tubular membranes from normal and dystrophic skeletal muscle, *Arch. Biochem. Biophys.* **215**:628–637.
Vergara, J., Delay, M., and Tsien, R. J., 1985, Inositol (1,4,5) trisphosphate. A possible chemical link in excitation–contraction coupling in muscle, *Proc. Natl. Acad. Sci. U.S.A.* **82**:6352–6356.
Walaas, O., Walaas, E., Lystad, E., Alertsen, A. R., Horn, R. S., and Forsum, S., 1977, A stimulatory effect of insulin on phosphorylation of a peptide in sarcolemma-enriched membrane preparation from rat skeletal muscle, *FEBS Lett.* **80**:417–422.

Chapter 7

Methodologies to Study Channel-Mediated Ion Fluxes in Membrane Vesicles

Ana Maria Garcia

1. INTRODUCTION

The use of techniques such as voltage clamp or patch clamp has provided a great deal of information about channel properties and about ion movement across ion channels in biological membranes. However, neither of these techniques can be used to study the presence and properties of ion channels in very small cells or in internal membranes such as the sarcoplasmic reticulum of skeletal muscle. A clever technique developed by Miller and Racker (1976) has made it possible to overcome these problems: the membrane of interest, once isolated and purified, is allowed to fuse to an artificial planar lipid bilayer under appropriate conditions. Thus, ion channels and probably other membrane components are incorporated into the artificial membrane (for a review on the technique, see Miller, 1983a,b). Using this approach, various channels have been studied, including a voltage-gated K^+ channel from sarcoplasmic reticulum (Miller, 1978), a Cl^- channel from *Torpedo* electroplax (White and Miller, 1979), and a Ca^{2+}-activated K^+ channel from skeletal muscle

---

ANA MARIA GARCIA • Boston Biomedical Research Institute, Department of Muscle Research, Boston, Massachusetts 02114; *present address:* Whitehead Institute, Cambridge, Massachusetts 02142.

transverse tubule (Latorre *et al.*, 1982). None of these channels was known to exist before these experiments were done.

The biggest advantage of incorporating channels into artificial planar bilayers is the great accessibility of both sides of the bilayer to manipulation, both in terms of changing electrical parameters and in varying the composition of the medium at either or both sides of the membrane. It is also possible to manipulate the lipid environment in which the channel is embedded and thus gain information on how lipid–protein interactions affect channel behavior.

However, despite the apparent power of this technique, there are questions that remain unanswered. Since most of the channels studied with this technique so far have not previously been characterized, it is impossible to know whether their properties have been modified by the insertion into a foreign membrane or if the channel under study is present in only a very small population of vesicles, perhaps a contamination: of 10^{10} vesicles added to the solution, only a few (from 1 to 10^3 at most) are fused to the planar bilayer (Miller, 1983b). Furthermore, there is evidence that certain channels are incorporated selectively into the planar bilayer under particular conditions (Miller, 1983b). Does this mean that different channels are segregated in separate vesicles or that under certain conditions some channels are inactivated and therefore, even if they are present, cannot be studied? Even more, is it that only a few specific channels may be incorporated into the artificial planar membrane?

The purpose of pointing out these problems is to stress the need of an alternate way to study channel behavior. This might be especially important from the point of view of the biochemist, whose ultimate goal is to purify and reconstitute the protein under study. The understanding of channel structure and function will come ultimately from both biochemical and electrophysiological studies.

In order to study the behavior of channels in their native membranes or in reconstituted liposomes, we have to consider that the rate of ion movement through the channel is on the order of 10^5–10^7 ions/sec. Native membranes in solution spontaneously form closed vesicles with an average diameter of 0.1 μm, which corresponds to an internal volume of 4 $\times$ 10^{-15} ml per vesicle (under controlled conditions, unilamellar liposomes of the same size can be formed). If the vesicles or liposomes are filled with a 0.1 M salt solution, only 10^4 ions are trapped inside. The conclusion is obvious: if ions are moving through channels, the vesicles can exchange their content within a few milliseconds. This imposes the necessity of using techniques with high time resolution for measuring ion fluxes. The remainder of this chapter describes two procedures to measure ion fluxes on a fast time scale: fluorescence quenching and light scattering. The application of these techniques is explained by using an example: the study of the properties of the K^+ channel from sarcoplasmic reticulum (SR) isolated from rabbit skeletal muscle. As indicated above, the questions to answer are about the presence and distribution of the channel in the isolated vesicles and how its properties compare with the ones described in the planar bilayer.

2. CHANNEL-MEDIATED Tl+ FLUX MEASURED BY FLUORESCENCE QUENCHING

2.1. Rationale

This ingenious technique, first used by Moore and Raftery (1980) to measure acetylcholine-receptor-mediated ion fluxes, makes use of the property of some molecules of quenching the fluorescence of a fluorescent molecule on collision. Some examples of quencher molecules are heavy atoms like Tl^+, Cs^+, acrylamide, methionine, and Br^- (Eftink and Ghiron, 1981). The extent of fluorescence quenching depends on the concentration of quencher and is described by the Stern–Volmer relationship as follows (Eftink and Ghiron, 1981):

$$F_0/F = 1 + K_{SV}[Q] \qquad (1)$$

where F_0 corresponds to the total fluorescence, F is the fluorescence in the presence of a concentration [Q] of the quencher, and K_{SV} is the Stern–Volmer constant for quenching. A plot of [Q] versus F_0/F gives a straight line whose slope corresponds to K_{SV}. Thus, once K_{SV} is known for a given set of conditions, this relationship can be used to determine unknown concentrations of quencher.

When a fluorescent probe is trapped inside vesicles, and these vesicles are mixed with a solution of quencher, the rate of fluorescence decrease will depend on the rate of quencher influx. Let us consider the case in which the quencher molecule is a salt. In order to have a net flux of the quencher molecule across the membrane, both species, cation and anion, must move together in order to maintain electroneutrality; alternatively, the quencher ion can be exchanged for other ion of the same charge from inside the vesicle. In both cases, the rate of movement will be determined by the less permeable of the two species moving across the membrane. Therefore, if the rate of quencher influx is limited by the movement of the compensating ion, the rate of fluorescence decrease will reflect the permeability of the latter. This is better illustrated by the following example.

As was pointed out above, by fusing SR vesicles into planar bilayers the presence of a cation-selective channel has been demonstrated (Miller, 1978). Its cation selectivity is $K^+ > Tl^+ > Na^+ > Li^+ \gg$ choline, N-methylglucamine (Coronado et al., 1980). If we load SR vesicles with a fluorescent probe in the presence of Li^+ glutamate and then mix them with a solution containing Tl^+ glutamate (Fig. 1), Tl^+ (quencher molecule) will move into the vesicles according to its concentration gradient, but it can do so only in exchange for Li^+. The anion, glutamate, is only slowly permeable, and its concentration is the same on both sides of the membrane. Since Tl^+ permeates the channel better than Li^+ (or Na^+ or choline), the rate of Tl^+ influx and therefore the rate of fluorescence quenching will be limited by the rate at which Tl^+ is exchanged by the ion (Li^+ or other) trapped inside the vesicles.

2.2. Methodology

The fluorescent probe to be used in this type of experiment must be hydrophilic, so that it will be trapped inside the vesicles and not bound to the membranes. Two examples found in the literature are 8-aminoaphthalene-1,3,6-trisulfonic acid (ANTS) (Moore and Raftery, 1980) and 1,3,6,8-pyrene tetrasulfonic acid (PTS) (Garcia and Miller, 1984). The probe, in the medium of choice, is loaded by either freeze–thawing (Moore and

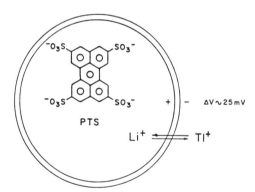

	In (mM)	Out (mM)
PTS^{-4}	10	–
Choline	40	–
Li^+	100	75
Tl^+	–	25
$Glutamate^-$	100	100
Glucose	100	150

Figure 1. Ionic conditions at the time of mixing. The diagram shows a SR vesicle loaded with the fluorescent probe PTS and the permeable cation Li^+, which is exchanged by external Tl^+. The table indicates the concentrations at the time of mixing in the stopped-flow apparatus. Also indicated in the diagram is the calculated membrane potential under these conditions.

Raftery, 1980) or prolonged incubation of a vesicle suspension in the medium containing the fluorescent probe (Garcia and Miller, 1984). Before the experiment, the external probe is removed by passing an aliquot of the suspension through an ion-exchange resin, by filtration through a molecular sieve resin, or by any other appropriate technique. The vesicles are then mixed in a stopped-flow apparatus with a solution of the same composition as the loading medium except that the cation has been replaced by Tl^+. The ionic conditions at the time of mixing are depicted in Fig. 1. The influx of Tl^+ is followed as the decrease in fluorescence of the trapped probe.

To eliminate the possibility that some of the probe is bound to the membranes or that it has not been completely removed or that it leaks out of the vesicles, it is advisable to determine the externally available fluorescence by mixing the vesicles with a slowly permeant quencher such as methionine or acrylamide (Garcia and Miller, 1984). This test will also tell us whether or not the permeability barrier of the vesicles has been damaged by the loading of the probe.

2.3. Results

When SR vesicles loaded with PTS in the presence of different cations are mixed with a solution containing Tl^+, a decrease in fluorescence is observed whose rate depends on the ion trapped inside the vesicles. The results are shown in Fig. 2. The fluorescence values have been converted into internal Tl^+ concentration using the Stern–Volmer relationship as described above, and the data have been fitted with a double exponential. The external Tl^+ concentration is 25 mM in all experiments. It is observed that when the vesicles have been preloaded with K^+, 70% of the Tl^+ is exchanged within the mixing time of the apparatus (3 msec), and the remaining 30% is exchanged more slowly (Fig. 2A). In the case of Li^+-loaded vesicles, a somewhat similar kinetic of Tl^+ exchange is observed: a fast phase with a measurable rate that corresponds to approximately 70% of the internal space and a slow phase. If the vesicles have been preloaded with choline instead, the rate of Tl^+ entry is much slower. When Li^+-loaded vesicles are treated with a cation ionophore (gramicidin A) prior to the experiment, Tl^+ is instantaneously equilibrated in all the vesicle population. Figure 2B shows a longer time course of Tl^+ entry into Li^+- or choline-loaded vesicles in which it can be clearly seen that there are fast and slow components for Tl^+/Li^+ exchange, whereas more than 80% of the exchange in the presence of choline is slow. The rate constants for Tl^+ exchange in these experiments are summarized in Table I.

The results previously shown strongly suggest that there are two

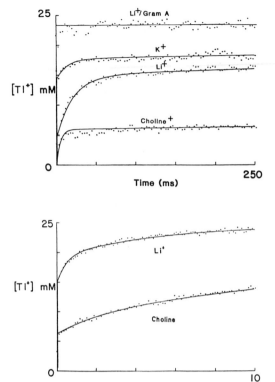

Figure 2. Time course of Tl$^+$ influx into SR vesicles loaded with the indicated cations. The concentration of Tl$^+$ inside the vesicles was calculated from the fluorescence decay according to equation 1. The experimental points were fitted to two exponentials. A: Fast time course. B: Long time course.

Table I. Rate Constants of Tl$^+$ Exchange[a]

	k_{fast}	f_{fast}	k_{slow}	f_{slow}
K$^+$	>300[b]	0.71	0.51	0.28
Na$^+$	>300[b]	0.70	0.52	0.29
Li$^+$	42	0.65	0.40	0.34
TEA	50	0.29	0.10	0.70
Choline	54	0.21	0.18	0.78
N-Methylglucamine	41	0.22	0.20	0.77

[a] The first-order rate constants (sec^{-1}) of the fast and slow exchange, k_{fast} and k_{slow}, respectively, were determined by computer fitting the Tl$^+$ influx data with a double exponential. Each value represents the mean of at least 10 determinations. The standard error of the mean was less than 10%. The fraction of Tl$^+$ exchanged at each rate is indicated as f_{fast} and f_{slow}.
[b] Faster than time resolution of measurement.

compartments for Tl^+ movement and that these compartments correspond to two populations of vesicles with different permeability properties. If we compare the rates of Tl^+ exchange in all the cases studied (Table I), it is clear that the fast fraction of exchange shows some ion selectivity, whereas the slow fraction is independent of the ion that was trapped inside the vesicles. The fact that the addition of gramicidin A increases the rate of exchange of the slow fraction supports this idea. Moreover, the ions that are rapidly exchanged correspond to the ions that are permeable through the K^+ channel in the planar bilayer and follow approximately the same selectivity sequence. Therefore, it is tempting to postulate that we are measuring ion fluxes through K^+ channels, which are present in 70% of the SR vesicles.

If the measured ion fluxes are indeed channel mediated, we would expect that the presence of a blocker of the K^+ channel in the planar bilayer would reduce the rate of exchange. When vesicles loaded with PTS were incubated with 1,10-bis-guanidino-n-decane (bisG10) (Garcia and Miller, 1984) before mixing them with Tl^+, the rate of Tl^+ exchange decreased. The results are shown in Fig. 3 for two concentrations of the channel blocker. The effect of the blocker was only observed in the fast fraction of exchange and was eliminated if the vesicles were pretreated with gramicidin A (data not shown). Moreover, the concentration of blocker necessary to reduce the rate of exchange of the fast fraction to half was 75 μM, a value comparable to the concentration needed to reduce to half the channel conductance in the planar bilayer (50 μM) (Fig. 4). This evidence strongly supports the idea that the fast fraction of Tl^+ exchange corresponds to a channel-mediated transport.

It is important to stress that in order for bisG10 to have any effect, it was necessary to preincubate the vesicles with the blocker. Mixing the vesicles with bisG10 in the stopped-flow apparatus resulted in no effect on the rates of Tl^+ exchange. Independent observations suggested to us

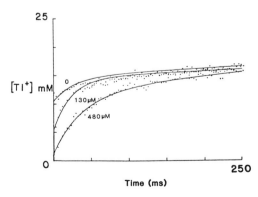

Figure 3. Effect of channel blocker on Tl^+ influx. Sarcoplasmic reticulum vesicles in K^+ medium were incubated with the indicated concentrations of bisG10 for 10 min before mixing with Tl^+ in the stopped-flow apparatus.

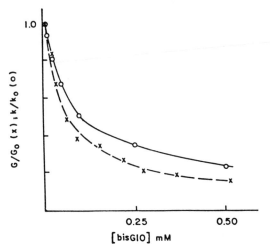

Figure 4. Effect of the channel blocker bisG10 in planar bilayers and vesicles. Circles: Relative macroscopic conductance when bisG10 is added to the *trans* chamber. Ionic conditions: *cis* chamber, 75 mM Li$^+$ glutamate, 25 mM Tl$^+$ glutamate, 10 mM MOPS, pH 7.0; *trans* chamber, 100 mM Li$^+$ glutamate, 10 mM MOPS, pH 7.0. Crosses: Relative rate of Tl$^+$ influx into Li$^+$-loaded vesicles preincubated with the indicated concentrations of bisG10.

that this procedure allows bisG10 to diffuse inside the vesicles. In other words, the effect of the blocker is asymmetric, in agreement with the planar bilayer results: a strong blocking is observed when bisG10 is added *trans*, i.e., the side opposite to which the vesicles are added. Several pieces of evidence from the planar bilayer work have suggested that the channel is asymmetric (Miller and Rosenberg, 1979; Coronado and Miller, 1982). Despite this fact, the channel always inserts into the planar bilayer with the same orientation. The results shown here allow us to say that the *trans* side of the bilayer corresponds to the interior of the vesicle and, therefore, to the lumen of the SR *in situ*.

The evidence presented so far leaves no doubt about the presence of a K$^+$ channel in 70% of the SR vesicles whose ionic selectivity and blocking characteristics are comparable in the native and artificial membranes. It remains to be determined whether the ionic conductance is the same in both cases. Also, it is important to calculate the permeability of the native membrane to monovalent cations, which can give us some idea about the role of the channel in the membrane *in situ*.

The rate constant for Tl$^+$ exchange into a homogeneous vesicle population in which each vesicle has n channels can be calculated from the following relationship:

$$k = RT\, g_i n / F^2\, V\, C_i \qquad (2)$$

where R, T, and F are the gas constant, absolute temperature, and Faraday constant, respectively, g_i is the single-channel conductance of the ith ion, n is the number of open channels, V is the vesicle volume, and C_i is the concentration of the ith ion. This relatioship is valid under conditions of steady-state current and zero voltage (Sten-Knudsen, 1978). Under these conditions, the membrane permeability is proportional to the rate constant k and to the ratio of vesicle volume to area (A) by:

$$P_i = k\ V/A \tag{3}$$

Using these two relationships, we can calculate the measured rate constants for Tl^+ exchange by using the rate constants for ion flux calculated from the single-channel conductance (Coronado et al., 1980). In this case, we assume a homogeneous population of vesicles of 0.1 μm diameter and with one open channel. The results are summarized in Table II. The agreement between the theoretical and experimental values is remarkable. Therefore, we can conclude that the Tl^+ exchange has the characteristics of a channel-mediated process and that these correlate very closely to the properties of the SR K^+ channel in the planar bilayer. In other words, the channel does not seem to have been extensively modified by the insertion into the planar bilayer. Considering this, the membrane permeability was calculated from the flux data, and the results are also shown in Table II. The very high permeability of the SR membrane to monovalent cations (10^{-3} cm/sec) suggests that most probably this membrane is always clamped to zero voltage. That is to say, the role of the K^+ channel would be to compensate for charge movements across the SR membrane.

2.4. Discussion

The results previously shown clearly demonstrate that it is possible to measure channel-mediated ion fluxes by means of fluorescence quench-

Table II. Rate Constants of Tl^+ Exchange and Membrane Permeability[a]

	g (pS)	k (sec^{-1}) Calc	k (sec^{-1}) Exp	P (cm/sec)
K^+	130	325	>300	>1 × 10^{-3}
Na^+	55	137	>300	>1 × 10^{-3}
Li^+	6	15	42	4 × 10^{-4}

[a] The theoretical rate constants (Calc) were calculated according to equation 2, using the indicated single-channel conductance at 100 mM salt (g) (see text for ref.). The experimental rate constant (Exp) correspond to the fast component of Tl^+ exchange indicated in Table I. The membrane permeability (P) was calculated according to equation 3, using the values of the experimental rate constant.

ing. The measured flux rates are compatible with a channel-mediated process. However, the rates as well as the permeability values calculated from them should be considered with caution. We assumed a vesicle population of homogeneous size, which is not the real situation. The rate constant is proportional to the vesicle volume, which means that small differences in the vesicle diameter will produce large changes in the rate of ion exchange and therefore in the calculated permeability. Moreover, in order to calculate the rate constant of ion flux and the SR membrane permeability from the single-channel conductance (eq. 2), we assumed conditions of steady-state flux and zero voltage. This is not the case in our experiments. The approximation to zero-voltage conditions is not unreasonable, since the calculated diffusion potentials are rather small (25 mV for Li^+, which is the least permeable of the three ions studied that permeate the channel); therefore, the values obtained will be on the lower limit of the real value. On the other hand, the flux will never be at steady state because the concentration of the internal cation, as well as the potential across the membrane, will be decreasing with time. Therefore, by using the ionic conditions at the time of mixing, we will be calculating the maximum rate of ion movement, but the measured rates will probably be lower than the real value. All in all, the conclusion is that we can consider the obtained values of flux rate and membrane permeability as order-of-magnitude measurements only. Despite this limitation, this technique allows the estimation of ionic permeabilities several orders of magnitude higher than other techniques such as isotope exchange (McKinley and Meissner, 1978).

This method could also be used to determine anion permeability, using anions such as Br^- as quenchers of a positively charged fluorescent molecule. In general, the main difficulties in using fluorescence quenching to measure channel-mediated ion fluxes will be determined by the loading of the fluorescent probe and by the selectivity of the channel studied: the molecule used as quencher should be the most permeable of the molecules to be studied.

3. CHANNEL-MEDIATED ION FLUXES MEASURED BY LIGHT SCATTERING

3.1. Rationale

The scattering of light by a suspension of particles can give information about the molecular weight of the particles as well as their size and shape (Stacey, 1956). For small particles, i.e., smaller than 1/20 of the wavelength, the amount of light scattered is directly proportional to the molecular weight of the particle. For particles of about the same size as the wavelength, the amount of light scattered is not directly proportional to the molecular weight, because larger particles have more than one scattering center and, therefore, destructive interference occurs as a result of the differences in phase of the light scattered from each center. In the case of large particles, measurements of the amount of light scattered at different angles give information about their size and shape.

The scattering properties of large particles have been used to determine membrane permeability. When a suspension of red blood cells (Sha'afi *et al.*, 1967) or *Torpedo* electroplax vesicles (White and Miller, 1981) is transferred from a medium of low osmolarity to a medium of higher osmolarity, the cells or vesicles shrink. Water, which is more permeable than solutes, moves out of the cell or vesicle to compensate for the difference in osmolarity, and therefore a decrease in volume takes place. On shrinking, the scattering centers of the particles move closer together, and thus, the destructive interference between them decreases. As a result, the amount of light scattered by the suspension increases. When the vesicles reswell as a result of influx of solute and water, the amount of light scattered returns to the initial value.

The rate of light scattering change is given by the rate of volume change; i.e., it depends on the permeability of the membrane to water and solutes. Therefore, if water is more permeable than the solutes, the vesicles will shrink; from the rate of shrinking we can calculate the permeability of water. Reswelling occurs when solute followed by water enters the vesicles. Obviously, the rate of reswelling will be determined by the rate of solute influx. If the solute is a dissociated salt, the rate of solute influx will be limited by the movement of the least permeable ionic species. Thus, using different salts in which either the cation or the anion is kept constant, we should be able to determine the membrane permeability to a large number of ions.

3.2. Experimental

The use of this technique is illustrated with the same example as in the case of fluorescence quenching. Isolated SR vesicles at approximately 1 mg protein/ml are equilibrated in 10 mM buffer (MES), pH 7.0, by overnight incubation on ice. The equilibrated vesicles are diluted to 0.2 mg/ml and degassed for 15–30 min. The vesicles at room temperature are mixed in a stopped-flow apparatus with a solution of higher osmolarity, and the changes in the amount of light scattered at 90° are recorded. All the solutions used should be degassed and filtered, preferably two or three times through a filter that does not release particles (Nuclepore™ filters of polycarbonate, 0.2-μm pore size, seem to be a good choice).

3.3. Results

A typical experiment is shown in Fig. 5A. Sarcoplasmic reticulum vesicles in 10 mM MES, pH 7.0, 1 mM EGTA, and 1 mM $CaCl_2$ were mixed with either 100 mM KCl or 100 mM glucose. As expected, the amount of light scattered by the vesicles increased, reflecting shrinking, followed by a slower return to the base line as the vesicles reswelled. From the rate of reswelling we can say that the permeability of the vesicles to glucose is much lower than that to KCl. However, the extent of shrinking in glucose was much larger than in KCl, even though the osmolarity of the KCl solution was twice that of glucose.

An even more striking result is observed in Fig. 5B. The same SR suspension used in the previous experiment was treated with melittin, a small peptide from bee venom that is a nonspecific ionophore. Independent experiments using radioactive tracers showed us that this treatment renders the SR vesicles permeable to solutes such as Ca^{2+}, glucose, and

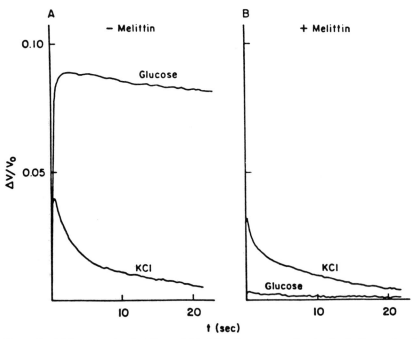

Figure 5. Light scattering by SR vesicles following an osmotic shock. A: Sarcoplasmic reticulum vesicles (0.2 mg/ml) in 1 mM EGTA, 1 mM $CaCl_2$, 10 mM MES, pH 6.5, were mixed in a stopped-flow apparatus with a solution supplemented with 100 mM KCl or glucose, as indicated in the figure. The increase in light scattering corresponds to vesicle shrinking. B: Experiment done under the same conditions except that the vesicles were preincubated with 40 µg/ml melittin for 30 min.

methionine. As expected, the osmotic response of the vesicles to glucose is abolished. However, the response to KCl is almost unaffected.

Our interpretation of these results is that the observed light scattering changes by an SR suspension are not caused solely by volume changes of the vesicles. It is possible that other membrane properties are affected by changes in ionic strength rather than by changes in osmolarity and that this is reflected in the light-scattering behavior. Therefore, since the results obtained with this type of experiment depend on the conditions of the medium, we conclude that it can not be validly used to measure permeabilities of SR vesicles.

3.4. Discussion

The main conclusion of the results shown here is that this technique cannot be widely applied to study membrane permeability. The fact that this method was successfully used in red blood cells (Sha'afi *et al.*, 1967) and *Torpedo* electroplax vesicles (White and Miller, 1981) does not mean that its use can be generalized. Moreover, whenever this technique has been used, including measurements in SR vesicles (Kometani and Kasai, 1978), the assumption is made that water is more permeable than any solute. This might not be true. The water permeability of membranes in general is on the order of 10^{-3} cm/sec (Andersen, 1978). If we consider the previously calculated SR K^+ permeability ($>10^{-3}$ cm/sec), and the fact that from planar bilayer experiments Cl^- is only slightly less permeable than K^+, the result would be that KCl can move as fast as water across the SR membrane. In other words, we would not expect to find SR vesicle shrinking when a KCl osmotic shock is given. This conclusion is supported by the experiments previously shown.

In summary, I have described here two techniques that might be used to measure channel-mediated ion fluxes in isolated vesicles or liposomes. Both techniques have advantages and disadvantages. The main characteristic of these techniques is their high time resolution. However, a method to measure channel-mediated ion fluxes by conventional radioisotope techniques has also been described (Garty *et al.*, 1983), which has the advantage that no expensive equipment is needed.

REFERENCES

Andersen, O., 1978, Permeability properties of unmodified lipid bilayer membranes, in *Membrane Transport in Biology*, Vol. I (D. Tosteson, ed.), Springer-Verlag, Berlin, Heidelberg, New York, pp. 369–439.

Coronado, R., and Miller, C., 1982, Conduction and block by organic cations in a K^+-

selective channel from sarcoplasmic reticulum incorporated into planar bilayers, *J. Gen. Physiol.* **79**:529–547.

Coronado, R., Rosenberg, R., and Miller, C, 1980, Ionic selectivity, saturation and block in a K^+-selective channel from sarcoplasmic reticulum, *J. Gen. Physiol.* **76**:425–446.

Eftink, M., and Ghiron, C., 1981, Fluorescence quenching studies with proteins, *Anal. Biochem.* **114**:199–227.

Garcia, A. M., and Miller, C., 1984, Channel-mediated monovalent cation fluxes in isolated sarcoplasmic reticulum vesicles, *J. Gen. Physiol.* **83**:819–839.

Garty, H., Rudy, B., and Karlish, S., 1983, A simple and sensitive procedure for measuring isotope fluxes through ion specific channels in heterogeneous populations of membrane vesicles, *J. Biol. Chem.* **258**:13094–13099.

Kometani, T., and Kasai, M., 1978, Ionic permeability of sarcoplasmic reticulum vesicles measured by light scattering method, *J. Membr. Biol.* **41**:295–308.

Latorre, R., Vergara, C., and Hidalgo, C., 1982, Reconstitution in planar bilayers of a Ca^{2+}-dependent K^+ channel from transverse tubule membranes isolated from rabbit skeletal muscle, *Proc. Natl. Acad. Sci. U.S.A.* **79**:805–809.

McKinley, D., and Meissner, G., 1978, Evidence for a K^+, Na^+ permeable channel in sarcoplasmic reticulum, *J. Membr. Biol.* **44**:159–186.

Miller, C., 1978, Voltage-gated cation conductance channel from fragmented sarcoplasmic reticulum: Steady-state electrical properties, *J. Membr. Biol.* **40**:1–23.

Miller, C., 1983a, Integral membrane channels: Studies in model membranes, *Physiol. Rev.* **63**:1209–1242.

Miller, C., 1983b, First steps in the reconstruction of ionic channel function in model membranes, in *Current Methods in Cellular Neurobiology* Vol. 3, (J. Barber, ed), pp. 1–37, John Wiley & Sons, New York.

Miller, C., and Racker, E., 1976, Ca^{2+}-induced fusion of fragmented sarcoplasmic reticulum with artificial planar bilayers, *J. Membr. Biol.* **30**:283–300.

Miller, C., and Rosenberg, R., 1979, Modification of a voltage-gated K^+ channel from sarcoplasmic reticulum by a pronase-derived endopeptidase, *J. Gen. Physiol.* **74**:457–478.

Moore, H.-P., and Raftery, M., 1980, Direct spectroscopic studies of cation translocation by *Torpedo* acetylcholine receptor on a time scale of physiological relevance, *Proc. Natl. Acad. Sci. U.S.A.* **77**:4509–4513.

Sha'afi, R., Rich, G., Sidel, V., and Solomon, A., 1967, The effect of unstirred layer on human red cell water permeability, *J. Gen. Physiol.* **50**:1377–1399.

Stacey, K., 1956, *Light Scattering in Physical Chemistry*, Academic Press, New York.

Sten-Knudsen, O., 1978, Passive transport processes, in *Membrane Transport in Biology*, Vol. I (D. Tosteson, ed.), Springer-Verlag, Berlin, Heidelberg, New York, pp. 5–112.

White, M., and Miller, C., 1979, A voltage-gated anion channel from electric organ of *Torpedo californica*, *J. Biol. Chem.* **254**:10160–10166.

White, M., and Miller, C., 1981, Chloride permeability of membrane vesicles isolated from *Torpedo californica* electroplax, *Biophys. J.* **35**:455–462.

Chapter 8

Optical Studies on Ionic Channels in Intact Vertebrate Nerve Terminals

Brian M. Salzberg, Ana Lia Obaid, and Harold Gainer

1. INTRODUCTION

Optical techniques for the detection of transmembrane electrical events have found a variety of applications over the past decade because they offer certain advantages over more conventional measurements. Because the membranes of interest are not mechanically violated, the methods may be relatively noninvasive. Spatial resolution is limited only by microscope optics and noise considerations; it is possible to measure changes in membrane potential from regions of a cell having linear dimensions on the order of 1 μm. Temporal resolution is limited by the response time of the probes and by the bandwidth imposed on the measurement, again by noise considerations, and response times faster than any known membrane time constant may be achieved. Because mechanical access is not required, unusual latitude is possible in the choice of preparation, and voltage changes may be monitored in membranes that are otherwise inaccessible. Finally, since no recording electrodes are employed, and the measurement is actually made at a distance from the preparation—in the image plane of an optical apparatus—it is possible to record changes in potential simultaneously from a large number of discrete sites. Several

---

BRIAN M. SALZBERG, AND ANA LIA OBAID • University of Pennsylvania, Philadelphia, Pennsylvania 19104.   HAROLD GAINER • National Institutes of Health, Bethesda, Maryland 20205.

recent reviews have surveyed the literature on optical measurement of membrane potential (Cohen and Salzberg, 1978; Salzberg, 1983; Waggoner, 1979; Freedman and Laris, 1981), and they should be consulted by the reader interested in these techniques *per se*.

Here, we begin by reviewing briefly the evidence that optical methods are equivalent, at least in a limited sense, to electrode measurements, but we do not consider technical details concerned with signal-to-noise enhancement or with the selection of appropriate molecular indicators. These are treated at length elsewhere (Salzberg, 1983; Cohen *et al.*, 1974; Ross *et al.*, 1977; Gupta *et al.*, 1981; Salzberg *et al.*, 1977; Grinvald *et al.*, 1982). We then proceed to illustrate the application of optical techniques to a problem that has proven difficult to approach by alternative means, i.e., direct recording of the action potential from the nerve terminals of a vertebrate, and the study of its ionic basis.

2. EQUIVALENCE OF OPTICAL AND ELECTRICAL MEASUREMENTS OF MEMBRANE POTENTIAL

An electrode measurement can provide continuous information about the absolute value and time course of a varying transmembrane voltage difference. Under the circumstances encountered most frequently, noise in the electrical recording may be reduced to negligible proportions, and the bandwidth is generally broad enough so as to introduce little or no distortion of the time course or frequency spectrum of the biological events of interest. The degree to which an optical measurement of membrane potential approximates this ideal is best examined in a preparation that permits simultaneous recordings by both techniques and that allows for both spatial and temporal control of the membrane potential. This is, of course, readily achieved in the giant axons of the squid, where the large diameter of the fibers permits the insertion of electrodes that reduce the longitudinal resistance of the cell and that, through a feedback circuit, uniformly control the transmembrane voltage over the large membrane area that is monitored optically.

The apparatus used to study the relationship of the optical changes to electrical events occurring in and across the membrane of the squid axon is illustrated schematically in Fig. 1 (Cohen *et al.*, 1974; Conti, 1975; Ross *et al.*, 1977; Gupta *et al.*, 1981). This apparatus could be used to study changes in extrinsic (i.e., dye-related) absorption, fluorescence, light scattering, intrinsic and extrinsic birefringence, and dichroism in cleaned intact axons. A modified version of this arrangement, in which the axon is mounted horizontally in a voltage-clamp chamber that is incorporated into the stage of a compound microscope, is used in our laboratory to study optical changes in perfused as well as intact giant axons.

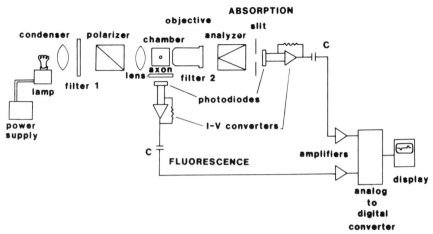

Figure 1. Schematic drawing of the apparatus used to obtain simultaneous optical and electrode measurements from giant axons of the squid *Loligo pealei*. Filter 1, a multicavity interference filter, has a full width at half height of about 30 nm. Heat filters are positioned between the condenser and filter 1. For wavelengths shorter than 720 nm, KG-1 heat filters (Schott Optical Glass Co., Duryea, PA) were employed; for longer wavelengths, the heat filters were Schott RGN-9 glass. The interference filters were usually blocked to 1.3 μm. The barrier filters (filter 2) were long-wavelength-pass filters from Schott Optical Glass Co. For absorption, linear dichroism, and birefringence measurements, an image of the axon was formed with a 10×, 0.22 n.a. objective lens, and variable slits in the objective image plane were used to block the light that did not pass through the axon. For polarized light measurements, the polarizer was usually a calcite Glan–Thompson, and the analyzer was a calcite Ahrens polarizer. The calcite polarizers (Karl Lambrecht Corp., Chicago, IL) provided better extinction than sheet polarizers, particularly at wavelengths longer than 800 nm. The photodetectors were either PV 444 or SGD 444, manufactured by EG&G, Salem, MA. The coupling capacitors, C, block the steady-state and low-frequency components of the optical signal; to measure resting light levels, the capacitors are bypassed. The coupling time constant was set to be greater than 10 times the sweep duration. The high-frequency response time constants were limited by means of the high-frequency cutoff on Tektronix 3A9 or 5A22 amplifiers. The amplifier outputs were converted to digital form for storage and subsequent processing. (After Gupta *et al.*, 1981.)

Light from a tungsten–halogen lamp is collimated, passed through interference and heat filters (filter 1), and focused to a spot centered on the space-clamped region of the axon. A barrier filter (filter 2) transmits the emitted fluorescence wavelengths but blocks the scattered incident light from reaching the photodetector in fluorescence experiments. In absorption, dichroism, and birefringence experiments, a microscope objective is used to form an image of a region of the axon, and a variable slit is positioned in the image plane, so that all of the light reaching the detector has passed through the axon. The slit can also be used to restrict the optical measurement to a particular region of the axon (i.e., edge or center) for birefringence studies in which calcite crystal polarizers are inserted in the incident beam and between the objective and the photodetector. Typically, the polarizer axis (electric vector of the incident light) is set at +45° to the longitudinal axis of the fiber, and the analyzer (second polarizer) axis is crossed at −45°. These polarizing components are removed from the light path for fluorescence and absorption measurements, and for dichroism studies, only the analyzer is removed. For absorption experiments of the kind shown in Fig. 2, the objective lens confines the field of view to a 1.5-mm length of space-clamped giant axon.

The cleaned giant axon is incubated in a seawater solution of some probe, and the fluorescence emission, absorption, linear dichroism, or birefringence can be measured during a voltage-clamp step. (Linear dichroism experiments require two measurements, one with the polarizer perpendicular and one with it parallel to the longitudinal axis of the axon.) The results of such an experiment using a merocyanine–rhodanine dye (Salzberg et al., 1976; dye XVII of Ross et al., 1977) are shown in Fig. 2. The absorption change (dotted trace) recorded during a membrane action potential in a space-clamped region of a squid giant axon (Fig. 2A) has a time course that closely resembles the change in membrane potential recorded with conventional platinized platinum electrodes (thin solid trace). The result that the optical signal is related to the membrane potential itself, rather than to the associated changes in conductance or current, is established unambiguously only by voltage-clamp experiments of the kind illustrated in Fig. 2B (Davila et al., 1974; Cohen et al., 1974; Ross et al., 1977; Gupta et al., 1981). The optical changes (top trace) in Fig. 2B that accompany the hyperpolarizing and depolarizing voltage-clamp steps (middle trace) are similar in size and time course and closely follow the membrane potential, although the depolarization is associated with large conductance changes and the ionic currents recorded in the bottom trace and the hyperpolarization was not.

When four voltage-clamp steps of different size were imposed on the membrane during a single sweep, and the resulting optical changes were plotted against the change in membrane potential, the data shown in Fig. 2C were obtained, indicating that this dye behaves as a linear potentiometric probe. This dye and its analogues also respond rapidly, as suggested by Fig. 2A; the absorption at 750 nm of stained axon membrane follows changes in potential with a delay of less than 2 μsec at room temperature (F. Bezanilla, A. L. Obaid, and B. M. Salzberg, unpublished observations). Many of the more than 2000 dyes that have now been studied (Cohen et al., 1974; Ross et al., 1977; Gupta et al., 1981; Grinvald et al., 1982) exhibit fast linear optical responses to changes in transmembrane potential, and, in some instances, the response can be measured with excellent signal-to-noise ratio. The linearity of the voltage dependence and the accuracy of the temporal response imply that molecular indicators of membrane potential such as merocyanine and styryl dyes (Salzberg et al., 1972; Davila et al., 1973; Cohen et al., 1974; Ross et al., 1977; Gupta et al., 1981; Grinvald et al., 1982; Loew and Simpson, 1981)

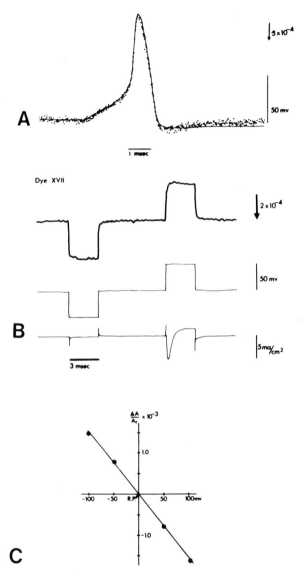

Figure 2. Equivalence of optical and electrode measurement of membrane potential. A: Changes in absorption (dots) of a squid giant axon stained with a merocyanine–rhodanine dye (XVII of Ross et al., 1977) during a membrane action potential. The thin solid trace is the electrode measurement recorded simultaneously. The vertical arrow at the right shows the fractional change in absorption for a single sweep. 750 nm; 32 sweeps; response time constant was 5 μsec. B: Changes in absorption (top trace) of a stained squid giant axon during 50-mV voltage-clamp steps (middle trace). The potential steps are accompanied by the ionic currents shown in the bottom trace. The axon had been stained with the same merocyanine–rhodanine dye. 750 nm; 128 sweeps; 20 μsec optical time constant. C: The absorption change exhibited by this dye is a linear function of membrane potential. (After Ross et al., 1977.)

are capable of monitoring synaptic and electrotonic potentials as well as action potentials, and they have now found wide application in a variety of systems (Salzberg *et al.*, 1983; Cohen and Salzberg, 1978; Waggoner, 1979; Freedman and Laris, 1981). Indeed, the fidelity of the optical response to the true transmembrane potential recently provided the basis for an optical determination of the series resistance, by Ohm's law, in the squid giant axon (Salzberg and Bezanilla, 1983).

Some caveats are required, however. In certain respects, optical measurement of membrane potential by means of these molecular transducers is not equivalent to an electrode measurement. Although the optical changes are frequently linear functions of the transmembrane potential, they depend on a multitude of additional factors that are difficult to estimate, and, therefore, the optical signals can provide no measure of the absolute magnitude of the potential change without a separate calibration. This is an important limitation on the equivalence of optical and electrical measurements of membrane potential. For example, the size of a potential-dependent absorption change will be proportional to the product of the extinction change of the dye, $\Delta\epsilon$, and the number of dye molecules that participate in the extinction change, Δn. Fluorescence changes depend, in addition, on changes in the quantum efficiencies of the dye molecules as well as on nonspecific fluorescence from dye molecules free in solution and bound to sites where the potential does not change (Waggoner and Grinvald, 1977).

Thus, the size of the optical signal depends on the membrane area from which it is measured, possibly the membrane topology, the amount and state of aggregation of bound probe, the sites of probe binding, and the sensitivity of the probe to other changes that may accompany the change in potential as well as on the collection efficiency and quantum efficiency of the apparatus (Salzberg *et al.*, 1977; Salzberg, 1983). Even if an electrode calibration of the absolute magnitude of the potential change associated with a given optical signal can be obtained, it will remain valid only transiently, since photolysis and changes in bound dye concentration may occur rapidly. [Probes that are truly electrochromic (Platt, 1961) might offer the possibility of an absolute voltage calibration if the wavelength shift in a given membrane system could be demonstrated to depend solely on potential, but no sensitive potentiometric probe has yet been shown unequivocally to respond electrochromically to membrane potential changes (see Loew *et al.*, 1985).]

In many circumstances, however, a faithful optical transcription of the time course of a transmembrane potential change may provide useful information that is otherwise obtainable only with difficulty or not at all. This seems clearly the case in the experiments on amphibian nerve terminals that are illustrated in the sections that follow.

3. OPTICAL RECORDING OF ACTION POTENTIALS FROM NERVE TERMINALS OF THE FROG *XENOPUS*

Our understanding of the physiology of synaptic transmission in the vertebrates has been limited by our inability to monitor directly the action potential in the nerve terminals themselves. The giant synapse of the squid has provided a very successful model for the study of the release of transmitter substances (Bullock and Hagiwara, 1957; Katz and Miledi, 1967; Llinas *et al.*, 1976) because in this preparation, both the postsynaptic and presynaptic membranes are accessible for electrophysiological investigation including voltage clamp. Much of our present knowledge of vertebrate synaptic physiology is derived from the study of the frog neuromuscular junction (Katz, 1969), but in this instance "the small size of the nerve terminal prevents a direct measurement of the presynaptic potential change" (Katz, 1969).

The vertebrate hypothalamo–neurohypophyseal system represents an alternative model for the study of excitation–secretion coupling at nerve terminals. Magnocellular neurons located in the hypothalamus (supraoptic and paraventricular nuclei in mammals, preoptic nucleus of lower vertebrates) project their axons as bundles of fibers through the median eminence and infundibular stalk to terminate in the neurohypophysis, where the neurohypophysial peptides and proteins (neurophysins) are secreted into the circulation. Indeed, the neurohypophysis has been a classical *in vitro* preparation for measuring the calcium-dependent release of peptide hormones under various conditions of stimulation (Douglas, 1963; Douglas and Poisner, 1964), and extracellularly recorded compound action potentials are readily measured in this organ (Dreifuss *et al.*, 1971). The analysis of excitation- secretion coupling in the neurohypophysis has been hindered, however, by the circumstance that these nerve terminals are also too small (0.5–1.0 μm) for intracellular measurement of the electrophysiological events that affect release.

In the remainder of this chapter, we demonstrate the use of optical probes to monitor rapid changes in membrane potential from a population of nerve terminals in the posterior pituitary (neurohypophysis) of *Xenopus*. Resting and slowly changing transmembrane voltages in synaptosomes prepared from rat brain homogenates have already been measured using cyanine (Blaustein and Goldring, 1975) and merocyanine (Kamino and Inouye, 1978) dyes, and action potentials have been recorded from the growth cones of tumor cells in tissue culture using an oxonol dye (Grinvald and Farber, 1981). We describe here the recording of action potentials from a population of nerve terminals in the intact neurohypophysis of *Xenopus* and the manipulation of the shape of the action potential by extracellular calcium and other agents known to alter the release of neurohormones and neurotransmitters. We also show that when voltage-sensitive Na^+ channels are blocked by tetrodotoxin (TTX) and K^+ channels are blocked by tetraethylammonium (TEA), direct electric field stimulation of the nerve terminals evokes large active responses that probably arise from an inward Ca^{2+} current associated with hormone release from these nerve terminals.

Figure 3B shows an optical recording that represents the intracellular potential changes during the action potentials in a population of synchronously activated nerve terminals in the neurohypophysis of the frog, *Xenopus laevis*. This recording was photographed from an oscilloscope screen

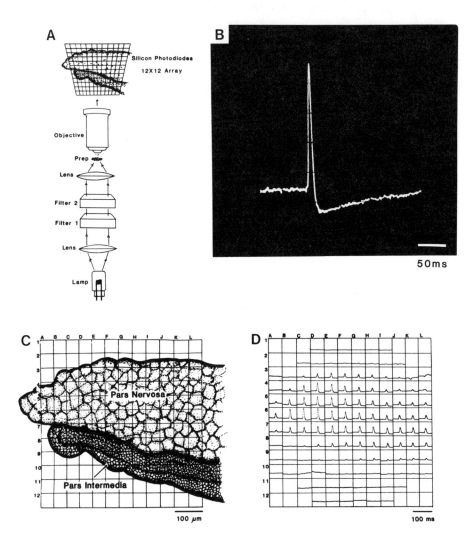

without signal averaging. The technique used here, multiple-site optical recording of transmembrane voltage (MSORTV), is an extension of the method used to record extrinsic absorption changes in the squid giant axon and in many other preparations (Salzberg et al., 1977; Grinvald et al., 1981; Senseman et al., 1983; Salzberg, 1983). The MSORTV system is illustrated schematically to the left of the optical spike (Fig. 3A). Light from a tungsten–halogen lamp was collimated, made quasimonochromatic with a heat filter (KG-1, Schott Optical Co., Duryea, PA) and an interference filter (700 nm; 70-nm full width at half-maximum), and focused by means of a bright-field condenser onto the pars nervosa of the posterior pituitary. The pars nervosa had been excised together with the hypothalamus and the infundibulum, and the entire preparation was vitally stained by incubating it for 25 min in a 100 µg/ml solution of the merocyanine–rhodanine dye, NK 2761 (Kamino et al., 1981; Gupta et al., 1981) (Nippon Kankoh Shikiso Kenkyusho, Inc., Okayama, Japan) in *Xenopus* Ringer's solution (112 mM NaCl; 2 mM KCl; 2 mM $CaCl_2$; 15 mM HEPES; 6 mM glucose; pH adjusted to 7.35).

Light transmitted by the stained preparation was collected by a high numerical aperture water-immersion objective, which formed a real image on a 12 × 12-element silicon photodiode matrix array (MD 144-O; Integrated Photomatrix, Inc., Mountainside, NJ) located in the image plane of a compound microscope (UEM, Carl Zeiss, Inc., Oberkochen, West

←

Figure 3. Multiple-site optical recording of transmembrane voltage (MSORTV) from nerve terminals in the neurohypophysis of *Xenopus*. A: Schematic diagram of the optical portion of the MSORTV system. Collimated light from an incandescent source is made quasimonochromatic with interference and heat filters and focused on the preparation by means of a bright-field condenser with numerical aperture matched to that of the objective. A high-numerical-aperture water-immersion objective projects a real image of a portion of the preparation onto the central 124 elements of a 144-element photodiode matrix array, whose photocurrent outputs are converted to voltages, AC coupled and amplified, multiplexed, digitized, and stored in a PDP 11/34A computer under direct memory access (DMA). A full frame is recorded, with an effective resolution of 18 bits, every 800 µsec. B: A photograph of an oscilloscope recording of the change in the transmitted intensity monitored by one channel (element E5) of the MSORTV system following a single brief shock to the hypothalamus. This element monitored intensity changes from a region of the posterior pituitary entirely within the pars nervosa. The fractional change in intensity during the action potential was approximately 0.25%. Single sweep; 722 ± 21 nm; rise time of the optical system (10–90%) 1.1 msec; AC-coupling time constant 1.0 sec. C: Drawing of the region of the posterior pituitary imaged on the photodiode matrix array, showing the positions of the individual detector elements with respect to the tissue. Drawn from a photomicrograph of the preparation and a transparent overlay representing the detector array. Objective 20×, 0.33 n.a. D: Extrinsic absorption changes obtained in a single sweep, following stimulation of the hypothalamus, superimposed on corresponding elements of the photodiode matrix array. The five elements in each corner of the array were not connected. The largest signals represent fractional changes in transmitted intensity of approximately 0.3%. (After Salzberg et al., 1983.)

Germany). The photocurrents generated by the central 124 array elements were separately converted to voltages and amplified as described previously by Salzberg et al. (1977) and Grinvald et al. (1981). In most of the experiments, all of the amplifier outputs were passed to a data acquisition system based on a PDP 11/34A computer (Digital Equipment Corp., Maynard, MA) capable of acquiring a complete frame every 0.8 msec with an effective resolution of 18 bits. This data acquisition system is similar to that described previously by Grinvald et al. (1981) and employed by Grinvald et al. (1982), Senseman et al. (1983), and Salzberg et al. (1983). In some experiments, the temporal resolution was markedly improved when selected outputs of the current-to-voltage converters were also passed in parallel to a 16-channel signal averager (TN 1500; Tracor Northern, Inc., Middleton, WI). These digitized output signals could be stored on magnetic tape for later display and analysis. A map of the preparation was obtained by superimposing a transparent overlay representing the photodiode array elements on a photograph taken through the trinocular tube of the microscope after removing the photodiode array housing.

The experiment shown in Fig. 3A employed a 20×, 0.33 numerical aperture water-immersion objective (Nikon) to image part of the neurointermediate lobe (pars nervosa plus pars intermedia) onto the photodetector array. Figure 3C is a drawing of the projected region of the gland, prepared from the photomicrograph. Figure 3D shows the MSORTV display of the action potentials recorded optically, in a single sweep, from different areas of the pars nervosa following a single brief stimulus applied to the hypothalamus. It is observed that the optical signals are confined to elements of the grid that correspond to the neurohypophysis; note that a small portion of the neurohypophysis lies under the pars intermedia (row 9). The optical signal illustrated in Fig. 3B represents the analogue output from channel E5. This record, which is typical, illustrates the large signal-to-noise ratio that may be achieved in these experiments. A variety of control experiments showed that these signals represent voltage changes; e.g., no signals could be recorded in the absence of illumination or in white light, and the signals could be inverted at shorter wavelengths (Senseman and Salzberg, 1980).

Evidence that the optical signals shown in Fig. 3 do, in fact, represent the membrane potential changes associated with the invasion of the nerve terminals of the neurohypophysis by the action potential is provided indirectly by some morphometric observations. In the rat, it is estimated that the 18,000 magnocellular neurons in the hypothalamus give rise to approximately 40,000,000 terminal endings in the pars nervosa of the posterior pituitary, and morphometric data suggest that here, secretory endings and swellings comprise 99% of the excitable membrane area, with axons making up the remaining 1% (Nordmann, 1977). Comparably detailed morphometric studies do not yet exist for the neurohypophysis of *Xenopus,* but the ultrastructural resemblance of the anuran neurohypophysis to that of the rat (Rodriguez and Dellman, 1970; Dellman, 1973; Gerschenfeld *et al.,* 1960) suggests that, in *Xenopus,* it is also very heavily enriched with nerve terminals. Thus, we infer that the optical signals shown in Fig. 3, originating in the lateral regions of the pars nervosa, are dominated by the membrane potential changes in the terminals of the magnocellular neurons. (For additional arguments, see Salzberg *et al.,* 1983.)

4. PROPERTIES OF THE ACTION POTENTIAL IN THE NERVE TERMINALS

The nerve terminal action potentials shown in Fig. 3 exhibit two rather unexpected features. First, they have relatively long durations (i.e., full width at half height of about 6 msec) compared with those recorded from vertebrate axons (e.g., <2 msec in the frog; Stampfli and Hille, 1976). There must, of course, be some temporal dispersion in the population recorded by a single photodetector element, and this represents, therefore, an upper bound on the spike width. However, internal evidence suggests that there is very little temporal dispersion, since the spike width is constant over a considerable range of objective magnifications. Also, the value for the width of the action potential is in good agreement with the 5- to 6-msec duration reported for the action potentials of magnocellular neuron somata (Poulain and Wakerley, 1982) and the compound action potential in the neurohypophysis (Dreifuss *et al.,* 1971). The more remarkable characteristic of the nerve terminal action potential illustrated in Fig. 3 is its conspicuous afterhyperpolarization, which is reminiscent of the afterhyperpolarization frequently associated with a calcium-mediated potassium conductance [e.g., snail neurons (Meech and Strumwasser, 1970), rat myotube (Barrett *et al.,* 1982), guinea pig Purkinje cells (Llinas and Sugimori, 1980a,b)]. We found that it was possible to alter the afterhyperpolarization recorded optically by changing the ionic composition of the bath.

Figure 4 shows the results of a series of experiments designed to examine the calcium dependence of this long-lasting hyperpolarization. The traces shown here represent typical outputs from single channels of the 124-channel MSORTV system; each trace was recorded in a single sweep. In each row, the optical signal on the left (NR) was recorded in normal Ringer's solution, which contained 2 mM Ca^{2+}. These control records are normalized to the same peak height. The middle trace in Fig. 4A (High Ca) shows the effect of a 10-min exposure to elevated (5 mM)

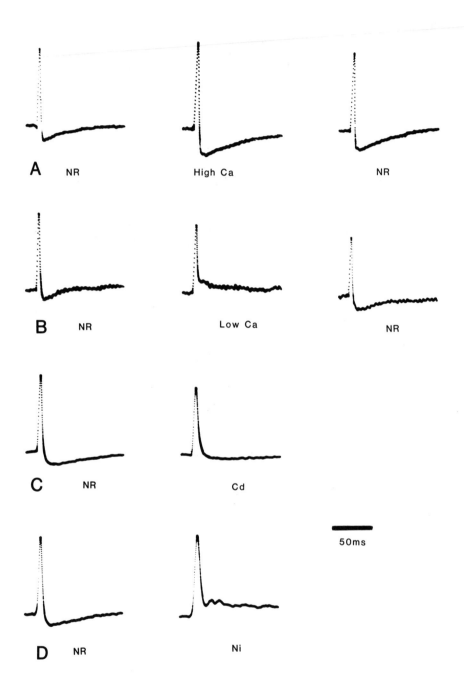

extracellular Ca^{2+} concentration; the afterhyperpolarization increased by approximately 90%. This probably represents a lower limit, since it is likely that some dye was bleached during the measurements, and some diminution of the signal is almost inevitable. We always found a large increase in the size of the afterhyperpolarization, although the magnitude of the effect was variable. An increase in the height (14% in Fig. 4A) was often noted; this finding suggests that an inward Ca^{2+} current might contribute to the rising phase of the action potential (see below). The record on the right in Fig. 4A (NR) shows the optical signal from the same region of the neurohypophysis 10 min after the return to 2 mM Ca^{2+}. The optical traces in Fig. 4B show the effect of reducing extracellular Ca^{2+} to 0.1 mM by replacing the Ca^{2+} with Mg^{2+}. The middle trace (Low Ca) shows that after 10 min, the undershoot is completely absent, replaced by a long-lasting depolarization. The magnitude of this effect also was variable from one experiment to the next; however, a reduction of extracellular calcium to 0.1 mM or less never failed to reduce the afterhyperpolarization. Further lowering of the extracellular Ca^{2+} occasionally eliminated the action potential entirely, but we consider this to be a consequence of steady-state depolarization to which the optical measurements are insensitive. The right-hand trace in Fig. 4B (NR) again illustrates the recovery (10 min following return to 2 mM Ca^{2+}). In this experiment, reduction of the magnitude of the signal resulting from bleaching was more apparent.

Although the effects of changes in extracellular Ca^{2+} were always

←
───

Figure 4. Calcium dependence of the long-lasting afterhyperpolarization of the action potential recorded from nerve terminals of the neurohypophysis. A: Effect of elevated Ca^{2+} on the nerve terminal action potential. Left-hand trace shows the nerve terminal action potential recorded in normal Ringer's solution (NR; 2 mM Ca^{2+}). This, and all the records that follow it, were obtained in single sweeps from an element of the photodiode array monitoring potential changes from nerve terminals in the neurohypophysis. The action potential shown in the middle trace (High Ca) was recorded after 10 min in a Ringer's solution containing 5 mM Ca^{2+}. The right-hand trace (NR) shows the nerve terminal action potential 10 min following the return to normal (2 mM Ca^{2+}) Ringer's solution. B: Effect of lowered extracellular Ca^{2+} on the shape of the nerve terminal action potential. The left-hand trace (NR) shows the control action potential. The action potential in the middle trace (Low Ca) was recorded 10 min after replacement of the bathing solution with one containing 0.1 mM Ca^{2+} and 2 mM Mg^{2+}. The right hand trace (NR) shows the nerve terminal action potential 10 min following return to control (2 mM Ca^{2+}) Ringer's solution. Some bleaching is evident from the reduction in overall signal size. C: Effect of 0.2 mM Cd^{2+} on the shape of the nerve terminal action potential. Left-hand trace (NR) shows the action potential in normal Ringer's solution. Right-hand trace (Cd) shows the action potential 2 min following the addition of 0.2 mM Cd^{2+} to the normal Ringer's solution. This effect could not be reversed. D: Effect of 1 mM Ni^{2+} on the shape of the nerve terminal action potential. Left trace (NR) shows the control action potential. Right-hand trace (Ni) shows the action potential 2 min following the addition of 1 mM Ni^{2+} to normal Ringer's solution. This effect was only partially reversible. (After Salzberg et al., 1983.)

reversible, this was not the case with other ionic substitutions. Cadmium ion is known to block calcium currents in a number of preparations (e.g., Standen, 1981), and Fig. 4C shows that after 2 min in 200 μM Cd^{2+} (trace on the right, Cd), the afterhyperpolarization was almost entirely eliminated. This effect was only partially reversible. Longer exposure to cadmium ion eliminated the afterhyperpolarization entirely. The peak height was also considerably reduced, but, in the absence of recovery, it is impossible to know whether this is, at least partially, an effect of bleaching. Row D (Fig. 4) shows that nickel ion, another inorganic calcium blocker (Kaufmann and Fleckenstein, 1965) at millimolar concentrations, also blocks the spike afterhyperpolarization. In some experiments, Ni^{2+} produced a small but significant increase in the spike height. The origin of this effect is not clear, and although a decrease in sodium inactivation is possible, a change in resting potential has not been ruled out. All of the experiments illustrated in Fig. 4 suggest that calcium plays a significant role in shaping the action potential in the vertebrate nerve terminals studied here. Experiments with other Ca^{2+} antagonists such as Mn^{2+} and Co^{2+} are consistent with this idea, although their potency seems to be less than that of cadmium.

5. IONIC BASIS OF THE DEPOLARIZING PHASE OF THE ACTION POTENTIAL

Because the entry of calcium seems to play a role in excitation–secretion coupling, we attempted to determine whether an inward calcium current contributed measurably to the depolarizing phase of the action potential in these terminals. *A priori,* calcium is unlikely to be the dominant carrier of the inward current in the nerve terminals of the neurohypophysis because the very large surface-to-volume ratio in these small secretory endings would then result in the imposition of a large, probably intolerable calcium load.

Figure 5 illustrates an experiment designed to examine this point. Because calcium antagonists leave the depolarizing phase of the action potential largely intact, sodium may be carrying the bulk of the inward current. If so, it may be blocked by tetrodotoxin (TTX), and a positive result would provide evidence that the rising phase of the action potential is caused primarily by the opening of voltage-dependent sodium channels. In order to test this possibility, it was necessary to change the stimulation conditions, since in the presence of TTX, block of the sodium action potential in the axons of the infundibulum would prevent the excitation of the nerve terminals in the neurohypophysis by blocking spike conduction. We were able to circumvent this difficulty by resorting to a form of

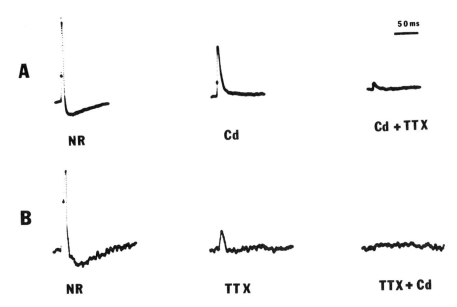

Figure 5. Effects of sodium and calcium channel blockers on the rising phase of the nerve terminal action potential elicited by direct field stimulation of the neurohypophysis. A: Left-hand trace (NR) shows the nerve terminal action potential in normal Ringer's solution. The middle trace (Cd) shows the action potential 18 min after the addition of 0.2 mM Cd^{2+} to the normal Ringer's solution. The right-hand trace (Cd + TTX) shows the optical signal recorded from the same population of nerve terminals 6 min after the addition of 5 μM TTX to normal Ringer's solution containing 0.2 mM Cd^{2+}. B: Left-hand trace (NR) shows the nerve terminal action potential in normal Ringer's solution. The middle trace (TTX) shows the residual potential change 6 min following the addition of 2 μM TTX to the Ringer's solution bathing the preparation. The right-hand trace (TTX + Cd) shows the elimination of this TTX-resistant response when calcium channels are blocked by Cd^{2+}. For this record, the preparation had been bathed in a Ringer's solution containing 2 μM TTX and 0.2 mM Cd^{2+} for 5 min. All traces are single sweeps, 700 nm. Rise time of the light-measuring system (10–90%) was 1.1 msec.

field stimulation in which the lateral tips of the pars nervosa were held by suction electrodes, and very brief (200- to 300-μsec) currents were passed between the two electrodes. In this fashion, the terminals could be excited directly, without the requirement of axonal transmission.

This technique was employed in the experiment shown in Fig. 5. The MSORTV record on the left in Fig. 5A shows the control, in 2 mM Ca^{2+} Ringer's solution. The amplitude of the optical spike increased with field strength, as expected, since a large number of nerve terminals were stimulated and monitored. A maximal response was obtained typically with stimuli of 100 to 200 V and was independent of stimulus polarity. As with the optical signals elicited by stimulation of the hypothalamus, the amplitude of the response was approximately 0.3% of the transmitted light intensity at 700 nm. The middle trace (Cd) was recorded from the same population of nerve terminals following 18 min exposure to 200 μM Cd^{2+} to block any inward calcium current. The reduction in the magnitude of the action potential is obvious, along with a considerable broadening of the spike and the complete loss of the afterhyperpolarization, the latter effects most likely resulting from the elimination of the large calcium-mediated potassium conductance. The extent of the reduction in the height of the action potential may be exaggerated, however, by factors other than blockage of calcium channels (e.g., dye bleaching). The right-hand trace in Fig. 5A (Cd + TTX) shows that 6 min after application of 5 μM TTX, the Cd^{2+}-resistant portion of the terminal action potential was completely eliminated. The residual signal seen in this record is the passive depolarization produced by the field stimulation; it was reversed with reversal of stimulus polarity, and it exhibited the same wavelength dependence as the optical recording of the normal action potential. (It should be noted, however, that the size of the passive electrotonus was very variable from element to element.)

The complementary experiment is shown in Fig. 5B. Here, the sodium current was first blocked with TTX (middle trace, Fig. 5B), and the small residual active response was eliminated by Cd^{2+} (right-hand trace, TTX + Cd) and substantially decreased in size by 1 mM Ni^{2+} (not shown). Thus, the experiments illustrated here suggest that the rising phase of the nerve terminal action potential in the neurohypophysis of *Xenopus* is mediated predominantly by sodium, although these data suggest at least a small calcium component. The magnitude of this calcium component is difficult to assess quantitatively, particularly because the TTX blockade of the normal regenerative sodium depolarization may limit the opening of voltage-dependent calcium channels. The important role for calcium entry in excitation–secretion coupling is fully consistent with its apparently small contribution to the terminal action potential (Katz and Miledi, 1969). In adrenal chromaffin cells, where calcium influx is the critical event in excitation–secretion coupling (Douglas, 1968), the Ca^{2+} component of the action potential is also small (Brandt et al., 1976).

Nonetheless, under conditions in which the terminal membrane depolarization is prolonged, sufficient inward calcium current may develop to give rise to a relatively large calcium action potential. The records shown in Fig. 6 are from one of 24 posterior pituitaries in which TEA was added, in addition to TTX, in order to block the voltage-dependent potassium conductance and thereby prolong the depolarization of the nerve terminal membrane.

Figure 6A shows a control action potential from a population of nerve terminals. The action potential exhibits the characteristic rapid upstroke that we attribute primarily to the effect of a fast inward sodium current,

Figure 6. Cadmium sensitivity of active responses (calcium spikes) elicited by direct field stimulation of the neurohypophysis in the presence of TTX and TEA. Optical recording of action potentials from nerve terminals of the *Xenopus* neurohypophysis following staining for 25 min in 0.1 mg/ml NK 2761. Outputs of a single channel from the photodiode matrix array in the MSORTV system. Top: Action potential recorded in normal Ringer's solution. Middle: Active response of the nerve terminals following 20 min exposure to 5 mM TEA plus 2 μM TTX in normal Ringer's solution. Bottom: Passive response (electrotonus) remaining 15 min after the addition of 0.5 mM Cd^{2+} to the TEA–TTX Ringer's solution bathing the preparation on stimulation with normal and reversed polarity. 700 nm. Rise time of the light-measuring system (10–90%) was 1.1 msec. Temperature 18–22°C. (After Obaid *et al.*, 1985.)

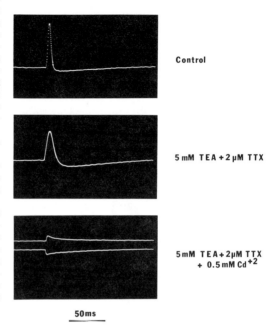

a rapid repolarizing phase resulting from K^+ efflux, and the typical afterhyperpolarization resulting from the Ca^{2+}-mediated increase in K^+ conductance. When 5 mM tetraethylammonium (TEA) was added to the Ringer's solution (not shown), the spike was prolonged, and, in most experiments, a hump appeared on the falling phase, possibly because of repetitive firing of some of the terminals. The addition of a high concentration of TTX, sufficient to block voltage-sensitive sodium channels, abolished the hump and unmasked a large afterhyperpolarization that presumably results from the enhanced activation of a Ca^{2+}-mediated K^+ conductance. Panel B shows the optical response recorded from the same population of terminals represented in Fig. 6A following 20 min exposure to 5 mM TEA and 2 μM TTX. The regenerative response illustrated is smaller and slower than the normal action potential and appears to depend entirely on the influx of calcium and the subsequent rise in a TEA-insensitive K^+ conductance. The further addition of 0.5 mM Cd^{2+} to the bath (Fig. 6C) completely blocked the remaining active responses. (We have already seen in Figs. 4 and 5 that such low levels of cadmium do not block the action potentials in the nerve terminals of the neurohypophysis in the absence of TTX.) The residual optical signals shown in Fig. 6C represent the purely passive electrotonus, as indicated by their symmetric response to stimulus polarity. The combined effects of TEA and Cd^{2+} could generally be reversed, but it was never possible to reverse completely the effect of TTX.

Although the optical signal recorded in Fig. 6B has the appearance of a calcium action potential, it is possible that the major contribution to the rising phase of the response in TTX comes from the influx of Na^+ through Ca^{2+} channels, which are not blocked by the toxin. It was possible, in these experiments, to assess qualitatively the relative contributions of Ca^{2+} and Na^+ to the active response by recording optical signals at different $[Ca^{2+}]_o$ at high and low $[Na^+]_o$, and at different $[Na^+]_o$ at high and low $[Ca^{2+}]_o$. These experiments were all carried out in the combined presence of 5 mM TEA and 1 μM TTX.

Figure 7 illustrates the principal results. Panel A shows that at normal $[Na^+]_o$, the amplitude and rate of rise of the upstroke, as well as the

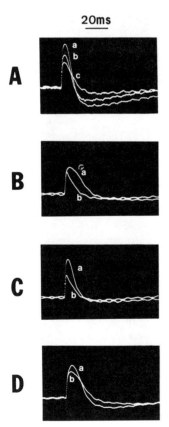

Figure 7. Calcium and sodium dependence of the active response elicited by direct field stimulation of nerve terminals after blocking the voltage-sensitive sodium and potassium channels with TTX and TEA. A: Optical response in the presence of 1 μM TTX and 5 mM TEA at normal sodium concentration (120 mM) and the following Ca^{2+} concentrations (for the times indicated): (a) 10 mM (19 min); (b) 2 mM (20 min); (c) 0.1 mM (1.9 mM Mg^{2+} substitution, 20 min). B: Optical responses in the presence of 1 μM TTX and 5 mM TEA, at low sodium concentration (8 mM, sucrose substitution) and the following calcium concentrations (for the times indicated): (a) 2 mM (16 min); (b) 0.2 mM (1.8 mM Mg^{2+} substitution, 21 min). C: Optical responses in the presence of 1 μM TTX and 5 mM TEA at low calcium concentration (0.2 mM, 1.8 mM Mg^{2+}) and the following sodium concentrations (for the times indicated): (a) 120 mM (normal, 21 min); (b) 8 mM (sucrose substitution, 21 min). D: Optical responses in the presence of 1 μM TTX and 5 mM TEA, at normal calcium concentration (2 mM) and the following sodium concentrations (for the times indicated): (a) 120 mM (normal, 16 min); (b) 8 mM (sucrose substitution, 21 min). The rise time (10–90%) of the light-measuring system was 1.1 msec. Single sweeps. Temperature 18–20°C. (After Obaid et al., 1985.)

amplitude of the afterhyperpolarization, increased as $[Ca^{2+}]_0$ increased. This is, of course, consistent with the idea that Ca^{2+} carries inward current in the presence of both TTX and TEA. At lowered extracellular concentrations of Na^+ (sucrose replacement, Fig. 7B), a decrease in $[Ca^{2+}]_0$ from normal 2 mM (trace a) to 0.2 mM (trace b) eliminated virtually all of the active response. A comparison of trace b in Fig. 7B (low Ca^{2+}, low Na^+) with trace c in Fig. 7A (low Ca^{2+}, normal Na^+) suggests that at low $[Ca^{2+}]_0$, Na^+ might contribute significantly to the inward current conducted by TTX-insensitive channels. On the other hand, at normal $[Ca^{2+}]_0$ (Fig. 7D), varying the $[Na^+]_0$ had little effect on the rate of rise of the optical signal, as if sodium ions are unable to compete effectively with normal concentrations of Ca^{2+}. However, at low $[Ca^{2+}]_0$ (Fig. 7C), the active component of the response was markedly $[Na^+]_0$ sensitive. Thus, it appears that in low $[Ca^{2+}]_0$, sodium ions are able to carry depolarizing membrane current through these cadmium-sensitive (TTX- and TEA-sensitive) calcium channels.

The calcium-mediated active responses recorded optically and described here appear to share some properties with the local subthreshold action potentials described by Hodgkin (1938) in unmyelinated crab nerve; with TTX blocking any propagated action potential in the axons of the infundibulum, the small calcium currents cannot excite a sufficient length of nerve to produce a propagated action potential. Nonetheless, their dependence on $[Ca^{2+}]_0$ and sensitivity to Cd^{2+} block indicates that they probably result from a voltage-sensitive calcium influx into the terminals (Hagiwara and Byerly, 1981; Kostyuk *et al.*, 1977; Obaid *et al.*, 1985), which probably mediates hormone release.

6. CONCLUDING REMARKS

The anatomy of the posterior pituitary provides a unique opportunity for the study *in vitro* of excitation–secretion coupling in the vertebrate: there is no postsynaptic excitable membrane to confound the interpretation of the optical signals (Salzberg *et al.*, 1983; Obaid *et al.*, 1985) and no barrier to the collection of its secretory products. In the mammalian neurohypophysis, large and rapid intrinsic optical changes appear to reflect both the arrival of the action potential at the terminals and some event intimately related to the secretory process itself (Salzberg *et al.*, 1985). We expect that direct optical measurement, using MSORTV, of the transmembrane potential changes from the nerve terminals of the vertebrate neurohypophysis, when correlated with secretion of neurohypophyseal peptides monitored both optically and biochemically, will provide new insights into the electrophysiology of transmitter and hormone release in higher animals.

ACKNOWLEDGMENTS. We are grateful to Professors C. M. Armstrong, L. B. Cohen, M. Morad, and R. K. Orkand for valuable discussions and to Professors Orkand and D. M. Senseman for their participation in some of the experiments discussed. We also thank L. B. Cohen and S. Lesher for providing the software used for data acquisition and display. Supported by USPHS grant NS 16824.

REFERENCES

Barrett, J. N., Magleby, K. L., and Palotta, B. S., 1982, Properties of single calcium activated potassium channels in cultured rat muscle, *J. Physiol. (Lond.)* **331**:211–230.

Blaustein, M. P., and Goldring, J. M., 1975, Membrane potentials in pinched-off presynaptic nerve terminals monitored with a fluorescent probe: Evidence that synaptosomes have potassium diffusion potentials, *J. Physiol. (Lond.)* **247**:589–615.

Brandt, B. L., Hagiwara, S., Kidokoro, Y., and Miyazaki, S., 1976, Action potentials in the rat chromaffin cell and effects of acetylcholine, *J. Physiol. (Lond.)* **263**:417–439.

Bullock, T. H., and Hagiwara, S., 1957, Intracellular recording from the giant synapse of the squid, *J. Gen. Physiol.* **40**:565–577.

Cohen, L. B., and Salzberg, B. M., 1978, Optical measurement of membrane potential, *Rev. Physiol. Biochem. Pharmacol.* **83**:33–88.

Cohen, L. B., Salzberg, B. M., Davila, H. V., Ross, W. N., Landowne, D., Waggoner, A. S., and Wang, C.-H., 1974, Changes in axon fluorescence during activity: Molecular probes of membrane potential, *J. Membr. Biol.* **19**:1–36.

Conti, F., 1975, Fluorescent probes in nerve membranes, *Annu. Rev. Biophys. Bioeng.* **4**:287–310.

Davila, H. V., Salzberg, B. M., Cohen, L. B., and Waggoner, A. S., 1973, A large change in axon fluorescence that provides a promising method for measuring membrane potential, *Nature (New Biol)* **241**:159–160.

Davila, H. V., Cohen, L. B., Salzberg, B. M., and Shrivastav, B. B., 1974, Changes in ANS and TNS fluorescence in giant axons from *Loligo, J. Membr. Biol.* **15**:29–46.

Dellman, H. D., 1973, Degeneration and regeneration of neurosecretory systems, *Int. Rev. Cytol.* **36**:215–315.

Douglas, W. W., 1963, A possible mechanism of neurosecretion-release of vasopressin by depolarization and its dependence on calcium, *Nature* **197**:81–82.

Douglas, W. W., 1968, Stimulus–secretion coupling. The concept and clues from chromaffin and other cells, *Br. J. Pharmacol.* **38**:451–474.

Douglas, W. W., and Poisner, A. M., 1964, Stimulus secretion coupling in a neurosecretory organ and the role of calcium in the release of vasopressin from the neurohypophysis, *J. Physiol. (Lond.)* **172**:1–18.

Dreifuss, J. J., Kalnins, I., Kelley, J. S., and Ruf, K. B., 1971, Action potentials and the release of neurohypophysial hormones *in vitro, J. Physiol. (Lond.)* **215**:805–817.

Freedman, J. C., and Laris, P. C., 1981, Electrophysiology of cells and organelles: Studies with optical potentiometric indicators, *Int. Rev. Cytol. Suppl.* **12**:177–246.

Gerschenfeld, H. M., Tramezzani, J. H., and De Robertis, E., 1960, Ultrastructure and function in neurohypophysis of the toad, *Endocrinology* **66**:741–762.

Grinvald, A., and Farber, I. C., 1981, Optical recording of Ca^{2+} action potentials from growth cones of cultured neurons using a laser microbeam, *Science* **212**:1164–1169.

Grinvald, A., Cohen, L. B., Lesher, S., and Boyle, M. B., 1981, Simultaneous optical monitoring of activity of many neurons in invertebrate ganglia, using a 124 element photodiode array, *J. Neurophysiol.* **45**:829–840.

Grinvald, A., Hildesheim, R., Farber, I. C., and Anglister, L., 1982a, Improved fluorescent probes for the measurement of rapid changes in membrane potential, *Biophys. J.* **39**:301–308.

Grinvald, A., Manker, A., and Segal, M., 1982b, Visualization of the spread of electrical activity in rat hippocampal slices by voltage-sensitive optical probes, *J. Physiol. (Lond.)* **333**:269–291.

Gupta, R. K., Salzberg, B. M., Grinvald, A., Cohen, L. B., Kamino, K., Lesher, S., Boyle, M. B., Waggoner, A. S., and Wang, C.-H., 1981, Improvements in optical methods for measuring rapid changes in membrane potential, *J. Membr. Biol.* **58**:123–137.

Hagiwara, S., and Byerly, L., 1981, Calcium channel, *Annu. Rev. Neurosci.* **4**:69–125.

Hodgkin, A. L., 1938, The subthreshold potentials in a crustacean nerve fibre, *Proc. R. Soc. Lond. (Biol.)* **126**:87–121.

Kamino, K., and Inouye, A., 1978, Evidence for membrane potential changes in isolated synaptic membrane ghosts monitored with a merocyanine dye, *Jpn. J. Physiol.* **28**:225–237.

Kamino, K., Hirota, A., and Fujii, S., 1981, Localization of pacemaking activity in early embryonic heart monitored using voltage sensitive dye, *Nature* **290**:595–597.

Katz, B., 1969, *The Release of Neural Transmitter Substances,* Charles C. Thomas, Springfield, Illinois.

Katz, B., and Miledi, R., 1967, A study of synaptic transmission in the absence of nerve impulses, *J. Physiol. (Lond.)* **192**:407–436.

Katz, B., and Miledi, R., 1969, Tetrodotoxin-resistant electrical activity in presynaptic terminals, *J. Physiol. (Lond.)* **203**:459–487.

Kaufmann, R., and Fleckenstein, A., 1965, Ca^{++}-kompetitive elektro-mechanische Entkoppelung durch Ni^{++}- und Co^{++}-Ionen am Warmblütermyokard, *Pfluegers Arch.* **282**:290–297.

Kostyuk, P. G., and Krishtal, O. A., 1977, Separation of sodium and calcium currents in the somatic membrane of mollusc neurons, *J. Physiol. (Lond.)* **270**:545–568.

Llinas, R., and Sugimori, M., 1980a, Electrophysiological properties of *in vitro* Purkinje cell somata in mammalian cerebellar slices, *J. Physiol. (Lond.)* **305**:171–195.

Llinas, R., and Sugimori, M., 1980b, Electrophysiological properties of *in vitro* Purkinje cell dendrites in mammalian cerebellar slices, *J. Physiol. (Lond.)* **305**:197–213.

Llinas, R., Steinberg, I. Z., and Walton, K., 1976, Presynaptic calcium currents and their relation to synaptic transmission: Voltage clamp study in squid giant synapse and theoretical model for the calcium gate, *Proc. Natl. Acad. Sci. U.S.A.* **73**:2918–2922.

Loew, L. M., and Simpson, L., 1981, Charge shift probes of membrane potential. A probable electrochromic mechanism for ASP probes on a hemispherical lipid bilayer, *Biophys. J.* **34**:353–365.

Loew, L. M., Cohen, L. B., Salzberg, B. M., Obaid, A. L., and Bezanilla, F., 1985, Charge shift probes of membrane potential. Characterization of aminostyrylpyridinium dyes in the squid giant axon, *Biophys. J.* **47**:71–77.

Meech, R. W., and Strumwasser, F., 1970, Intracellular calcium injection activates potassium conductance in *Aplysia* nerve cells, *Fed. Proc.* **29**:834.

Nordmann, J. J., 1977, Ultrastructural morphometry of the rat neurohypophysis, *J. Anat.* **123**:213–218.

Obaid, A. L., Orkand, R. K., Gainer, H., and Salzberg, B. M., 1985, Active calcium responses recorded optically from nerve terminals of the frog neurohypophysis, *J. Gen. Physiol.* **85**:481–489.

Platt, J. R., 1961, Electrochromism, a possible change in color producible in dyes by an electric field, *J. Chem. Phys.* **34**:862–863.

Poulain, D. A., and Wakerley, J. B., 1982, Electrophysiology of hypothalamic magnocellular neurons secreting oxytocin and vasopressin, *Neuroscience* **7**:773–808.

Rodriguez, E. M., and Dellman, H. D., 1970, Hormonal content and ultrastructure of the disconnected neural lobe of the grass frog (*Rana pipiens*), *Gen. Comp. Endocrinol.* **15**:272–288.

Ross, W. N., Salzberg, B. M., Cohen, L. B., Grinvald, A., Davila, H. V., Waggoner, A. S., and Wang, C.-H., 1977, Changes in absorption, fluorescence, dichroism and birefringence in stained giant axons: Optical measurement of membrane potential, *J. Membr. Biol.* **33**:141–183.

Salzberg, B. M., 1983, Optical recording of electrical activity in neurons using molecular probes, in *Current Methods in Cellular Neurobiology*, Vol. 3: *Electrophysiological Techniques* (J. L. Barker and J. E. McKelvey, eds.), John Wiley & Sons, New York.

Salzberg, B. M., and Bezanilla, F., 1983, An optical determination of the series resistance in *Loligo*, *J. Gen. Physiol.* **82**:807–818.

Salzberg, B. M., Davila, H. V., Cohen, L. B., and Waggoner, A. S., 1972, A large change in axon fluorescence, potentially useful in the study of simple nervous systems, *Biol. Bull.* **143**:475.

Salzberg, B. M., Cohen, L. B., Ross, W. N., Waggoner, A. S., and Wang, C.-H., 1976, New and more sensitive molecular probes of membrane potential: Simultaneous optical recordings from several cells in the central nervous system of the leech, *Biophys. J.* **16**:23a.

Salzberg, B. M., Grinvald, A., Cohen, L. B., Davila, H. V., and Ross, W. N., 1977, Optical recording of neuronal activity in an invertebrate central nervous system: Simultaneous monitoring of several neurons, *J. Neurophysiol.* **40**:1281–1291.

Salzberg, B. M., Obaid, A. L., Senseman, D. M., and Gainer, H., 1983, Optical recording of action potentials from vertebrate nerve terminals using potentiometric probes provides evidence for sodium and calcium components, *Nature* **306**:36–40.

Salzberg, B. M., Obaid, A. L., and Gainer, H., 1985, Large and rapid changes in light scattering accompany secretion by nerve terminals in the mammalian neurohypophysis, *J. Gen. Physiol.* **86**:395–411.

Senseman, D. M., and Salzberg, B. M., 1980, Electrical activity in an exocrine gland: Optical recording using a potentiometric dye, *Science* **208**:1269–1271.

Senseman, D. M., Shimizu, H., Horwitz, I. S., and Salzberg, B. M., 1983, Multiple site optical recording of membrane potential from a salivary gland: Interaction of synaptic and electrotonic excitation, *J. Gen. Physiol.* **81**:887–908.

Stampfli, R., and Hille, B., 1976, Electrophysiology of the peripheral myelinated nerve, in *Frog Neurobiology* (R. Llinas and W. Precht, eds.), Springer-Verlag, Berlin, pp. 3–32.

Waggoner, A. S., 1979, Dye indicators of membrane potential, *Annu. Rev. Biophys. Bioeng.*, **8**:47–68.

Waggoner, A. S., and Grinvald, A., 1977, Mechanisms of rapid optical changes of potential sensitive dyes, *Ann. N.Y. Acad. Sci.* **303**:217–242.

Chapter 9

Optical Detection of ATP Release from Stimulated Endocrine Cells
A Universal Marker of Exocytotic Secretion of Hormones

Eduardo Rojas, Rosa Maria Santos, Andres Stutzin, and Harvey B. Pollard

1. INTRODUCTION

In many types of secretory cells the secretory granules and vesicles contain adenosine-5'-triphosphate (ATP) in addition to the specific transmitters or hormones. Examples include chromaffin cells (Winkler and Westhead, 1980), pancreatic B cells (Sussman and Leitner, 1977), nerve cells (Dowdall et al., 1974; Wagner et al., 1978; Pollard and Pappas, 1979), platelets (Code, 1952; Born, 1958; Weil-Malherbe and Bone, 1958), and mast cells (Uvnas, 1971). It should therefore be possible, at least in principle, to obtain quantitative, real-time measurements of the kinetics of secretion of vesicle contents by monitoring the ATP release (Leitner et al., 1975; Ito et al., 1980; Livett and Boksa, 1984) using the luciferase-catalyzed luminescent oxidation of luciferin.

We present here the results obtained with two different types of endocrine cells using a highly purified luciferase preparation. Using this technique, we have measured the kinetics of the Ca^{2+}-dependent acetyl-

---

EDUARDO ROJAS, ROSA MARIA SANTOS, ANDRES STUTZIN, AND HARVEY B. POLLARD • Laboratory of Cell Biology and Genetics, NIADDK, National Institutes of Health, Bethesda, Maryland 20205.

choline-induced ATP release from chromaffin cells. In addition, some of the properties of stimulus–secretion coupling through the cholinergic receptor were determined. The concentration of ATP in the cytosol was estimated by measuring the ATP liberation after digitonin permeabilization (Wilson and Kirshner, 1983; Dunn and Holz, 1983) or after high-electric-field cell permeabilization (Knight and Baker, 1982) in a low-Ca^{2+} medium. The granular and vesicular compartments of ATP were evaluated by solubilizing the intracellular organelles with Triton.

Although maximal stimulation of pancreatic islets releases about 1% of the total cellular insulin and the ATP content of the B-cell granules is smaller (Leitner *et al.*, 1975) than that in chromaffin cells (Winkler and Westhead, 1980), the sensitivity of the technique allows quantitative measurements with about 50 collagenase-isolated islets. We show here that the luciferin–luciferase method can be used to monitor the early events taking place during glucose-stimulated insulin release.

2. METHODOLOGICAL CONSIDERATIONS

2.1. The Luminescent Instrument

A diagram of the instrument used to stimulate the cells (or islets) and to measure the light is presented in Fig. 1. The reaction chamber (upper part) consisted of a small plastic tube with hydrophobic surfaces so as to keep the Krebs solution used confined to the bottom of the tube. Various secretagogues were added in small volumes (10 μl) to the cell (or islet) mixture with a microliter pipetter (Gilson) through a light-tight holder. The temperature of the solution in the tube was controlled at 36°C. Emitted light from the reaction chamber was detected using a photomultiplier (lower part) with the anode held at virtual-ground voltage by means of an amplifier with a negative-feedback resistance (either 10^8 or 10^9 Ω).

2.2. The Luciferin–Luciferase Reaction

Bioluminescence may be defined as the emission of visible radiation by an enzyme-catalyzed reaction. The efficiency of the reaction, i.e., number of photons emitted divided by the number of molecules of substrate oxidized ranges from 0.1 to nearly 1.0 as for the luciferin–luciferase mixture (Eley *et al.*, 1970). During the oxidation of luciferin, 50 to 72 kcal/mol quanta is released. This energy is utilized to create an electron excited state either directly or by energy transfer to fluorescent molecules formed or present during the reaction. The chemical requirements for light emission can be expressed by the following scheme (Cormier *et al.*, 1973):

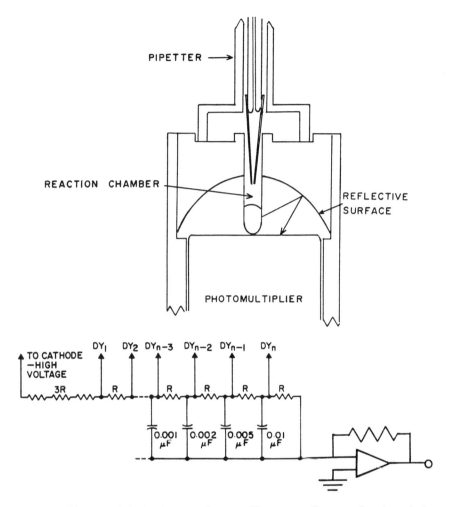

Figure 1. Diagram of the luminescent detector. Upper part: Cross section through the reaction chamber. Lower part: Selected low-dark-current photomultiplier and current/voltage converter (operational amplifier 52K from Analog Devices).

$$\text{Luciferin} + \text{ATP·Mg}^{2+} + O_2$$
$$= CO_2 + \text{oxyluciferin} + \text{AMP} + \text{PP}_i\text{·Mg}^{2+} + h\nu \quad (1)$$

From this general equation we get the expression that actually represents the bioluminescent reaction, namely,

$$\text{Luciferin} + \text{ATP·Mg}^{2+} = \text{luciferyl-adenylate} + \text{PP}_i \quad (2)$$

called the activation step, and

Luciferyl-adenylate + O_2 = AMP + oxyluciferin + CO_2 + $h\nu$ (3)

which is called the redox step.

The highest energy emission corresponds to a wavelength maximum of 546 nm. The optimum pH is 7.75, and the optimum temperature is about 25°C (Biggley et al., 1967).

The technique as used here measures the rate at which light is emitted by the ongoing reaction (not the amount of ATP present). However, it is easy to infer from the rate of reaction how much ATP is present. Figure 2 illustrates the calibration of the detector and the reaction.

It may be seen (Fig. 2, left side) that the light output follows the amount of ATP present in the reaction chamber. For this calibration curve, the concentration of the luciferin–luciferase mixture was kept constant. Similar levels of emitted light were obtained either by single additions of ATP or by consecutive additions. Depending on the concentration of the luciferin–luciferase mixture (Fig. 2, right side), the order of the reaction

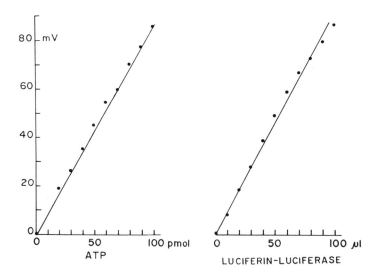

Figure 2. Calibration curve for the luciferin–luciferase reaction. A: Light output (mV) as the function of ATP (pmol) in the reaction chamber. Each vial with the enzyme mixture was dissolved in 5 cm^3 of Krebs solution (stock). Luciferin–luciferase in the reaction medium was kept constant, i.e., 50-μl aliquot from the stock solution. The final volume of the reaction mixture in the chamber was 200 μl. To obtain each point, 10-μl samples of a given ATP solution were added. B: Light output as a function of luciferin–luciferase. All points were obtained in the presence of a constant quantity of ATP (50 pmol) in the reaction medium.

ranges between 1.5 and 2. This implies that the light emitted is proportional to the concentration of ATP added and not to the amount of substrate present.

Emitted light levels during the experiments with chromaffin cells or islets of Langerhans were constant and linearly related to ATP levels up to 1400 pmol in the reaction medium. Light levels, although reduced by the dilution of the ATP during the additions of small volumes, were unaffected by Triton X-100 (up to 1%) and digitonin (up to 200 μM). Furthermore, the reaction was inhibited by K^+ (25% at 100 mM K^+) and Cd^{2+} (50% at 1 mM). The assays with the cells were performed either at 20 or 35°C with similar results. Since glucose-induced insulin release is blocked below 30°C, the assays with the islets were carried out at 36°C.

2.3. Preparation of Chromaffin Cells

Chromaffin cells were prepared from bovine adrenal medulla by collagenase digestion as described previously in detail (Greenberg and Zinder, 1982). The cells were placed in Eagle's essential medium at a concentration of 50 million cells per flask (30 cm^3 of medium) and allowed to recover from the isolation procedure for a 3-day period in a CO_2 incubator at 37°C. After this incubation, the cells were collected by centrifugation and resuspended in a modified Krebs solution (135 mM NaCl, 10 mM Hepes-NaOH at pH 7.2, 5 mM KCl, 2 mM $CaCl_2$, 2 mM $MgCl_2$, and 10 mM D-glucose). The cell preparation was kept at room temperature prior to initiation of experiments at a concentration of 10^7 cells/cm^3 of Krebs solution.

2.4. Collagenase Isolation of Mouse Islets of Langerhans

Islets were isolated by collagenase digestion (Lacy and Kostianovsky, 1967) in the presence of soybean trypsin inhibitor. After three washes in control Krebs solution (135 mM NaCl, 10 mM Na-Hepes, 10 mM $NaHCO_3$, 1 mM $MgCl_2$, 2.6 mM $CaCl_2$, 5 mg/ml bovine serum albumin, pH 7.2), the islets were preincubated at 37°C for 30 min in control Krebs solution plus bovine serum albumin (1 mg/cm^3) and glucose (2.8 mM).

2.5. Luciferin/Luciferase Preparation

The enzyme preparation (from Analytical Luminescence Laboratory, San Diego, CA) consisted of a mixture of highly purified luciferase, purified bovine serum albumin (Calbiochem-Behring), and luciferin. The contents of a sealed vial were dissolved in 5 cm^3 of Krebs solution. The reaction mixture (130 μl Krebs solution, 15 to 25 μl cells, and 50 μl of the luciferin–luciferase mixture) was placed in the plastic tube. Cell density in the reaction mixture was kept below 2.5×10^6 cells/cm^3 in order to maintain ecto-ATPases at low activity. Under resting conditions, the light output of the reaction mixture containing cells was constant at about 0.25% of the maximum signal recorded in each experiment. The digitonin (Calbiochem-Behring) solution used was passed through a cationic exchanger column (Dowex-300).

3. ACETYLCHOLINE-INDUCED ATP RELEASE FROM CHROMAFFIN CELLS: CALCIUM DEPENDENCE

In chromaffin cells, the secretion of catecholamines is evoked by stimulation of a nicotinic cholinergic receptor and depends on calcium in the medium (Douglas and Rubin, 1961; Douglas, 1968; Kidokoro and Ritchie, 1980). As shown in Fig. 3, this is also the case for stimulation of ATP release. For the experiment illustrated in Fig. 3A, 10 μM acetylcholine was applied to cells in the luciferin–luciferase medium, and prompt secretion of ATP was observed. The vertical axis represents the time integral of the ATP in the medium. The time course of the rate of secretion, which is the time derivative of the record shown in Fig. 3, indicated that the ATP secretory activity was complete in less than 1 min, consistent with data for secretion of catecholamines (Oka et al., 1980). In the presence of 1.9 mM Ca^{2+}, nearly 10% of the cellular ATP was released with acetylcholine stimulation (10 μM). In a modified Krebs medium with no added calcium (1 μM Ca^{2+}), the response of the cells to acetylcholine was abolished (Fig. 3B). This was also true for higher concentrations of acetylcholine (up to 100 μM).

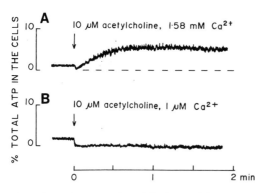

Figure 3. Calcium dependence of acetylcholine-induced ATP release. Chart records of the photomultiplier current output with 250,000 cells in the reaction mixture. These represent the time integral of the ATP release. Vertical calibration: 10% of maximum signal obtained by addition of Triton X-100 (1% final concentration) at the end of the experiment. Arrows indicate the addition of acetylcholine (10 μM). A: Free calcium concentration (measured with a calcium electrode) was 1.58 mM. B: Free calcium concentration was 1 μM. The rapid decrease of the light signal to below the basal level on addition of 10 μl of Krebs solution with acetylcholine (200 μM) was caused by the dilution of the ATP present in the medium.

4. NICOTINIC RECEPTOR DESENSITIZATION

Acetylcholine receptors in chromaffin cells can be desensitized by prolonged exposure to high concentrations of agonists; i.e., further addition of agonist does not elicit further catecholamine release. As shown in Fig. 4A, this property of the nicotinic acetylcholine receptor is also expressed in the case of the ATP release. Submaximal doses of acetylcholine (10 μM) can evoke substantial ATP release (up to 15% of the cellular ATP). Addition of a further identical dose 2 min later (20 μM for record A) resulted in no further ATP release. Control experiments showed that the application of a single 20 μM dose elicited 30 to 50% more ATP release, and, thus, this refractoriness of ATP secretion did not represent depletion of ATP. Furthermore, a sudden elevation of $[K^+]_o$ (keeping the sum $[Na^+]_o + [K^+]_o$ in the modified Krebs solution constant) after stimulation of the nicotinic acetylcholine receptor did elicit a substantial additional ATP release response (see Fig. 5).

Additional experiments showed that the nicotinic acetylcholine receptor was primarily involved in the acetylcholine stimulation of ATP release. As shown in Fig. 4B, pretreatment of the cells with d-tubocurarine induced a substantial inhibition of the acetylcholine-evoked ATP release. Similar results were obtained with another specific inhibitor of nicotinic acetylcholine receptors, hexamethonium.

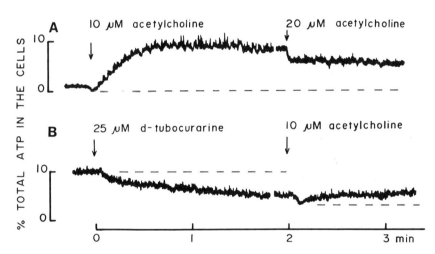

Figure 4. Desensitization of the nicotinic acetylcholine receptor for ATP release. Upper part: Records represent the time integral of the ATP released from 250,000 chromaffin cells in response to two consecutive applications of acetylcholine. The arrow on the left indicates the time for the first addition (concentration in the reaction mixture 10 μM); arrow on the right for the second addition (20 μM). Lower part: Blockade by D-tubocurarine. Dashed line represents the estimated level of ATP before the addition of the agonist or blocker.

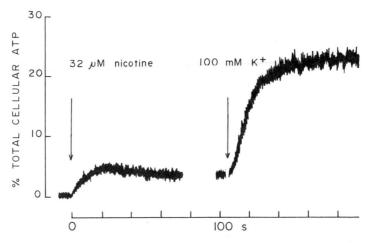

Figure 5. ATP release induced by membrane depolarization after nicotinic stimulation. Records represent the time integral of the ATP released from 250,000 chromaffin cells. Responses to consecutive applications of nicotine (62 μM) and K$^+$ (100 mM).

On the other hand, the acetylcholine-evoked ATP release in the presence of atropine, a blocker of the muscarinic acetylcholine receptor, was 20 to 30% smaller than in the absence of atropine. Application of muscarine alone (up to 50 μM) did not elicit ATP secretion.

5. GRANULAR NATURE OF THE SECRETED ATP

ATP is distributed both in the secretory granules and in the cytosol (Winkler and Westhead, 1980; Ungar and Phillips, 1983) of the adrenal medullary cells. To establish the origin of the secreted ATP, we permeabilized the cells either by application of digitonin (20–40 μM) or by brief exposures (2 μsec) to an electrical field (2000 V/cm) in a medium with isethionate in place of chloride.

Figure 6 represents a diagram of a chromaffin cell between the electrodes used to apply the high-voltage discharges together with the expressions used to calculate the transmembrane voltage at the two poles of the cell where dielectric breakdown occurs.

When chromaffin cells suspended in a medium mimicking the intracellular environment (low Ca^{2+} and Cl$^-$) were subjected to high-voltage discharges, a fraction of the cell ATP was rapidly released to the medium (Fig. 7, arrow on the left). This may represent the true cytosolic levels at ATP. Increasing Ca^{2+} to 200 μM induced a further release of ATP, and application of 40 μM digitonin in the presence of 200 μM Ca^{2+} (Fig.

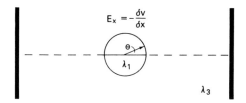

TRANSMEMBRANE POTENTIAL AT
THE ONSET OF THE DISCHARGE

$V_o(\Theta) = k \cdot r \cdot E_x \cdot \cos \Theta$

$k = f(r, \delta, \lambda_1/\lambda_2, \lambda_2/\lambda_3)$

$V_o(\Theta=0) = 1.5\ V$

$\lambda_1 = \lambda_2 = 0.01\ \Omega^{-1} cm^{-1}$ (conductivity internal and external medium)

$\lambda_3 = 10^{-10}\ \Omega^{-1} cm^{-1}$ (conductivity of membrane)

$r = 5 \times 10^{-4}\ cm$ (radius)

$\delta = 10^{-6}\ cm$ (membrane thickness)

$E_x = 2 \times 10^3\ V/cm$

Figure 6. The high-electric-field permeabilization technique. Upper part represents a diagram of the electrode arrangement with a spherical cell at the center. The conductivity of the cytosol (λ_1) and the extracellular medium (λ_2) is assumed to be the same and equal to 0.01 $\Omega^{-1}cm^{-1}$. The conductivity of the membrane (λ_3) was taken as $10^{-10}\ \Omega^{-1}cm^{-1}$. With these values, $\lambda_1/\lambda_2 = 1$ and $\lambda_2/\lambda_3 - 10^8$. (Taken with modifications from Knight and Baker, 1982.)

7, arrow in the middle) gave rise to a further ATP liberation of 30 to 35% of the total ATP in the cells. This ATP may represent some granule-bound component releasable by calcium (Wilson and Kirshner, 1983; Dunn and Holz, 1983). Finally, treatment of the reaction mixture with 1% Triton released the remainder of the ATP in the cells (Fig. 7, arrow on the right), probably that within still intact chromaffin granules. The ATP level reached with the Triton treatment gave the total ATP as 1200 pmol (Fig. 7). Since the reaction mixture contained *ca.* 150,000 cells, the measured ATP content per chromaffin cell was 8 fmol. Under these experimental conditions, namely, isethionate in place of chloride and low $[Ca^{2+}]_o$ (0.1–1 μM) in the external medium, we calculated that 4 to 5% of the ATP in the cells is liberated with both permeabilization methods (average of nine experi-

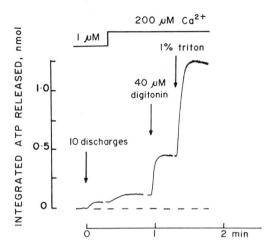

Figure 7. ATP release from permeabilized chromaffin cells. Records of the time integral of the ATP release from 150,000 cells in Cl^--free, isethionate Krebs solution. Arrow on the left: application of 10 high-voltage discharges (frequency 0.25 sec^{-1}). The concentration of free calcium was then augmented from 1 to 200 μM as indicated above the records. Arrow in the middle: application of digitonin. Arrow on the right: application of 1% Triton.

ments). Increasing $[Ca^{2+}]_o$ to 200 μM induces a further liberation of ATP in the range from 9 to 10% of the total ATP in the cells.

The ATP released after permeabilization of the cells in the experiment illustrated in Fig. 7 (low Ca^{2+} and low Cl^-) amounted to 4.8% of the total ATP.

One other interesting observation from the data in Fig. 7 was that digitonin permeabilization seemed to provoke further ATP release when added after high-voltage permeabilization. Since no comparison of the two methods has been reported, we exposed two aliquots of the same cell preparation to either 20 μM digitonin, a level found optimal by ourselves and others (Wilson and Kirshner, 1983; Dunn and Holz, 1983), or five high-voltage discharges, both in a Cl^--free, 1 μM Ca^{2+}, isethionate medium. A careful comparison of ATP release evoked by different numbers of discharges revealed that, with our apparatus, no further release could be elicited by more than five discharges. As shown in Fig. 8, we measured ATP release from chromaffin cells induced by either 20 μM digitonin (upper record) or by 5 high-electric-field discharges (lower record). Clearly, in the presence of 1 μM Ca^{2+} and in a Cl^--free medium, digitonin evoked twice as much release as did an optimal number of high-voltage discharges. On the other hand, when the $[Ca^{2+}]_o$ was adjusted to 0.1 μM, both methods released similar amounts of ATP from the cytosol (about 5%).

Catecholamine release from permeabilized cells is known to depend on the presence of ATP in the medium (Knight and Baker, 1982; Dunn and Holz, 1983). Because of the nature of our assay, we could not increase the $[ATP]_o$ to reach the levels reportedly required (5 mM) for optimal exocytotic catecholamine release without complete oxidation of the lu-

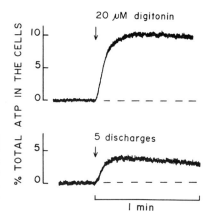

Figure 8. Comparison between ATP release induced by 20 μM digitonin (upper record) and by five high-voltage discharges (lower record). Experiments with 250,000 cells in Cl⁻-free, isethionate medium in presence of 1 μM Ca^{2+}. High-voltage discharges were applied at a frequency of 0.2 sec^{-1}.

ciferin in the reaction mixture. Therefore, the measured stimulation of the ATP release from permeabilized cells by a sudden increase in $[Ca^{2+}]_o$ from 1 to 200 μM, as in Fig. 6, or the quantitative differences between digitonin and high-voltage permeabilization in Fig. 8 may only represent fractions of the release possible if the optimal $[ATP]_o$ could be included in the reaction mixtures. Very little more can be said at present because the action of ATP during secretion remains unknown.

6. ATP RELEASE EVOKED BY MEMBRANE DEPOLARIZATION IS MEDIATED BY ACTIVATION OF VOLTAGE-GATED CALCIUM CHANNELS

Cadmium ions block voltage-gated Ca^{2+} channels in many excitable cells (Hagiwara and Byerly, 1981). In the case of the chromaffin cells, a complete blockade of the ATP release associated with sudden depolarizations of the chromaffin cell membrane could be demonstrated with cadmium. For example, in the presence of 1 mM Cd^{2+} an elevation of $[K^+]_o$, keeping $[Na^+]_o$ plus $[K^+]_o$ constant in the modified Krebs solution, failed to induce the ATP release observed in control experiments in the absence of Cd^{2+}. The partial inhibition of the luciferin–luciferase assay by Cd^{2+} was easily corrected for by a separate standard curve constructed in the presence 1 mM Cd^{2+}.

Consistent with the cadmium results were our studies with nifedipine, a potent blocker of the voltage-gated calcium channel (Nayler and Horowitz, 1983) and of catecholamine release (Garcia et al., 1984). In the micromolar concentration range, this drug strongly inhibited ATP release evoked by K^+-induced membrane depolarization (see Fig. 9).

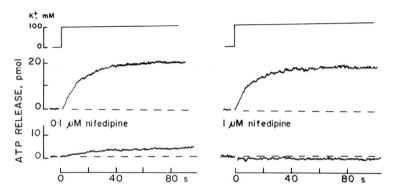

Figure 9. Effects of nifedipine on K^+-induced ATP release. Time course of the quantity of ATP in the reaction chamber during the control experiments (upper records) and in the presence of 0.1 μM (lower, left side) or 1 μM nifedipine. About 50,000 cells in the reaction medium.

7. ATP RELEASE FROM COLLAGENASE-ISOLATED ISLETS OF LANGERHANS

Recently, glucose-stimulated insulin release from single mouse islets has been correlated to the electrical activity recorded from a B cell within the same islet. The time lag between the addition of glucose and the appearance of the insulin in the samples, although dependent on the glucose level prior to stimulation, was found to be of the order of minutes (Scott *et al.*, 1981). On the other hand, the time constant for the rise to peak of insulin release elicited by 50 mM $[K^+]_o$ is only about 30 sec. This time lag is caused, at least in part, by the restricted diffusion of insulin along the tortuous clefts between the islet cells and across the unstirred layer around the islet. Comparing the time course of the appearance of ATP with that of insulin in the perfusate from a chamber with collagenase-isolated islets revealed that the time to peak of ATP release was much smaller than that for insulin, consistent with the idea that the rate-limiting step was diffusion (Leitner *et al.*, 1975). Placing the luciferin–luciferase ATP detector in the medium, outside the cells, should give an even smaller time constant.

Figure 10 shows that this is the case for glucose-induced ATP release from islets of Langerhans (upper record). It can be seen that a sudden elevation of the glucose from 2.8 to 33mM gave rise to a rapid liberation of ATP. Figure 10 also shows that when the B-cell membrane was depolarized by increasing $[K^+]_o$, maintaining constant $[Na^+]_o + [K^+]_o$, a rapid release of ATP occurred. ATP secretion was blocked if the secretagogues (glucose and K^+) were applied in a low-Ca^{2+} medium (lower record), in agreement with the insulin release data. These results suggest

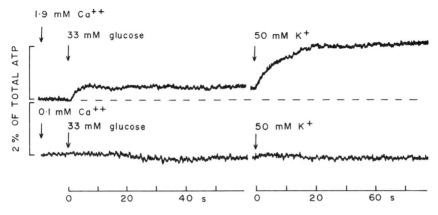

Figure 10. ATP release from collagenase-isolated islets of Langerhans. Fifty pancreatic islets from obese *ob/ob* mice in 50 μl of modified Krebs solution containing the luciferin–luciferase at 36°C. Islets were preincubated for 25 min in the presence of theophylline (2 mM) and glucose (2.8 mM).

that the secretory response of the islet is faster than previously thought and that the time resolution achieved with the luciferin–luciferase technique is suitable for the measurement of the kinetics of ATP release, presumably from the granular store. Thus, the stimulus–secretion process in islets of Langerhans can be followed with high time resolution.

8. CONCLUSIONS

The main conclusion from our study is that Ca^{2+}-dependent ATP secretion from two different endocrine cells can be quantitatively measured using a highly purified luciferase preparation to assay released ATP. It is also clear that stimulated ATP release shares many important properties with the processes of hormone secretion of the cell systems used here. One particularly attractive feature of this assay of ATP is that it can allow us to monitor exocytosis with a time resolution in the millisecond range.

ATP is distributed in both secretory granules and cytosol, as mentioned above, and thus the origin of secreted ATP would continue to be hypothetical without recourse to the quantitative information provided by the cell permeabilization techniques used here. Taking the volume of the granule as 4.6×10^{-15} cm^3, the ATP concentration in the granule as 130 μmol/cm^3, and 10^4 granules per cell (Ungar and Phillips, 1983), the granular ATP content per cell is estimated as 6 fmol. Taking the cytosolic cell volume as 50% of the cell volume, i.e., 2.5×10^{-10} cm^3 and the cytosolic concentration of ATP as 1.5 μmol/cm^3, the ATP in this compartment is estimated as 0.4 fmol per chromaffin cell, or less than 7% of the total ATP that can be extracted from the cell. We have seen that in low Ca^{2+} ($\leq$ 1 μM), 7 to 10% of the cellular ATP is liberated after treatment of the cells with 20 μM digitonin.

Digitonin, for reasons that are not well understood, appears to solubilize the plasma membrane of the chromaffin cell, leaving the organelles such as chromaffin granules, mitochondria, and endoplasmic reticulum intact (Wilson and Kirshner, 1983; Dunn and Holz, 1983). The permeabilized cell loses large molecules such as lactate dehydrogenase (LDH), but not DBH or catecholamines, which are found in the granules. The high-electric-field permeabilization step, on the other hand (Zimmerman et al., 1974; Knight and Baker, 1982), under the same experimental conditions induced the release of 4 to 5% of the cellular ATP. Application of high-electric-field pulses, a few microseconds in duration makes holes in the plasma membrane estimated to be 4 to 5 Å in diameter. These holes do not allow LDH to escape (Knight and Baker, 1982), and thus, externally applied luciferase cannot enter the cells. Therefore, the detection of ATP in the case of high-voltage-permeabilized cells takes place outside the cell boundary, unlike the situation with digitonin-permeabilized cells. Therefore, the greater percentages obtained with digitonin, namely 7 to 10% compared with 4 to 5% obtained with the high-voltage pulses, may be attributable to the more efficient membrane removal by digitonin.

Stimulation of the islet of Langerhans with 33 mM glucose induced 1% of the total islet ATP to be released. This figure is in perfect agreement with data on glucose-induced insulin release (Scott et al., 1981), supporting the concept that the released ATP was from the granular store.

9. SUMMARY

The luminescent oxidation of luciferin catalyzed by luciferase has been used to follow the ATP release from freely suspended bovine chromaffin cells and collagenase-isolated mouse islets of Langerhans. ATP release from medullary chromaffin cells can be induced by acetylcholine and other secretagogues, and the process can be followed quantitatively by monitoring the luminescent oxidation of luciferin. In adrenal medullary chromaffin cells, acetylcholine (1–100 μM) induced ATP release of up to 30% of the total cellular ATP. This secretion required external free calcium and could also be elicited by K$^+$-induced membrane depolarization. Blockers of the voltage-gated Ca^{2+} channel effectively blocked the ATP release induced by depolarization of the chromaffin cell membrane. Furthermore, the acetylcholine antagonist d-tubocurarine inhibited the acetylcholine-evoked release of ATP.

The size of the cytosolic ATP compartment was estimated by either solubilizing the cell membrane using digitonin (20–40 μM) or by application to the cells of brief pulses (2 μsec) of high electric field (2,000 V/cm) in a low-Ca^{2+} ($< 10^{-6}$ M) and Cl^--free medium. The vesicular compartment was exposed using either 200 μM digitonin or 1% Triton X-100. Glucose, the physiological secretagogue for insulin release in the islet of Langerhans, and elevated K^+ induced ATP release from the islet B cells in a Ca^{2+}-dependent manner. These data support the concept that ATP is released together with the insulin by exocytosis of B-granule contents.

ACKNOWLEDGMENTS. The authors are pleased to thank Drs. I. Atwater and R. Ornberg for comments and discussion. Thanks are also given to Diane Seaton for preparation of the chromaffin cells.

REFERENCES

Biggley, W. H., Lloyd, J. E., and Seliger, H. H., 1978, The spectral distribution of firefly light II, *J. Gen. Physiol.* **50**:1681–1697.

Born, G. V. R., 1958, Changes in the distribution of phosphorus in platelet-rich plasma during clotting, *Biochem. J.* **68**:695–704.

Code, C. F., 1952, Histamine in blood, *Physiol. Rev.* **32**:47–65.

Cormier, M. J., Wampler, J. E., and Hori, K., 1973, Bioluminescence: Chemical aspects, *Fortschr. Chem. Org. Naturst.* **30**:1–60.

Douglas, W. W., 1968, Stimulus–secretion coupling: The concept and clues from chromaffin and other cells, *Br. J. Pharmacol.* **34**:451–474.

Douglas, W. W., and Rubin, R. P., 1961, The role of calcium in the secretory response of the adrenal medulla to acetylcholine, *J. Physiol. (Lond.)* **159**:40–57

Dowdall, M. J., Boyne, A. F., and Whittaker, V. P., 1974, Adenosine triphosphate: A constituent of cholinergic synaptic vesicles, *Biochem. J.* **140**:1–12.

Dunn, L. A., and Holz, R. W., 1983, Catecholamine secretion from digitonin-treated adrenal medullary chromaffin cells, *J. Biol. Chem.* **258**:4989–4993.

Eley, M., Lee, J., Lhoste, J. M., Lee, C. Y., Cormier, M. J., and Hemmerich, P., 1970, Bacterial bioluminescence, comparisons of bioluminescence emission spectra, the fluorescence of luciferase reaction mixture and the fluorescence of flavin cations, *Biochemistry* **9**:2902–2922.

Garcia, A. G., Sala, F., Reig, J. A., Viniegra, S., Frias, J., Fonteriz, R., and Gandia, L., 1984, Dihydropyridine BAY-K-8644 activates chromaffin cell calcium channels, *Nature* **309**:69–71.

Greenberg, A., and Zinder, O., 1982, λ- and β-receptor control of catecholamine secretion from isolated adrenal medulla cells, *Cell Tissue Res.* **226**:655–656.

Hagiwara, S., and Byerly, L., 1981, Calcium channel, *Annu. Rev. Neurosci.* **4**:69–125.

Ito, S., Nakazato, Y., and Ohga, A., 1980, Comparison of catecholamine, dopamine β-hydroxylase and adenine nucleotides secretion between perfused adrenal gland and isolated chromaffin cells, *Adv. Biosci.* **36**:87–94.

Kidokoro, Y., and Ritchie, A. K., 1980, Chromaffin cell action potentials and their possible role in adrenaline secretion from rat adrenal medulla, *J. Physiol. (Lond.)* **307**:199–216.

Knight, D. E., and Baker, P. F., 1982, Calcium-dependence of catecholamine release from bovine adrenal medullary cells after exposure to intense electric fields, *J. Membr. Biol.* **68**:107–140.

Lacy, P. E., and Kostianovsky, M., 1967, Method for the isolation of intact islets of Langerhans from the rat pancreas, *Diabetes* **16**:35–39.

Leitner, J. W., Sussman, K. E., Vatter, A. E., and Schneider, F. H., 1975, Adenine nucleotides in the secretory granule fraction of rat islets, *Endocrinology* **95**:662–677.

Livett, B. G., Boksa, P., 1984, Receptors and receptor modulation in cultured chromaffin cells, *Can. J. Physiol. Pharmacol.* **62**:467–476.

Naylor, W. G., and Horowitz, J. D., 1983, Calcium antagonists: A new class of drugs, *Pharmacol. Ther.* **20**:203–262.

Oka, M., Isosaki, M., and Watanabe, J., 1980, Calcium flux and catecholamine release in isolated bovine adrenal medullary cells: Effects of nicotinic and muscarinic stimulation, *Adv. Biosci.* **36**:29–36.

Pollard, H. B., and Pappas, G., 1979, Veratridine-activated release of adenosine-5'-triphosphate from synaptosomes: Evidence for calcium dependence and blockade by tetrodotoxin, *Biochem. Biophys. Res. Commun.* **88**:1315–1321.

Scott, A. M., Atwater, I., and Rojas, E., 1981, A method for the simultaneous measurement of insulin release and B-cell membrane potential in single mouse islet of Langerhans, *Diabetologia* **21**:470–475.

Sussman, K. E., and Leitner, J. W., 1977, Conversion of ATP into other adenine nucleotides within isolated islet secretory vesicles. Effect of cyclic AMP on phosphorus translocation, *Endocrinology* **101**:694–701

Ungar, A., and Phillips, J. H., 1983, Regulation of the adrenal medulla, *Physiol. Rev.* **63**:787–843.

Uvnas, B., 1974, The molecular basis for the storage and release of histamine in rat mast cell granules, *Life Sci.* **14**:2355–2366.

Wagner, J. A., Carlson, S. S., and Kelly, R. B., 1978, Chemical and physical characterization of cholinergic synaptic vesicles, *Biochemistry* **17**:1199–1206.

Weil-Malherbe, H., and Bone, A. D., 1958, The association of adrenaline and noradrenaline with blood platelets, *Biochem. J.* **70**:14–22.

Wilson, S. P., and Kirshner, N., 1983, Calcium-evoked secretion from digitonin-permeabilized adrenal medullary chromaffin cells, *J. Biol. Chem.* **258**:4994–5000.

Winkler, H., and Westhead, E., 1980, The molecular organization of adrenal chromaffin granules, *Neuroscience* **5**:1803–1823.

Zimmermann, U., Pilwat, G., and Riemann, F., 1974, Dielectric breakdown of cell membranes, *Biophys. J.* **14**:881–899.

Part II

Channels in Biological Membranes

Chapter 10

Mechanotransducing Ion Channels

Frederick Sachs

1. INTRODUCTION

Mechanical transduction is the most widespread sensory modality in animals. Specialized organs of the cochlea (Hudspeth, 1983) and vestibular system provide our sense of hearing and local gravity. Our sense of touch and vibration, particularly in the skin, is mediated by a variety of specialized mechanosensors and possibly free nerve endings (Iggo and Andres, 1982). Muscle tension is transduced by Ia afferent nerves in muscle spindles, while tendon tension is transduced by Golgi tendon organs. Joint position is transduced by poorly understood receptors in the joint capsule. The inflation of hollow organs is reported by sensory nerves of the autonomic nervous system, although the location of the actual transducers is not known and may reside in nerve cells, muscle cells, or other types of supporting cells. Inflation of blood vessels is used to measure blood pressure. Inflation of the lung, gut, bladder, mammary glands, and probably cerebrospinal fluid all provides essential feedback for autonomic regulation. Osmoregulation at the systemic level and cell volume itself (Kregenow, 1981) are likely to be controlled by feedback from membrane tension.

FREDERICK SACHS • Department of Biophysics, State University of New York at Buffalo School of Medicine, Buffalo, New York 14214.

All of the above processes (with the possible exception of cell volume regulation) involve the transduction of mechanical stress into an electrical signal. Despite the ubiquitous nature of mechanical transduction, little is known of the mechanism(s). Some of the theories for mechanical transduction include piezoelectric effects (Naftalin, 1965), stress-induced changes in surface charge density coupling to voltage-sensitive ion channels (Gross et al., 1983), coupling of ion fluxes to water flows (O'Leary, 1970; Teorell, 1971), membrane stress releasing bound Ca^{2+}, Mg^{2+}, or other transmitter, which would then interact with an ion channel (Edwards, 1983), or, finally, ion channels themselves (Corey and Hudspeth, 1983).

Most of the above theories can be shown not to adequately explain the observed properties of mechanical transduction. For example, piezoelectric transduction cannot explain sustained current flow in the bullfrog sacculus (Corey and Hudspeth, 1983) or the crayfish stretch receptor (Edwards et al., 1981). For reasonable estimates of the sensitivity to stretch, the changes in membrane area associated with stretch are not sufficient to account for transduction by changes in capacity or surface charge.

One of the main difficulties in evaluating mechanisms has been the lack of knowledge of the actual membrane stress. If one actually knew the intrinsic stress sensitivity, then it would be possible to evaluate whether a given model could explain the observed sensitivity. However, the structures of sensory elements in the specialized structures that have been studied electrophysiologically—hair cells of the cochlea, pacinian corpuscles, muscle spindles, and invertebrate stretch receptors—are too complicated to be able to place useful limits on the stress at the transducing element.

Recently, Falguni Guharay and I have discovered an ion channel in the membrane of chick skeletal muscle that responds to membrane tension (Guharay and Sachs, 1984a,b). We have studied the channel using single-channel patch-clamp techniques, and because of the relatively well-defined geometry of the patch, we have been able to place absolute limits on the tension and hence on the sensitivity. It is not yet clear that the channel serves a physiological role in the muscle cells. But as I explain later on, most channels are not sensitive to mechanical stress, and specialized structures are necessary to produce the high levels of sensitivity that are observed. We must presume that the stretch-activated (SA) channels subserve a specific function. Similar channels have been reported to be present in muscle cultures from *Xenopus* (Brehm et al., 1984), red cells (Hamill, 1983) and hair cells (Ohmori, 1984). Anecdotal evidence suggests that they are present in tissue-cultured neurons and heart cells, denervated frog muscle, and the guard cells of plant leaves. Regardless of our lack of understanding of the physiological function of these stretch-activated channels in muscle, they serve as a useful prototype mechanotransducer.

2. RECORDING SA CHANNELS

To make tight, gigohm seals against the cell membrane, a patch pipette is placed against the cell membrane and then suction is applied to draw in a small bleb of membrane (Hamill *et al.*, 1981). The bleb of membrane is typically 2–5 μm in length and 2 μm in diameter (Sakmann and Neher, 1983). Some as yet mysterious forces hold the bleb of membrane in place and seal it tightly to the glass (Sachs, 1984). In tissue-cultured chick skeletal muscle cells (Guharay and Sachs, 1984), when suction is applied to the patch, bursts of single-channel activity appear (Fig. 1). The amount of activity is reversibly controlled by varying the

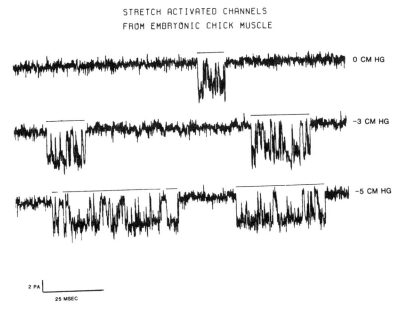

Figure 1. Currents from stretch-activated channels of a chick skeletal muscle in tissue culture. Data are shown at three different values of suction. The record was made with an excised, inside-out patch bathed with 150 mM K^+ saline in the pipette and 150 mM Na^+ saline in the bath, the inverse of the normal gradient. Pipette potential +50 mV, 23°C, 5-kHz bandwidth. Currents flowing out of the pipette (inward currents in conventional notation) are shown downward.

suction. In the early experiments we used a simple mercury manometer to measure pressure. We now use a syringe mounted on a linear motor with feedback from a solid state sensor (Motorola) to apply controlled pressure or suction.

3. GENERAL CHARACTERISTICS

The channels are activated by suction in the range of 1–5 cm Hg, whereas the membrane of normal cells tends to lyse in the range of 5–10 cm Hg. Patches of membrane can be excised from the cell, and the response is maintained with the same sensitivity although we generally cannot apply as much suction before the seal breaks down. The SA channels in cell-attached or "ripoff" patches can be activated by positive pressure, presumably by reversing the curvature of the dome of membrane. The SA channels in outside-out patches, where the extracellular face of the membrane is exposed to the bath, can be activated by positive pressure (J. Hidalgo, unpublished observations). There appears to be little if any adaptation, so that channel activity is maintained as long as suction is applied.

Because the channel conductance (see below) is similar to that of the low-conductance nicotinic acetylcholine-activated channel in these cells (Auerbach and Sachs, 1983), we made several checks to see that the channels were not a form of the nicotinic channel (Guharay and Sachs, 1985). When the patches are treated with 2 μg/ml of α-bungarotoxin, which is sufficient to block nicotinic channel activity, the SA channels remain active. When the patch is treated with high concentrations of acetylcholine sufficient to cause desensitization, SA channel activity persists. Combined with the differences in reversal potential, these data are sufficient to say that the SA channel is unlikely to be an altered nicotinic channel.

We also checked to see whether the SA channel gating could arise from stretch causing a nonspecific leak, which would include an influx of Ca^{2+} causing activation of a Ca^{2+}-sensitive channel. We filled the patch pipette with buffered Ca^{2+} at 10 nM and at 10 mM and found no difference in response. We excised a patch and held "cytoplasmic" Ca^{2+} at various concentrations between 10 nM and 10 mM and found no significant changes in activity. An additional argument against nonspecific leaks around the seal is that the seal resistance does not change during the application of suction.

We checked the generality of the stretch response by examining the kinetics of Ca^{2+}–K^+ channels, Ach channels, and the F channel, a very "flickery" K^+ channel that appears in these cells. None of these channels showed a measurable stretch response. In fact, as I show below, even for the SA channel, only one out of four distinguishable states is sensitive to mechanical distortion.

4. CONDUCTANCE PROPERTIES

The channel is cation selective, and in normal saline the channel conductance of cell-attached patches is 35 pS. The conductance is constant (at least in the hyperpolarizing quadrant) up to -160 mV (Fig. 2). The reversal potential is -30 mV, suggesting from the Goldman–Hodgkin–Katz equation that $P_K : P_{Na} = 4$. When the pipette contains isotonic K^+ saline and the channel carries inward K^+, the channel conductance is 70 pS with a reversal potential of 0 mV in cell-attached patches. From conductance measures, $P_K : P_{Na} = 2$. The conductance in symmetrical Cs^+ solutions is approximately the same as in symmetrical Na^+ solutions.

5. KINETIC PROPERTIES

Histograms of open and closed times were collected using an automated pattern recognition program (Sachs *et al.*, 1982). We were lucky enough to obtain one or two patches that contained only a single channel. In these records, the open-time histogram could be fitted with a single exponential, but the closed-time histogram required three exponentials (Fig. 3). This shows that any kinetic model of the SA channel must contain at least one open and three closed states.

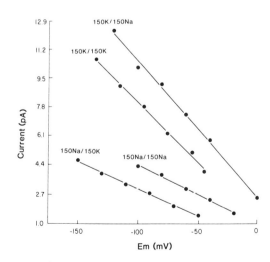

Figure 2. Current–voltage relationship of SA channels. The conductance was 70 pS with K^+ carrying the current and 35 pS with Na^+ carrying the current. The reversal potentials under biionic conditions indicated $P_K : P_{Na} = 4$.

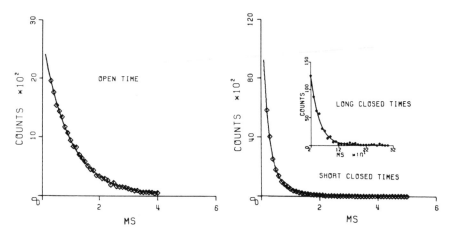

Figure 3. Open-time (panel A) and closed-time (panel B) histograms from a record containing a single SA channel. The open-time distribution can be fitted with a single exponential (solid line) with a time constant of 0.96 ± 0.01 msec. The closed-time distribution requires three exponentials. The distribution of short intervals has been fitted with a sum of two exponentials ($\tau_1 = 0.18 \pm 0.02$ and $\tau_2 = 0.59 \pm 0.12$ msec), and the distribution of long intervals (inset) has been fit with a single exponential beginning at 6 msec ($\tau_3 = 31 \pm 4$ msec). Suction −5 cm Hg, other conditions as per Fig. 1. Error limits on τs represent 90% confidence.

When we studied the effects of suction on the various time constants, it became clear that the only time constant that showed a consistent effect was the longest of the closed-time components. This time constant, τ_3, showed an exponential dependence on the square of the applied pressure (Fig. 4). A dependence on the square of pressure is suggestive of a simple mechanical mechanism, since energies of distortion, as in a spring, depend on the square of the force.

For multistep kinetics, the time constants are made up of various combinations of all the rate constants so that a knowledge of the time constants is not equivalent to a knowledge of the rate constants. Rate constants are model dependent. We decided to use the simplest possible scheme:

$$C_1 \underset{k_{21}}{\overset{k_{12}}{\rightleftarrows}} C_2 \underset{k_{32}}{\overset{k_{23}}{\rightleftarrows}} C_3 \underset{k_{43}}{\overset{k_{34}}{\rightleftarrows}} O_4$$

where the Cs represent closed states and O the open state (dubbed O_4 to simplify the numbering of the states). It is not trivial to derive rate constants from the time constants and their amplitudes. For the closed states, the time constants and amplitudes are related through the roots of five

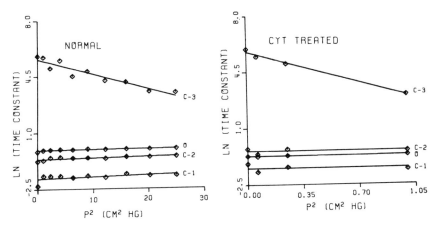

Figure 4. The pressure dependence of the four time constants in cell-attached mode with (panel A) and without (panel B) prior treatment with cytochalasin B (10^{-5} M overnight). The Cs represent the three closed-time distributions, and the unlabeled traces are the open-time distributions. Notice the linear dependence (in semilogarithmic format) of the longest-lived distribution, C_3, on the square of the pressure. Also note the lack of pressure dependence of the other time constants. Cytochalasin increased the pressure sensitivity of τ_3 from 10 cm^2 Hg/e-fold to 0.36 cm^2/e-fold.

cubic equations involving the rate constants (Colquhoun and Hawkes, 1981). Although the solutions can be obtained analytically, we found it simpler to write the predicted time constants and amplitudes in terms of the rate constants of the model and then solve the set of five equations (three time constants and two fractional amplitudes) using numerical techniques.

The rate constants we extracted from the data had the expected form, with k_{12} being the slowest rate and the one sensitive to pressure (Fig. 5). What was surprising was that only one rate constant was sensitive to pressure. More recent data (Guharay and Sachs, 1984b) have shown that of the six rate constants in the model, only k_{12} is sensitive to membrane potential, decreasing with hyperpolarization at a rate of about 50 mV/e at pH 7.4.

Considered as an Eyring rate process, k_{12} represents the crossing of a single energy barrier. In fact, since the reverse rate, k_{21}, does not change with pressure, we can interpret the effect of pressure as raising the energy of C_1 but not affecting the barrier height itself. The sensitivity of k_{12} to stretch can be interpreted as the ratio of the free energy produced by mechanical distortion to the mean thermal energy, $kT = 0.025$ eV. The problem is now to make a model to predict the sensitivity.

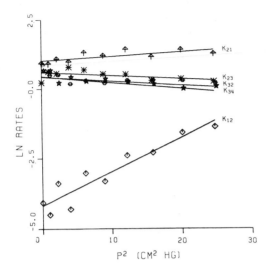

Figure 5. Rate constants of the linear sequential model (see text) as a function of pressure for a cell-attached patch of a normal cell. Notice that only one rate constant, k_{12}, is significantly sensitive to pressure and that it depends exponentially on the square of pressure, 0.12 cm² Hg/e-fold.

6. THE MODEL

How much energy is available to gate the channel? In the absence of any specific structural information about the channel, we modeled it as a cylindrical plug in the membrane (Fig. 6). Membrane tension will expand the diameter of the plug, and the work done will be the maximum available to gate the channel. It is a maximum because not all the energy may be available for gating. In analogy with Hooke's law, the increase in area, ΔA, produced by membrane tension is proportional to the tension via an area elasticity (Helfrich, 1973),

$$T = K_A \Delta A/A \tag{1}$$

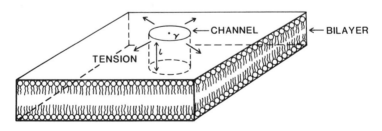

Figure 6. Model for calculating the energy of distortion. The channel is indicated as a cylindrical plug of radius r and thickness l embedded in the bilayer.

where T is tension (dyne/cm), A is area (cm^2), and K_A is the area elasticity (dyne/cm). For small distortions, $\Delta A/A \ll 1$, the energy derived from this distortion is

$$\Delta G = TdA = (A/2K_A)T^2 \qquad (2)$$

The relevant area is πr^2, where r is the radius of the cylindrical plug as shown Fig. 6.

The membrane tension can be calculated from Laplace's law assuming that the patch is hemispherical, which appears to be a reasonable approximation (Sakmann and Neher, 1983):

$$T = Pd/4 \qquad (3)$$

where P is the transmembrane pressure (dyne/cm^2) and d is the patch diameter. The patch diameter is in the range of 1–4 μm, and for convenience we have used a nominal value of 2 μm. To get a feel for the values of tension, a pressure difference of 1 cm Hg (1.33 × 10^4 dyne/cm^2) produces a tension of 0.67 dyne/cm in a 2-μm-diameter patch.

Combining Laplace's law (eq. 3) with equation 2 and the absolute reaction rate formalism for rate constant gives

$$k_{i,j} = k_0 \, e^{\theta P^2}$$
$$\theta = d^2 A/(32 K_A kT) \qquad (4)$$

where k_0 is the zero-pressure value of k_{12} and kT is Boltzmann's constant times the absolute temperature ($kT = 4.04 \times 10^{-14}$ dyne cm at 20°C).

7. COMPARING THE MODEL TO THE DATA

The prediction of equation 4 is clear that the rate constant should depend exponentially on the square of the applied pressure. Figure 5 shows that the data follow this prediction.

The scaling constant θ is a measure of the sensitivity. How does it match the data? The difficulty with making the calculation is that we have no hard information about either A or K_A. Using the maximal observed sensitivity of $\theta = 4.36$/cm Hg and our nominal patch diameter of 2 μm, we can constrain the ratio $r^2/K_A > 2.5 \times 10^3$, with r in Angstroms and K_A in dyne/cm. The inequality again reflects an assumption of 100% efficiency.

There are reasonable limits to the values for K_A and r. For red blood cells, K_A is about 250 dyne/cm (Evans *et al.*, 1976). Bilayers may be more elastic, with values ranging from 180 to as low as 3 dyne/cm (Lis *et al.*, 1981; Kwok and Evans, 1981; Evans and Hochmuth, 1978) with most biologically relevant lipids in the range of 100 dyne/cm. If the channel were as soft as 100 dyne/cm, the channel radius would have to be greater than 500 Å to provide the observed sensitivity. A protein 1000 Å in diameter and 100 Å thick, as a channel is likely to be (Kistler *et al.*, 1982), would have the incredible molecular weight of 37 million.

The situation gets even worse if we take more reasonable values for protein elasticity. Less energy is available from stiffer molecules. For the plug model, $K_A \approx El$, where E is Young's modulus and l is the thickness of the plug. There are no values for membrane proteins, but for silk and collagen the K_As are in the range of 10^9–10^{10} dyne/cm². For a 100-Å-thick plug, these numbers are equivalent to K_As of 10^3–10^4 dyne/cm. A channel with $K_A = 10^4$ would need a radius of 5000 Å to account for the observed sensitivity.

It is energetically inefficient to make a gigantic molecule simply to increase the perimeter. It is simpler to use a smaller molecule to form the channel and gather force from a large area by attaching "strings" between the channel and distant points on the membrane. The cytoskeleton is a possible anatomic substrate for the "strings." The strings should be attached to the membrane at distances of ≈1000 nm to account for the observed sensitivity to tension. It is not clear how many strings need to be attached to a channel, but two is a minimum, and the perimeter of 100-Å-diameter channel will only permit the attachment of five of the smallest (60 Å diameter) cytoskeletal microfilaments and fewer of the larger-diameter ones (Weber and Osborn, 1981). Preliminary histological studies have shown a submembranous network of filaments with interconnecting points approximately 1000 Å apart (Fig. 7).

Since we suspected the involvement of the cytoskeleton, we tested the cytochalasins A, B, and E for effects on the SA channel, expecting that if the drugs cut the strands we would lose sensitivity. In fact, the sensitivity increased by a factor of ≈30 (Fig. 4). We interpreted this to mean that the cytochalasins did not cut strands in series with the channel but cut strands in parallel that supported some of the membrane tension. The amount of suction required to break the patch decreased significantly after treatment with cytochalasins, a result that supports the notion of an increased tension in the bilayer.

Since the SA channel has never been studied before, we have little structural information about the channel. Recent studies on the effect of extracellular pH (Guharay and Sachs, 1984b) show that the conductance and reversal potential are unaffected by pH in the range of pH 7–10. However, both the stretch and voltage dependence do titrate over that range. The pH effect on both pressure and voltage are exerted through

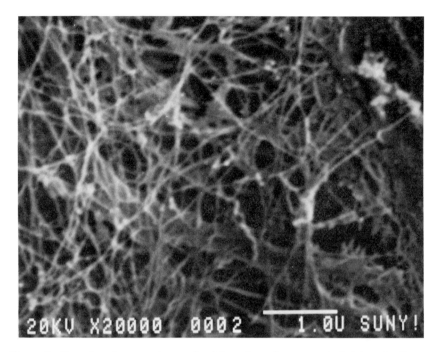

Figure 7. Scanning electron micrograph of the submembranous cytoskeleton of a tissue-cultured chick skeletal muscle. There is a random meshwork, which has an average strand length on the order of 0.1 μm. Since there were only two or three SA channels/patch, only a fraction of the strand junctions could possibly include SA channels. (The cells were fixed in 2% glutaraldehyde, the membrane was removed by a brief wash in 0.1% Triton X-100, and the cells were critical-point dried and then sputtered with gold.)

k_{12}, and both titrate at the same $pK = 9.1$. This suggests that a single group, either an N-terminal amino acid or a lysine in a hydrophobic environment, is responsible for the effects of pH on k_{12}. The pH effects can be summarized by the following equations:

$$k_{12} = k_{0_{12}} \exp[\alpha(pH)V_m + \theta(pH)\, P^2]$$

$$\alpha(pH) = 0.010 + 0.018/[1 + 10^{(9.1-pH)}]$$

$$\theta(pH) = 0.66 + 0.47/[1 + 10^{(9.1-pH)}]$$

$$k_{0_{12}} = 33/sec$$

where V_m is the membrane potential. The fact that $k_{0_{12}}$ is pH independent implies that there is no significant change in surface charge over the pH range of 7–10.

8. FUTURE PROSPECTS

We are currently attempting to evaluate the anatomic and physiological effects of additional cytoskeletal reagents in order to define the components of the cytoskeleton that are involved. We are also trying to find out in which types of cells the SA channels are present, how the sensitivity varies from one cell type to another, and how the SA channels change during development.

It is also important to study specialized mechanoreceptor cells (see Ohmori, 1984) to see what nature has done to make these cells better than most at detecting distortion; I think we now know which questions to ask.

Since the cytoskeleton is capable of contraction (Weber and Osborn, 1981), the SA channels may be under metabolic control; contraction of series strands would increase activity, whereas activation of parallel strands would decrease activity.

ACKNOWLEDGMENTS. I would like to thank Rich McGarrigle for taking the electron micrographs, Jim Neil for programming, Tony Auerbach and Jorge Hidalgo for helpful discussion, and particularly Falguni Guharay without whom this work would not have been done. I would also like to thank the *Journal of Physiology* for permission to use the figures that appear here. The research shown here was supported by NINCDS grant NS-13194.

REFERENCES

Auerbach, A., and Sachs, F., 1984, Kinetics and conductance properties of gaps within bursts of nicotinic ion channel activity, *Biophys. J.* **45**:187–198.

Brehm, P., Kullberg, R., and Moody-Corbett, F., 1984, Properties of nonjunctional acetylcholine receptor channels on innervated muscle of *Xenopus Laevis, J. Physiol. (Lond.)* **350**:631–648.

Colquhoun, D., and Hawkes, A. G., 1981, On the stochastic properties of single ion channels, *Proc. R. Soc. Lond. [Biol.]* **211**:205–235.

Corey, D., and Hudspeth, A. J., 1983, Kinetics of the receptor current in bullfrog saccular hair cells, *J. Neurosci.* **3**:962–976.

Edwards, C., 1983, The ionic mechanisms underlying the receptor potential in mechanoreceptors, in *The Physiology of Excitable Cells* (A. D. Grinnel and W. J. Moody, eds.), Alan R. Liss, New York, pp. 497–503.

Edwards, C., Ottoson, D., Rydqvist, B., and Swerup, C., 1981, The permeability of the transducer membrane of the crayfish stretch receptor to calcium and other divalent ions, *Neuroscience* **6**:1455–1460.

Evans, E. A., and Hochmuth, R. M., 1973, Mechano-chemical properties of membrane, in *Current Topics in Membrane Transport* (F. Bronner and A. Kleinzeller, eds.), Academic Press, New York, pp. 1–64.

Evans, E. A., Waugh, R., and Melnik, L., 1976, Elastic area compressibility modulus of red cell membrane, *Biophys. J.* **16**:585–595.

Gross, D., Williams, W. S., and Connor, J. A., 1983, Theory of electromechanical effects in nerve, *Cell. Mol. Neurobiol.* **3**:89–111.
Guharay, F., and Sachs, F., 1984a, Stretch activated single ion-channel currents in tissue-cultured embryonic chick skeletal muscle, *J. Physiol. (Lond.)* **352**:685–701.
Guharay, F., and Sachs, F., 1984b, Mechano-transducer ion channels in chick skeletal muscle: The effects of extracellular pH, *J. Physiol. (Lond.)* **363**:119–134.
Guharay, F., and Sachs, F., 1985, Mechano-receptor ion channels are not nicotinic, *Biophys. J.* **47**:2039.
Hamill, O. P., 1983, Potassium and chloride channels in red blood cells, in *Single-Channel Recording* (B. Sakmann and E. Neher, eds.), Plenum Press, New York, pp. 451–471.
Hamill, O. P., Marty, A., Neher, E., Sakmann, B., and Sigworth, F. J., 1981, Improved patch-clamp techniques for high resolution current recording from cells and cell-free membrane patches, *Pfluegers Arch.* **391**:85–100.
Helfrich, W., 1973, Elastic properties of lipid bilayers: Theory and possible experiments, *Z. Naturforsch.* **28C**:693–703.
Hudspeth, A. J., 1983, Mechanoelectrical transduction by hair cells in the acousticolateralis sensory system, *Annu. Rev. Neurosci.* **6**:187–215.
Iggo, A., and Andres, K. H., 1982, Morphology of cutaneous receptors, *Annu. Rev. Neurosci.* **5**:1–31.
Kistler, J., Stroud, R. M., Klymkowsky, M. W., Lalancette, R. A., and Fairclough, R. H., 1982, Structure and function of an acetylcholine receptor, *Biophys. J.* **37**:371–383.
Kregenow, F., 1981, Osmoregulatory salt transporting mechanisms: Control of cell volume in anisotropic media, *Annu. Rev. Physiol.* **43**:493–505.
Kwok, R., and Evans, E. A., 1981, Thermoelasticity of large lecithin bilayer vesicles, *Biophys. J.* **35**:637–652.
Lis, L. J., McAlister, M., Fuller, N., Rand, R. P., and Parsegian, V. A, 1982, Measurement of the lateral compressibility of several phospholipid bilayers, *Biophys. J.* **37**:667–672.
Naftalin, L., 1965, Some new proposals regarding acoustic transmission and transduction, *Cold Spring Harbor Symp. Quant. Biol.* **30**:169–180.
Ohmori, H., 1984, Studies of ionic currents in the isolated vestibular hair cell of the chick, *J. Physiol. (Lond.)* **350**:561–581.
O'Leary, D. P., 1970, An electrokinetic model of transduction in the semicircular canal, *Biophys. J.* **10**:859–875.
Sachs, F., 1984, The noise produced by patch electrodes: A model, *Biophys. J.* **45**:57a.
Sachs, F., Neil, J., and Barkakati, N., 1982, Automated analysis of single channel data, *Pfluegers Arch.* **395**:331–340.
Sakmann, B., and Neher, E., 1983, Geometric parameters of pipettes and membrane patches, in *Single-Channel Recording* (B. Sakmann and E. Neher, eds.), Plenum Press, New York, pp. 37–51.
Teorell, T., 1971, A biophysical analysis of mechano-electrical transduction, in *Principles of Receptor Physiology*, (W. R. Loewenstein, ed.), Springer-Verlag, New York, pp. 291–339.
Weber, K., and Osborne, M., 1981, Microtubule and intermediate filament networks in cells viewed by immunofluorescence microscopy, in *Cytoskeletal Elements and Plasma Membrane Organization*, (G. Poste and G. Nicholson, eds.), Elsevier/North-Holland, Amsterdam, pp. 1–55.

Chapter 11

Ionic Channels in Plant Protoplasts

Nava Moran, Gerald Ehrenstein, Kunihiko Iwasa, Charles Bare, and Charles Mischke

1. INTRODUCTION

Membrane transport in animal cells still keeps many a researcher intrigued; transport in plant cells, by comparison, is virtually *terra incognita* in spite of the increasingly intensifying research during the last few decades (Clarkson, 1984).

A major component of plant membrane transport is the electrogenic proton pump, a Mg^{2+}-dependent, K^+-stimulated ATPase transferring protons out of the cytoplasm through the external membrane, the plasmalemma (Spanswick, 1981; Sze and Churchill, 1981), and into the vacuole via the separating membrane, the tonoplast (Cretin *et al.*, 1982). The electrochemical gradient of protons then fuels the "uphill" transport of many other molecules, such as amino acids, sugars, or ions, via symports or antiports (Etherton and Rubinstein, 1978; Guy *et al.*, 1978).

In addition, there have been indications of the existence of conductive pathways for the movement of ions through the plasmalemma other than

---

NAVA MORAN, GERALD EHRENSTEIN, and KUNIHIKO IWASA • Laboratory of Biophysics, NINCDS, National Institutes of Health, Bethesda, Maryland 20205.   CHARLES BARE and CHARLES MISCHKE • Agricultural Research Service, Weed Science Laboratory, Agricultural Environmental Quality Institute, Beltsville, Maryland 20705. *Present address for N.M.:* Department of Neurobiology, The Weizmann Institute of Science, Center for Neurosciences and Behavioral Research, Rehovot, Israel 76100.

energized (primarily or secondarily) transport. In some cases these conductive pathways have been termed channels, after the much more studied and explicitly demonstrated phenomena in animal cells. At least three types of channels have been mentioned, for Cl^-, K^+, and Ca^{2+}.

Chloride channels, voltage gated and (in most of the cases) Ca^{2+} dependent, have been invoked in the explanation of the regenerative phase of the action potential in the giant cells of *Nitella* (Kishimoto, 1961, 1965) and the related alga, *Chara* (Findley and Hope, 1964).

Keifer and Lucas (1982) attributed the control of membrane potential in *Chara corallina* by different modifiers to potassium channels. In the alga *Hydrodictyon*, potassium currents have been described that are similar to the delayed recitifier in neurons in their voltage dependence and in their abolishment by the blocker tetraethylammonium (TEA) (Findley and Coleman, 1983).

Potassium and chloride fluxes through channels have been implicated in the osmotic changes underlying the function of pulvini or pulvinous tissue affecting the movement in plants (Satter, 1981) and the closing and opening of guard cells (Rashke, 1975).

The existence of Ca^{2+} channels in plants has been suggested on the basis of the interference of the Ca^{2+}-channel blocker verapamil with Ca^{2+}-dependent cytokinin-induced bud formation (Saunders and Hepler, 1983).

Other ions could use a channel route, for example, bicarbonate (HCO_3^-), the passive entry of which has been described as the preferred form of carbon entry into the photosynthesizing plant (Steemann, 1960).

The evidence for the existence of gated ionic channels in plants has been, until recently, only circumstantial at best. The lack of information on ion movements through plant cell membrane can be ascribed in large part to the difficulties in applying electrophysiological techniques to studies of this preparation. Some of the problems, such as the existence of plasmodesmata interconnecting all cells, resulting in a formation of a symplast, or the negatively charged polysaccharide cell wall, affecting ion concentrations near and voltage measurements across the plasmalemma, have been solved by isolating single wall-less protoplasts (Cocking, 1972).

A major problem plaguing plant electrophysiologists, particularly in studies of higher plant cells, is the uncertainty about the location of the tip of an inserted measuring electrode with respect to the two membranes: plasmalemma and tonoplast. This is because of the small size of the cells and the relatively large size of the vacuole. This problem has recently been circumvented by accessing the bare plasmalemma from the outside with a "patch" electrode Moran *et al.*, 1984; Schroeder *et al.*, 1984).

In this chapter we describe some results from applying this method to cells of wheat and carrot.

2. SOME METHODOLOGICAL CONSIDERATIONS

2.1. Preparation of Protoplasts

Carrot (*Daucus carota*, c. v. danvers) protoplasts were obtained from callus grown in Murashige and Scoog media (Reinert and Yeoman, 1982) with 0.05 mg/liter of 2,4-dichlorophenoxyacetic acid added.

For the preparation of protoplasts, callus clumps weighing approximately 1–2 g were agitated for 3 hr at room temperature in 25 ml of medium containing 1% (w/v) Drieselase™ (Plenum Science Research, Inc., USA), in solution 1 [0.5 M sorbitol, 1 mM $CaCl_2$, and 5 mM MES buffer (2-N-morpholineethenesulfonic acid)], pH 5.5.

Wheat (*Triticum aestivum* L. var. Arthur) was grown in commercially available Jiffy-mix for 7 days at 28°C day, 20°C night temperatures, with 18 hr day length.

For the preparation of protoplasts, fresh wheat leaves (weighing about 4 g) were cut into 1-mm-wide strips and soaked (without agitation) for 2–3 hr in light at 30°C in 25 ml of enzyme medium consisting of (w/v) 2.5% cellulase, 0.3% Macerase™ (both enzymes from Behring Diagnostic, USA), 0.5% bovine serum albumin, all in solution 1, pH 5.5 (Edwards *et al.*, 1978).

For both carrot and wheat, the digestive medium was removed, and 40 ml of solution 1, pH 6, was added and carefully mixed. The released protoplasts were filtered through a Monel 107-μm screen and centrifuged for 5 min at 100 r.p.m. with an IEC table-top centrifuge (setting 2). The protoplast pellet was gently resuspended in 10 ml of 0.5 M sucrose, 1 mM $CaCl_2$, 5 mM MES, pH 6. On top of this suspension a discontinuous gradient was then applied, layering first 2 ml of 0.4 M sucrose, 1 mM $CaCl_2$, 5 mM MES, pH 6, then 1 ml of solution 1, pH 6. This preparation was then centrifuged for 5 min at 200 r.p.m. (setting 3). The protoplast layer (appearing as a band between the two uppermost layers of the gradient) was collected, resuspended in 30 ml of solution 1, and centrifuged for 5 min at 200 rpm (setting 3). Multiple washing can be done with no apparent damage. Protoplasts were finally stored in solution 1 on ice until used, usually for 1 to 7 hr.

2.2. Electrophysiology

During the experiment, the protoplasts were transferred to a 35-mm tissue culture dish (Primaria, Falcon, USA). The experiments were performed at 21° to 24°C. A patch-clamp system (List Electronics, West Germany) was used to voltage-clamp the membranes. The amplified output current from the membrane patch was filtered through a four-pole Butterworth filter (Krohn–Hite) at 2.5 kHz low-pass RC mode and digitized at 10 kHz. The patch pipette was fabricated from soft (soda-lime) glass according to the method of Hamill et al. (1981) and coated with wax. The tip diameter of the electrode was <1 μm, and its resistance was 7–12 MΩ. The pipette potential was measured using Ag/AgCl electrodes from E. W. Wright (Guilford, CT).

The success rate of forming a tight seal (gigaseal) between the plasmalemma and the patch pipette was not larger than 10% of all attempted electrode–cell contacts. On-cell patches offered the least disturbed preparation, whereas excised patches permitted the control of the ionic environment on both sides of the membrane and a selective observation of ion transport independent of ATP or other metabolites (Hamill et al., 1981). Because of the fragility of the plasmalemma, even a very slight suction usually resulted in a rupture of the membrane under the electrode and consequently in the formation of either a pinched-off vesicle or a whole-cell configuration. We distinguished between these two configurations on the basis of the difference between the capacitative transients in the current elicited by a square voltage pulse applied to the electrode; when the settling time of the transient exceeded approximately 3 msec, the latter case was assumed. The whole-cell configuration was then converted, by gently pulling the electrode away, into an outside-out configuration. In the alternative case, the vesicle was briefly exposed to air to obtain an inside-out configuration (Hamill et al., 1981).

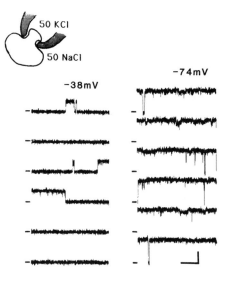

Figure 1. Single-channel currents recorded at the indicated hyperpolarizing potentials applied to an on-cell patch on carrot protoplast. Upward deflections indicate channel opening, and the horizontal bars to the left of each trace the level of current at channel closing. Note the transition to an almost all-open state at the larger hyperpolarization. Scale bars: 2 pA (vertical), 100 msec (horizontal). Inset: Schematic representation of the recording configuration, including the concentrations of the main ions (in mM). In addition, the bath medium contained 1 mM $CaCl_2$, 0.5 mM KCl, 0.56 M sorbitol, 20 mM MES, pH 6, and the pipette contained 1 mM $CaCl_2$, 0.5 mM NaCl, 0.4 M sorbitol, 20 mM MES, pH 7, and 5 mM EGTA.

3. VOLTAGE-DEPENDENT CHANNELS OPENED BY HYPERPOLARIZATION

Figure 1 depicts currents recorded from a carrot protoplast. This is a relatively rare case of a recording made in an on-cell configuration (see inset). Note the increased duration of open time and the shortening of closed time as the patch is hyperpolarized by 36 mV.

Figure 2 depicts fluctuations of current recorded from an outside-out excised patch from wheat protoplast held at various membrane potentials. The patch contained several channels, as indicated by fluctuations among several discrete levels. Figure 3A shows an amplitude histogram (continuous line) obtained from a 23-sec record from the data of Fig. 2 at a membrane potential of -65 mV. This histogram was subsequently fitted

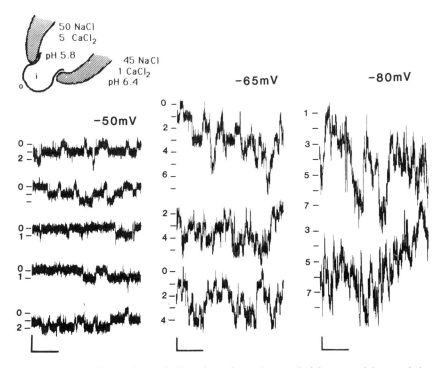

Figure 2. Current fluctuations at indicated membrane-hyperpolarizing potentials recorded from an outside-out excised patch from a wheat protoplast. Downward deflections result from channel opening. Note the multilevel jumps (levels marked by numbers), indicating the presence of several channels in the patch. Scale bars: 2 pA (vertical), 100 msec (horizontal). Inset: Schematic representation of the recording configuration. In addition to the concentrations indicated (in mM), the recording medium contained 0.4 M mannitol and 8 mM HEPES buffer.

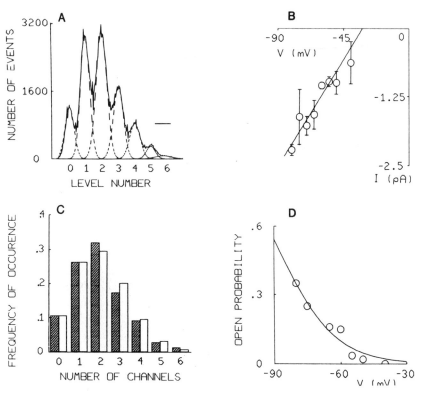

Figure 3. Characterization of the multichannel patch. A: Amplitude histogram using current fluctuations depicted in Fig. 2 (-65 mV). Continuous line: Experimental data. Dashed lines: Gaussian distributions of current amplitudes about each level. The sum of these distributions constitutes the best fit to the data. Horizontal bar: 1.5 pA. B: Current–voltage relationship for a single open channel. Symbols: Average separation between neighboring peaks of the amplitude histogram at each membrane potential. Vertical bars: Standard deviation. Line: Best fit to data: $I = 0.044 \cdot V + 1.41$ (I in pA and V in mV). C: Observed relative occupancy of each level versus calculated frequency of occurrence of a particular number of channels: shaded bars, "observed" relative area under each normal curve of Fig. 3A; open bars, calculated best-fitting binomial distribution of the number of channels open simultaneously, assuming 13 independent and identical channels in the patch and two conductance levels—conducting and nonconducting—for each channel. The fit resulted in the value of 0.16 for the open probability of a single channel (at -65 mV). D: Open probability of a single channel as a function of membrane potential. Symbols: Values estimated as in Fig. 3C. Line: Best fit to data: $P_O = [1 + \exp(V - 88)/12.6]^{-1}$ (V in mV).

with a sum of Gaussian distributions (dashed lines). This resulted in values for amplitudes of the levels of current fluctuations—the horizontal separation of the peaks—as well as in values of their relative occupancy—the area under each dashed curve relative to the total.

The average amplitudes were plotted versus membrane potential in

Fig. 3B, with standard deviations indicated by bars. The linearity of the current–voltage (I–V) relationship suggests a single type of channel. The slope of the I–V curve yields the single-channel conductance of 44 ± 6 pS, and its zero-current intercept the reversal potential of −34 ± 13 mV.

Figure 3C shows the observed relative occupancy of each level of current, i.e., the fraction of time any (n) number of channels is open at a membrane potential of −65 mV (shaded bars). Alongside we plotted the best-fitting binomial distribution of the relative occupancy, calculated assuming that the patch contained 13 identical and independent channels, each with the probability of being open of 0.16 (open bars).

The probability of being open (open probability) resulting from such a procedure is plotted as a function of membrane potential in Fig. 3D. For the range of potentials shown, the time spent in an open configuration increased as the membrane was increasingly hyperpolarized. This voltage dependence of the open probability (P_O) was fitted to

$$P_O(V) = \{1 + \exp[-(V - V_0)/K]\}^{-1} \quad (1)$$

resulting in the values of the midpoint voltage, V_0, of −88 ± 2 mV and in the parameter K of −12.6 ± 1.7 mV.

4. CHANNELS AFFECTED BY TEA

Figure 4 illustrates the effect of tetraethylammonium (TEA) on channels in an inside-out excised patch from another wheat cell. From the untreated patch we recorded three different single-channel amplitudes (Figs. 4A, 5): the largest, with conductance of 38 pS, occurred only at depolarizing potentials, increased their open probability with increased depolarization, and required more than 1 min after the change of voltage to reach steady state; the smallest, 8 pS, appeared at both polarities, and their voltage dependence could not be assessed; the medium-sized ones, 26 pS, were seen only at hyperpolarization and appeared to be voltage independent.

Ten millimolar TEA abolished the 38-pS and the 8-pS channels completely within a few seconds of application. The only channels that could then be observed were 17-pS channels that opened only at hyperpolarization and seemed to be independent of voltage (Figs. 4B, 5). This suggests that these are the 26-pS channels modified by TEA. Another effect of TEA was to raise the base-line resistance, measured when the channels appeared to be closed, from 16 GΩ to 27 GΩ (not shown).

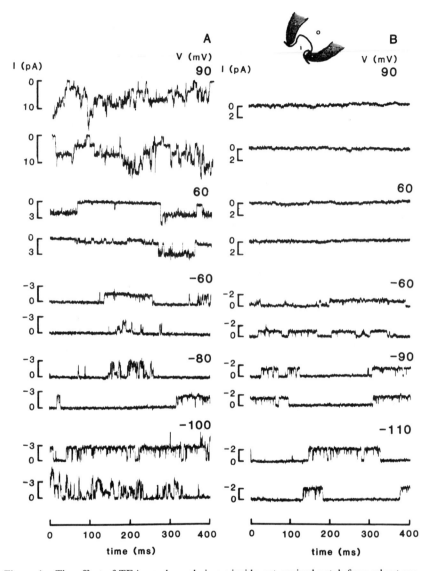

Figure 4. The effect of TEA on channels in an inside-out excised patch from wheat protoplast. Current fluctuations recorded at indicated depolarizing (positive) or hyperpolarizing (negative) membrane potentials; current polarity varies accordingly; zero current indicates channel closing. Inset: Schematic representation of the recording situation. The recording medium, symmetrical on both sides of the membrane, contained the following: 0.4 M sorbitol, 50 mM KCl, 1 mM $CaCl_2$, 20 mM MES, pH 6. A: Untreated membrane. B: A few seconds after application of 10 mM TEA to the bath.

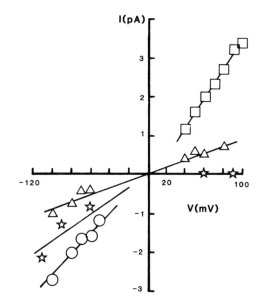

Figure 5. Current–voltage relationship for the channels illustrated in Fig. 4. □, 38-pS channel; ○, 26-pS channel; △, 8-pS channel, all before TEA treatment. ☆, after TEA treatment, 17-pS channel and null current.

5. CONCLUSIONS

From roughly 20–30% of the successful patches, we recorded voltage-dependent channels, some opening in response to depolarization and some—as in two of the cases described above—in response to hyperpolarization.

The fitted open probability–voltage (P_O–V) curve of Fig. 3D changes e-fold per -12.6 ± 1.7 mV change of membrane potential. This is only about half of the steepness of the P_O–V relation for sodium channel in neuroblastoma (e-fold change of open probability per 5.5 ± 2.0 mV; Huang et al., 1982) and is rather similar to that for the Ca^{2+}-dependent K^+ channel in pituitary cells (e-fold change in P_O per 9.1 mV; Wong et al., 1982).

Because the formation of the seal between the membrane and the electrode seems to occur in plant cells in a less controlled manner than in animal cells (see Section 2.2), the polarity of the patch could not be definitively ascertained with excised patches.

Although the polarity of the stimulation with respect to the membrane in the case of a voltage-dependent channel in an excised patch is thus ambiguous, there is, obviously, no ambiguity in the contents of solutions on either side of the membrane and the polarity with respect to the ground electrode of the reversal potential for the recorded currents. Therefore, given nonsymmetrical solutions, the selectivity of the channel may be defined.

For example, the single-channel currents illustrated in Figs. 2 and 3 reverse (the extrapolation based on the assumption of linearity of the I–V curve, Fig. 2B) when the potential in the pipette is -34 mV (negative with respect to the ground electrode). The dissimilar distribution of the concentrations of Ca^{2+} and of H^+ between the pipette and the bath (ratio of 5 : 1 and 4 : 1, respectively, see inset, Fig. 2A) fits the polarity and is the closest in magnitude to the observed reversal potential, whereas the concentration ratio of Na^+ is not sufficient (10 : 9), and that of Cl^- is both insufficient and in the wrong direction. This implies that the channels observed were selective to Ca^{2+} and/or H^+.

Another approach to channel identification is the use of channel-specific blockers. Tetraethylammonium, although one of the less specific drugs, has been found to selectively block cationic channels, in particular the K^+ channel (Armstrong, 1975; Latorre and Miller, 1983). The complete disappearance of the 38- and 8-pS channels on the addition of 10 mM TEA (Figs. 4, 5) suggests that both are types of K^+ channel (or substates of a K^+ channel). The 26-pS channel seen at hyperpolarization is probably also a K^+ channel, since it appears modified by TEA (Blatz and Magleby, 1984).

The presence of single channels has been demonstrated so far in algae (Krawczyk, 1978), in wheat (Moran *et al.*, 1984), in carrot (this study), and in broad bean (Schroeder *et al.*, 1984). In our work on wheat and carrot cells, representatives of monocots and dicots, respectively, we did not observe significant differences in properties of ion channels between the two groups.

If, indeed, some of the channels we recorded are K^+ channels, their widespread presence in the plant kingdom suggests some general function for these channels, which could be, for example, volume control of cells, and which may be reflected, in turn, in movement of some specialized cells or tissues such as guard cells or pulvini.

REFERENCES

Armstrong, C. M., 1975, K pores of nerve and muscle membranes, in *Membranes: A Series of Advances*, Vol. 3 (G. Eisenman, ed.), Marcel Dekker, New York, pp. 325–358.

Blatz, A. L., and Magleby, K. L., 1984, Ion conductance and selectivity of single calcium-activated potassium channels in cultured rat muscle, *J. Gen. Physiol.* **84**:1–23.

Clarkson, D. T., 1984, Ionic relations, in *Advanced Plant Physiology* (M. B. Wilkins, ed.), Pitman Publishing, Marshfield, Massachusetts, pp. 319–353.

Cocking, E. C., 1972, Plant cell protoplast—isolation and development, *Annu. Rev. Plant Physiol.* **23**:29–50.

Cretin, H., Martin, B., and D'Auzac, J., 1982, Characterization of a magnesium-dependent proton translocating ATPase on *Hevea latex* tonoplast, in *Plasmalemma and Tonoplast:*

Their Functions in the Plant Cell (D. Marmé, E. Marré, and R. Hertel, eds.), Elsevier, Amsterdam, pp. 201–208.

Edwards, G. E., Robinson, S. P., Tyler, N. J. C., and Walker, D. A., 1978, Photosynthesis by isolated protoplasts, protoplast extracts, and chloroplasts of wheat, *Plant Physiol.* **62**:313–319.

Etherton, B., and Rubinstein, B., 1978, Evidence for amino acid–H^+ co-transport in oat coleoptiles, *Plant Physiol.* **67**:59–63.

Findlay, G. P., and Coleman, H. A., 1983, Potassium channels in the membrane of *Hydrodictyon africanum*, *J. Membr. Biol.* **75**:241–251.

Findlay, G. P., and Hope, A. B., 1964, Ionic relations of cells of *Chara australis*. VII. The separate electrical characteristics of the plasmalemma and tonoplast, *Aust. J. Biol. Sci.* **17**:62–77.

Guy, M., Reinhold, L., and Laties, G. G., 1978, Membrane transport of sugars and amino acids in isolated protoplasts, *Plant Physiol.* **61**:593–596.

Hamill, O. P., Marty, A., Neher, E., Sakmann, B., and Sigworth, F. J., 1981, Improved patch-clamp techniques for high-resolution current recording from cells and cell-free membrane patches, *Pfluegers Arch.* **391**:85–100.

Huang, L. Y. M., Moran, N., and Ehrenstein, G., 1982, Batrachotoxin modifies the gating kinetics of sodium channels in internally perfused neuroblastoma cells, *Proc. Natl. Acad. Sci. U.S.A.* **79**:2082–2085.

Keifer, D. W., and Lucas, W. T., 1982, Potassium channels in *Chara carallina*, control and interaction with the electrogenic H^+ pump, *Plant Physiol.* **69**:781–788.

Kishimoto, U., 1961, Current–voltage relation in *Nitella*, *Biol. Bull.* **121**:370–371.

Kishimoto, U., 1965, Voltage clamp and internal perfusion studies on *Nitella* internodes, *J. Cell. Comp. Physiol.* **66**:43–54.

Krawczyk, S., 1978, Ionic channel formation in a living cell membrane, *Nature* **273**:56–57.

Latorre, R., and Miller, C., 1983, Conduction and selectivity in potassium channels, *J. Membr. Biol.* **71**:11–30.

Moran, N., Ehrenstein, G., Iwasa, K., Bare, C., and Mischke, C., 1984, Ion channels in plasmalemma of wheat protoplasts, *Science* **226**:835–838.

Rashke, K., 1975, Stomatal action, *Annu. Rev. Plant Physiol.* **26**:309–340.

Reinert, T., and Yeoman, M. M., 1982, *Plant Cell and Tissue Culture. A Laboratory Manual*, Springer-Verlag, New York.

Satter, R. L., and Galston, A. W., 1981, Mechanisms of control of leaf movements, *Annu. Rev. Plant Physiol.* **32**:83–110.

Saunders, M. J., and Hepler, P. K., 1983, Ca antagonists and calmodulin inhibitors block cytokinin-induced bud formation in *Funaria*, *Dev. Biol.* **99**:41–49.

Schroeder, J. I., Heidrich, R., and Fernandez, J. M., 1984, Potassium selective single channels in guard cell protoplasts of *Vicia faba*, *Nature* **312**:361–362.

Spanswick, R. M., 1981, Electrogenic ion pumps, *Annu. Rev. Plant Physiol.* **32**:267–289.

Steemann, N. E., 1960, Uptake of CO_2 by the plant, in *Encyclopedia of Plant Physiology*, Vol. V, *The Assimilation of Carbon Dioxide* (W. Ruhland, ed.), Springer-Verlag, New York, pp. 70–84.

Sze, H., and Churchill, K. A., 1981, Mg^{2+}/K^+-ATPase of plant plasma membranes is an electrogenic pump, *Proc. Natl. Acad. Sci. U.S.A.* **78**:5578–5582.

Wong, B. S., Lecar, H., and Adler, M., 1982, Single calcium-dependent potassium channels in clonal anterior pituitary cells, *Biophys. J.* **39**:313–317.

Chapter 12

Channels in Kidney Epithelial Cells

Sandra Guggino

1. INTRODUCTION

Since the first conclusive demonstration of the presence of channels in epithelia, in particular the demonstration of amiloride-sensitive sodium channels involved in active Na^+ reabsorption across the apical membrane of the frog skin (Lindemann and Van Driessche, 1976), the question arose as to how much of the conductive properties of epithelia are controlled by ion channels. The advent of the patch-clamp technique allowed a new way of investigating this problem in epithelial cells from kidney. Thus, by use of the patch-clamp technique on isolated kidney tubules and on cultured kidney cell lines, a comparison can be made between the conductance properties of the membranes and the properties of the ion channels found in the same segments. We can then begin to assess what components of the conductive properties of these cells are accounted for by channel activity.

Potassium secretion and the regulation of K^+ excretion in the kidney are known to occur in the distal nephron segments (Field and Giebisch, 1985). Figure 1A shows the model of a cell from the cortical collecting

---

SANDRA GUGGINO • Laboratory of Molecular Aging, Gerontology Research Center, National Institute on Aging, National Institutes of Health, Baltimore, Maryland 21224.

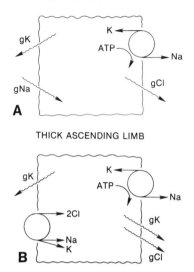

Figure 1. A: The cortical collecting duct (CCT) has a major role in the regulation of K^+ secretion and Na^+ reabsorption and is under the hormonal control of aldosterone. The apical membrane Na^+ conductance is inhibited by amiloride, and the apical K^+ conductance is inhibited by Ba^{2+} and Cs^+. B: The TAL is a segment in the distal nephron that exhibits a large net Cl^- reabsorption that is responsible for maintaining the high salt content in the kidney medulla. This segment can produce a net reabsorption or net secretion of K^+ depending on K^+ intake. Shown here are the major routes for ion movements in the cells. Note the presence of a Ba^{2+}-sensitive apical membrane K^+ conductance as well as an electroneutral cotransport of K^+, Na^+, and Cl^- that is furosemide sensitive.

tubule (CCT), a segment of the distal nephron that exhibits avid K^+ secretion and Na^+ reabsorption (Stoner et al., 1974). Figure 1B shows another distal segment, the thick ascending limb (TAL), which functions to reabsorb salt but not water in order to dilute the urine. In both segments, K^+ is pumped into the cell via the Na^+, K^+-ATPase and then is lost through an apical cell membrane conductance that is Ba^{2+} sensitive (Hebert and Andreoli, 1984a; O'Neil, 1983). In addition, K^+ is also taken up across the apical membrane of the TAL by an electrically silent cotransport with Na^+ and Cl^-. This results in a net secretion of K^+ by the CCT and a net reabsorption of K^+ by the TAL. The apical membrane of the CCT has a Na^+ conductance that, like that of other tight epithelia such as the frog skin, is stimulated by aldosterone (O'Neil and Helman, 1977) and inhibited by amiloride (O'Neil and Boulpaep, 1979).

A Ca^{2+}-activated K^+ channel that has properties common to the Ca^{2+}-activated K^+ channels found in other cells such as the secreting gland cells, muscle cells, and the T-tubule of skeletal muscle (see Latorre and Miller, 1983, for a review) is present on the apical membrane of a population of cultured chick kidney cells and on the apical membrane of a clone from medullary TAL of rabbit. The presence of such a K^+ channel on the apical cell membrane of kidney distal tubule cells can account for some of the properties of the distal nephron K^+ conductance, namely, (1) that the K^+ conductance is Ba^{2+} sensitive and (2) that the Ba^{2+}-sensitive K^+ conductance and net K^+ secretion are increased in the presence of antidiuretic hormone (ADH) (a hormone that increases levels of cAMP in the distal tubule of the kidney) (Morel, 1981). The Ca^{2+}-activated K^+ channel in the apical membrane of chick kidney cells possesses the properties of Ba^{2+} sensitivity and stimulation of activity by forskolin and ADH. These properties can account for the effects of these agents on the macroscopic membrane conductance in the apical membrane of distal tubule kidney cells.

2. CELL CULTURE

Isolation of cells for the primary culture of whole chick kidney basically followed a procedure modified from Henry (1979). Kidneys from 3-day-old white Leghorn chicks were perfused with Ringer to remove blood, quickly dissected, then minced to small pieces about 1 mm^3. Incubation at 37°C with collagenase (Sigma type I, 1.4 mg/ml) and hyaluronidase (Sigma type I, 1.0 mg/ml) with an occasional gentle stirring, followed by treatment with trypsin (Gibco, 0.125%) for 1 min, yielded a suspension of single cells and small tubular fragments. Cells were removed from undigested pieces of tissue by two gravity settlements and then plated at a density of 10^6 cells/ml on glass coverslips. The cells grew to confluency in approximately 3 days in McCoy's medium containing 10% fetal calf serum gassed with 5% CO_2. Antibiotics (1% penicillin, neomycin, and streptomycin, Gibco) were present for the first 24 hr; then antibiotic-free medium was used. The chick cells used in these experiments were elongated, had phase-bright edges and small indistinct nuclei, and comprised a small but characteristically distinct proportion of the total cell population. Patch recordings were made at 22–24°C within 1 hr of removal of the cells from the incubator.

Cells from pass 24 to pass 35 from a clone of medullary TAL from rabbit were also used for both cell-attached and excised-patch experiments. These cells were cultured as previously described by Green *et al.* (1985). The cell perimeter was polygonal in shape, about 10 μm in diameter, with phase-bright edges. Patch recordings on these cells were done at 37°C.

3. PATCH-CLAMP METHODOLOGY

3.1. Patch Pipets

Pipets were produced as described in detail by Hamill *et al.* (1981) using flint glass hematocrit capillaries. Pipets were pulled in two steps with a commercial vertical pipet puller (Kopf, Model 700C, Tujunga, CA).

The pipet tip was coated to within 100 μm of the opening with freshly prepared Sylgard™ (Dow Corning) and fire polished on a microforge (Stoelting, 600× magnification). Pipets routinely had an open-tip resistance of 5–8 MΩ in a standard Ringer's solution.

3.2. Electrical Recording and Data Analysis

Membrane currents were recorded with an extracellular patch-clamp amplifier (List L/M EPC-5, List Electronic, Darmstadt, FRG) set at a gain of 50 mV/pA and a filter setting of 10 kHz. The signal from the patch-clamp amplifier was recorded on an FM tape recorder (Gould Model 6500, Gould Electronics, Cleveland, OH) and later digitized at a sampling interval of 100 μsec with a Nicolet 4094 digital oscilloscope (Nicolet Instrument Co., Madison, WI) after postfiltering at 0.5 to 2 kHz low pass with an eight-pole Bessel filter (Frequency Devices, Inc., Haverhill, MA). Fractional open time and mean open and closed times were analyzed with a VAX 780 computer (Digital Equipment Corporation).

3.3. Cell-Attached Patches

Seals were made by gently lowering the patch pipet onto the surface of a cell until a small increase in patch resistance was observed; then gigaseals between 5 and 10 GΩ were obtained with gentle mouth suction. For cell-attached patches, the patch pipet and bath contained an identical Ringer solution composed of (in mM) 135 NaCl, 5 KCl, 1 $CaCl_2$, 1 $MgCl_2$, and 25 Na-HEPES at pH 7.4. The cell-attached patch-clamp recordings, as well as the apical cell membrane potential measurements using conventional intracellular microelectrodes, were performed in the presence of this Ringer in order to maintain the normal interior-negative resting membrane potential of the cell during experiments. Thus, in the cell-attached patch-clamp mode, only that small area across the patch was depolarized. The membrane potential across the patch, the patch potential, is the difference between the resting apical membrane potential and the applied pipet voltage; thus, a negative pipet voltage depolarized the patch.

Forskolin (Calbiochem) and ADH (arginine vasopressin, Sigma) were dissolved in the Ringer's solution described above with ethanol used as a carrier. The drugs were perfused by gentle back pressure from a broken microelectrode positioned near the surface of the cell-attached patch. The broken microelectrode was held above the water surface and then introduced into the solution and positioned near the cell once base-line channel activities were recorded; perfusion was ended by relieving back pressure and removing the pipet from the bath.

3.4. Inside-out Excised Patches

Gigaseals were made in the cell-attached mode, current–voltage relationships were obtained, and then the patch was excised by pulling up the micromanipulator. Excised patches were most successful when a Ringer with 10^{-7} M Ca^{2+} buffered with EGTA was perfused onto the cells just before excision, because this seemed to reduce vesiculation of the patch.

Experiments on excised patches that involved the substitution of gluconate$^-$ for Cl^-, K^+ for Na^+, addition of Ba^{2+}, or low Ca^{2+} were accomplished by perfusing the culture dish with the experimental Ringer's solution. During these experiments, the excised patch was moved away from the surface of the cells and near the inflow port.

4. POTASSIUM CHANNEL CHARACTERISTICS

A representative recording from a cell-attached patch experiment is illustrated in Fig. 2. At resting membrane potential, -49 ± 2 mV (mean + S.E., $n = 27$), zero pipet voltage, channel openings were infrequent, but depolarization of the patch increased the frequency of opening significantly, such that at -60, -80, and -100 mV pipet voltage, channels were in the open state 0.05%, 0.1%, and 0.3% of the time (Fig. 2). Estimates of fractional open time, $f(v)$, from other patches also showed a low frequency of opening at resting membrane potential, which increased with depolarization. For example, at pipet voltages of -90, -100, and -110

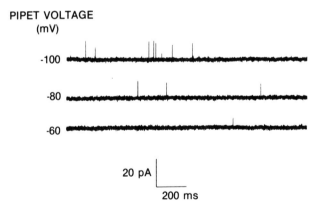

Figure 2. Single-channel recordings from a cell-attached patch measured on cultured kidney cells show an increased activation with voltage. Channel openings are upward deflections, indicating outward movement of charge. At resting membrane potentials, the channel is spontaneously active only infrequently; however, depolarization increases the frequency of opening of the channel.

mV, the $f(v)$s for another patch were 0.1, 0.2, and 0.8%, respectively. The mean open times, which were also voltage dependent, increased from 0.5 msec at -100 mV to 0.8 msec at -120 mV in the patch depicted in Fig. 2. The channels seen in cell-attached patches from these cultured kidney cells were characterized by very low probabilities of opening, very rapid open times, but with a relatively high density, because there were often more than two channels per patch.

The single-channel conductance can be estimated from the slope of the current–voltage relationship. The average single-channel or unit conductance for six cell-attached patches was 106 ± 9 pS (mean $\pm$ SE, $n = 6$) with a range of 80 to 140 pS. The single-channel conductance in these cultured chick kidney cells is similar to the conductance of Ca^{2+}-activated, Ba^{2+}-sensitive K^+ channels previously reported for the cortical collecting tubule of the rabbit (90 pS) (Hunter et al., 1984). The conductance for cell-attached MTAL cells was 133 ± 13 pS ($n = 3$).

In a cell-attached patch of chick kidney cells that showed 100-pS fluctuation, subsequent inside-out excision of the patch into a solution containing 87 mM K^+ and 10^{-7} M Ca^{2+} buffered with EGTA at pH 7.4 on the intracellular face of the patch yielded a current–voltage relationship with a slope of 75 pS. Because the single-channel conductance in the cell-attached and excised modes was similar and within the range measured for all the cell-attached channels, and the current flow followed the K^+ concentration gradient at zero pipet voltage, this suggested a K^+-selective channel. The calculated Nernst or reversal potential with 87 mM K^+ in the bath and 5 mM K^+ in the pipet was $+74$ mV pipet voltage. The current–voltage relationship was nonlinear in the range of $+60$ mV to $+70$ mV pipet voltages and did not reverse by $+70$ mV pipet voltage. This nonlinearity has been noted for the Ca^{2+}-activated K^+ channel in chromaffin cells (Yellen, 1984; see Suarez-Isla, Chapter 23) under bionic conditions with 0.1 M K^+ on the internal side of the patch and 0.1 M Na^+ on the other side.

In other experiments on excised patches made from cell-attached patches that also showed 100-pS outward current fluctuations, a manipulation of the Cl^- gradient by gluconate substitution was without effect on channel fluctuations. This provided additional evidence that the outward current fluctuations in the cell-attached patches were indeed the result of outward K^+ movements rather than inwardly-directed Cl^- movements.

In several cell types, including nonexcitable cells such as erythrocytes (Gardos, 1958), an increase in cytoplasmic Ca^{2+} activates a K^+ conductance. This Ca^{2+} sensitivity is expressed at the single-channel level as an increase in the fractional open time. The curve of the single-channel open-state probability (p_0) as a function of membrane potential shifts along the voltage axis with changes of internal [Ca^{2+}], so for a ten-fold change in [Ca^{2+}] the shift is about 65 mV (Latorre *et al.*, 1982).

We tested the Ca^{2+} activation of the K$^+$ channel in chick cells by lowering the Ca^{2+} on the intracellular face of an inside-out excised patch. The channel in Fig. 3A was activated in 1 mM Ca^{2+}, but as the free calcium concentration was lowered by the addition of buffered EGTA, the channel spent more time in the closed state. Such low Ca^{2+} concentrations also decreased open dwell times in the Ca^{2+}-activated K$^+$ channel of CCT cells (Hunter *et al.*, 1984). Similar experiments in medullary TAL cells revealed Ca^{2+}-sensitive current fluctuations.

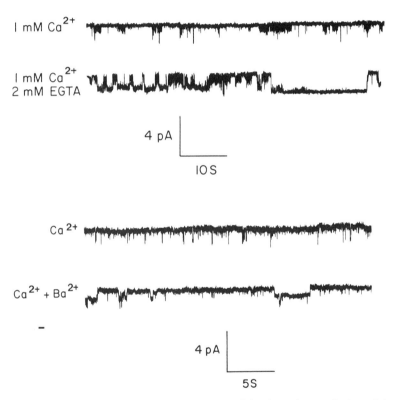

Figure 3. The effect of lowering Ca^{2+} on the intracellular face of an excised patch is to decrease the fractional open time. The addition of 2 mM EGTA and 1 mM Ca^{2+}, pH 7.4, gives a calculated concentration of free Ca^{2+} of 5×10^{-7} M, which decreases channel activity. Open channels are upward deflections. Another channel of 30 pS that is Ca^{2+} activated and that carries inward current at resting potential can be observed as longer openings of smaller amplitude in these traces. B: A demonstration that Ba^{2+} causes an increase in long closings in excised membrane patches. Bath solutions contain 1 mM Ca^{2+}. The addition of 5 mM Ba^{2+} (lower trace) is made at the intracellular face of the excised patch. Pipet voltage is -10 mV.

The influence of Ba^{2+}, a known inhibitor of K^+ conductance in epithelia (Nagel, 1979), on channel fluctuations is shown in Fig. 3B. When 5 mM Ba^{2+} was added to the intracellular face of an excised patch, the channels spent an increased time in the closed state. Other patches, excised into 10^{-7} M Ca^{2+}, show complete inhibition of channel opening when 5 mM Ba^{2+} was added to the intracellular side of the patch. Thus, channel opening in these cells is inhibited by either low Ca^{2+} or Ba^{2+} at the intracellular face. The K^+ channel from the apical membrane of cultured rabbit TAL cells also showed a Ba^{2+} block that is voltage dependent (Fig. 4). This provides additional evidence that the chick kidney cells as well as the medullary TAL cells possess on the apical surface a Ba^{2+}-sensitive, Ca^{2+}-activated K^+ channel. In addition, the apical membrane potential of both of these cells was reduced by Ba^{2+} (Guggino *et al.*, 1984, 1985). For example, the chick cells depolarized from 49 ± 2 to 38 ± 2 mV (n = 11) with 2 mM Ba^{2+}. Thus, it seems that the activity of this K^+ channel might be related to a Ba^{2+}-sensitive apical membrane poten-

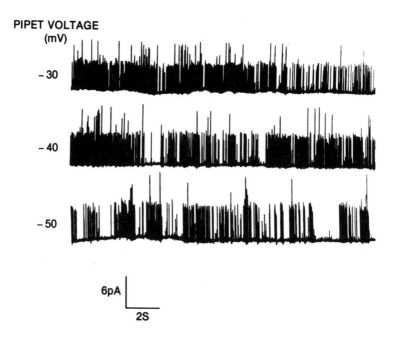

Figure 4. A demonstration of Ba^{2+} block on MTAL cells. Barium added to the intracellular face of a patch causes an increase in long closings as the voltage becomes more negative.

tial. The presence of a Ba^{2+}-sensitive K^+ channel in isolated tubules of the CCT or in cultured medullary TAL cells correlates with Ba^{2+}-sensitive K^+ conductances on these membranes (O'Neil, 1983; Hebert et al., 1984).

5. CHANNEL MODULATION

Forskolin, an activator of adenylate cyclase (Seamon and Daly, 1981), and ADH, which acts on the distal segments (see Handler and Orloff, 1981, for a review) through cAMP-mediated pathways, were chosen to determine whether channel activity is modulated by conditions that affect the K^+ conductance and K^+ secretion in the distal nephron.

Forskolin (10^{-5} M) added near a cell-attached patch greatly increased the frequency of opening and the mean open time of the K^+ channel. The increase in frequency of opening that occurred at all voltages is illustrated in Fig. 5. For example, at -40 mV pipet voltage, there was a 16-fold increase in fractional open time as well as a fivefold increase in mean open time. Even at zero pipet voltage, the normal resting membrane potential, the low incidence of channel opening significantly increased in the presence of forskolin. The single-channel conductance did not change after forskolin. Control experiments performed by applying the Ringer's solution containing carrier ethanol did not increase the frequency of channel opening.

Antidiuretic hormone has been shown to enhance K^+ secretion in the rat distal tubule (Field et al., 1984). Hebert et al. (1984) suggest that one stimulatory effect of ADH in the TAL is expressed as an increase in the apical K^+ conductance. In cultured kidney cells, ADH (10^{-6} M) perfused near a cell-attached patch increased fractional open time and mean open time of the K^+ channel (Fig. 6). The effects of forskolin and ADH, when taken together, suggest that the channel on the apical membrane of the cultured chick kidney cells is modulated through a cAMP-mediated process.

6. CONCLUSIONS

The results of experiments measuring channel fluctuations in a selected population of cells in a primary culture of whole chick kidney and from a clone of MTAL cells suggest that these cells possess a Ba^{2+}-sensitive, Ca^{2+}-activated K^+ channel. The properties of this channel resemble those of the Ca^{2+}-activated K^+ channel found in many other cell types including nerve (Adams et al., 1982) and muscle cells (Pallota et al., 1981) and chromaffin cells (Marty, 1981). In these cell types, the Ca^{2+}-

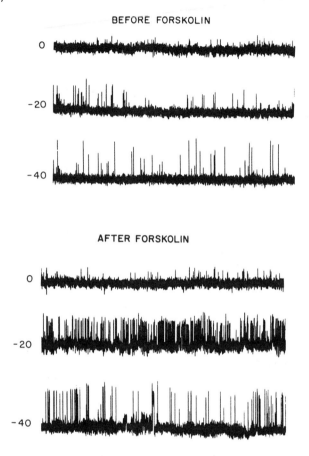

Figure 5. The activation of a K^+ channel before and after the addition of forskolin is illustrated at three voltages. Note that at each voltage there is an increase in the number of openings as well as an increase in the length of time each channel is open.

activated K^+ channel has a high conductance (~200 pS) in symmetrical solutions of 0.1 M K^+, and a high selectivity for K^+ over Na^+, such that under K^+/Na^+ biionic conditions, the current–voltage relationship approaches the voltage axis asymptotically. This channel also shows complicated opening and closing kinetics that can be modeled approximately by assuming two closed states and one open state of the channel (Latorre et al., 1982). In addition, the fractional open time is modulated by the internal Ca^{2+} concentration (Wong et al., 1982), although the Ca^{2+} sensitivity shows a large variability among channels, even within one cell type.

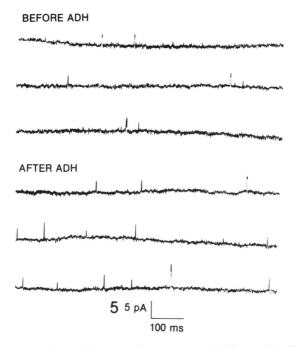

Figure 6. The application of ADH to a cell-attached patch held at -100 mV pipet voltage causes activation of this channel. The top trace shows channel activity before the addition of hormone; in the lower traces an increase in fractional open time is apparent.

At least one other type of kidney distal cell is known to possess Ba^{2+}-sensitive K^+ channels in the apical membrane—these are the CCT cells (Hunter et al., 1984). The function of these channels or their contribution to the underlying conductive properties of the apical membrane is just beginning to be determined. In other cells such as excitable cells, the Ca^{2+}-activated K^+ channel functions in adaptation of neurons to stimulation (Meech, 1978) or in the generation of pacemaker activity (Gorman et al., 1981). In pancreatic β cells (Atwater et al., 1983) and salivary glands (Maruyama et al., 1983), the channel acts to link hormonal stimulus and secretion through modulation of the membrane potential. In kidney cells, the Ca^{2+}-activated K^+ channel may contribute to the K^+ conductance of the apical membrane in the distal nephron.

The presence of a Ba^{2+}-sensitive K^+ conductance in the distal tubule, the site of K^+ secretion in the kidney, is well established. Potassium is also the primary conducting species across the apical membrane of MTAL cells of mouse (Hebert and Andreoli, 1984), the rabbit cortical TAL (Greger and Schlatter, 1983), and the early distal tubule of *Amphiuma* kidney (Guggino et al., 1982; Oberleithner et al., 1983). The substitution of Ba^{2+} for K^+ in all these segments produces a depolarization of the apical membrane potential and a large increase in the resistance ratio (apical resistance divided by basolateral resistance), suggesting that a K^+ conductance is inhibited by Ba^{2+}. The cortical collecting tubule also has a large K^+ conductance that is inhibited by Ba^{2+} (Koeppen et al., 1983; O'Neil and Sansom, 1984). A Ba^{2+}-induced depolarization of apical membrane potential occurs in cultured chick kidney cells and cultured rabbit medullary TAL cells. In three types of kidney cells, the presence of a Ba^{2+}-sensitive K^+ conductance correlates with the presence of Ba^{2+}-sensitive Ca^{2+}-activated K^+ channels, suggesting that this channel may contribute to K^+ conductive properties and K^+ secretion from these cells.

Hebert et al. (1984) have recently concluded that ADH increases the Ba^{2+}-sensitive component of the K^+ conductance in mouse medullary thick ascending limb cells. Recent evidence by Field et al. (1984) suggests that ADH stimulates net K^+ secretion by distal tubule segments. Thus, one would expect that if a Ba^{2+}-sensitive K^+ channel is a component of the apical K^+ conductance in the distal tubule, the channel current would be stimulated by ADH, as is the Ba^{2+}-sensitive K^+ conductance. It was found that ADH does increase the K^+ channel activity in these cultured chick kidney cells, an effect that would cause an increase in Ba^{2+}-sensitive K^+ conductance in kidney tubules in vivo.

The contribution of a K^+ channel to the total K^+ current across the apical cell membrane of renal cells depends on the number of channels, the probability of opening, and the current carried through the channel. The single-channel current, in turn, depends on the single-channel conductance and the electrochemical gradient for K^+, which in most renal cells favors a K^+ flux out of the cell. The Ca^{2+}-activated K^+ channel in chick kidney carries a small outward current at resting membrane potentials because in the absence of modulators, the probability of opening is very low. However, following application of forskolin or ADH, the fraction of the time the channels spend in the open state increases dramatically, causing an increase in the total K^+ current through these channels. Thus, taken together, the two aspects of channel activity that have been monitored—inhibition of channel activity by Ba^{2+} and stimulation of channel activity by ADH through cAMP-mediated mechanisms—can account for qualitatively similar effects on conductance properties when these agents are added to kidney cells in vivo or in vitro.

REFERENCES

Adams, P. R., Constantin, A., Brown, D. A., and Clark, R. B., 1982, Intracellular Ca^{2+} activates a fast voltage-sensitive K^+ current in vertebrate sympathetic neurons, Nature 296:746–749.

Atwater, I., Rosario, L., and Rojas, E., 1983, Properties of the Ca-activated K$^+$ channel in pancreatic β-cells, *Cell Calcium* **4**:451–461.

Field, M. J., and Giebisch, G. H., 1985, Hormonal control of potassium excretion, *Kidney Int.* **27**:379–387.

Field, M. J., Stanton, B. A., and Giebisch, G. H., 1984, Influence of ADH on renal potassium handling: A micropuncture and microperfusion study, *Kidney Int.* **25**:502–511.

Gardos, G., 1958, The function of calcium in the potassium permeability of human erythrocytes, *Biochem. Biophys. Acta* **30**:653–654.

Gorman, A. L. F., Herman, A., and Thomas, M. V., 1981, Intracellular calcium and the control of neuronal pacemaker activity, *Fed. Proc.* **40**:2233–2239.

Green, N., Algren, A., Hoyer, J., Triche, T., and Burg, M., 1985, Differentiated lines of cells from rabbit renal medullary thick ascending limb cells grown on amnion, *Am. J. Physiol.* **249**(18):C97–C104.

Greger, R., and Schlatter, E., 1983, Properties of the lumen membrane of the cortical thick ascending limb of Henle's loop of rabbit kidney, *Pfluegers Arch.* **396**:315–324.

Guggino, W. B., Stanton, B. A., and Giebisch, G., 1982, Electrical properties of isolated early distal tubule of the *Amphiuma* kidney, *Fed. Proc.* **41**:1597.

Guggino, S. E., Suarez-Isla, B. A., Guggino, W. B., Green, N., and Sacktor, B., 1984, Ba^{++} sensitive, Ca^{++} activated K$^+$ channels in cultured rabbit medullary thick ascending limb cells (MTAL) and cultured chick kidney cells (CK), *Kidney Intern.* **27**(1):309.

Guggino, S. E., Suarez-Isla, B. A., Guggino, W. B., Green, N., and Sacktor, B., 1985, The influence of barium on apical membrane potentials and potassium channel activity in cultured rabbit medullary thick ascending limb cells (MTAL). *Fed. Proc.* **44**(3):443.

Hamill, O. P., Marty, A., Neher, E., Sakmann, B., and Sigworth, F. I., 1981, Improved patch clamp technique for high resolution current recording from cells and cell free membrane patches, *Pfluegers Arch.* **391**:85–100.

Handler, J. S., and Orloff, J., 1981, Antidiuretic hormone, *Annu. Rev. Physiol.* **43**:611–624.

Hebert, S. C., and Andreoli, T. E., 1984a, Control of NaCl transport in the thick ascending limb, *Am. J. Physiol.* **246**:F745–F756.

Hebert, S. C., and Andreoli, T. E., 1984b, Effects of antidiuretic hormone on cellular conductive pathways in mouse medullary thick ascending limbs of Henle: 11. Determinants of ADH-mediated increases in transepithelial voltage and net Cl$^-$ absorption, *J. Membr. Biol.* **80**:221–233.

Hebert, S. C., Friedman, P. A., and Andreoli, T. E., 1984, Effects of antidiuretic hormone on cellular conductive pathways in mouse medullary thick ascending limbs of Henle: 1. ADH increases transcellular conductance pathways, *J. Membr. Biol.* **80**:201–219.

Henry, H. L., 1979, Regulation of the hydroxylation of 25-OH vitamin D$_3$ *in vivo* and in primary cultures of chick kidney cells, *J. Biol. Chem.* **254**(8):2722–2729.

Hunter, M., Lopes, A. G., Boulpaep, E. L., and Giebisch, G., 1984, Single channel recordings of calcium-activated potassium channels in the apical membrane of rabbit cortical collecting tubules, *Proc. Natl. Acad. Sci. U.S.A.* **81**:4237–4239.

Koeppen, B. M., Biagi, B. A., and Giebisch, G. H., 1983, Intracellular microelectrode characterization of the rabbit cortical collecting duct, *Am. J. Physiol.* **244**:F35–47.

Latorre, R., and Miller, C., 1983, Conduction and selectivity in potassium channels, *J. Membr. Biol.* **71**:11–30.

Latorre, R., Vergara, C., and Hidalgo, C., 1982, Reconstitution in planar lipid bilayers of a Ca^{2+}-dependent K$^+$ channel from transverse tubule membranes isolated from rabbit skeletal muscle, *Proc. Natl. Acad. Sci. U.S.A.* **79**:805–809.

Lindemann, B., and van Driessche, W., 1976, Sodium-specific membrane channels of frog skin are pores: Current fluctuations reveal high turnover, *Science* **195**:292–294.

Marty, A., 1981, Ca^{2+}-dependent K^+ channels with large unitary conductance in chromaffin cell membranes, *Nature* **291**:497–500.

Maruyama, Y., Gallacher, D. V., and Petersen, O. H., 1983, Voltage and Ca^{2+}-activated K^+ channels in baso-lateral acinar cell membranes of mammalian salivary glands, *Nature* **302**:827–829.

Meech, R. W., 1978, Calcium-dependent potassium activation in nervous tissues, *Annu. Rev. Biophys. Bioeng.* **7**:1–18.

Morel, F., 1981, Sites of hormone action in the mammalian nephron, *Am. J. Physiol.* **240**:F159–F164.

Nagel, W., 1979, Inhibition of potassium conductance by barium in frog skin epithelium, *Biochim. Biophys. Acta* **552**:346–357.

Oberleithner, H., Guggino, W. B., and Giebisch, G., 1982, Mechanism of distal tubular chloride transport in *Amphiuma* kidney, *Am. J. Physiol.* **242**:F331–F339.

O'Neil, R. G., 1983, Voltage-dependent interaction of barium and cesium with the potassium conductance of the cortical collecting duct apical cell membrane, *J. Membr. Biol.* **74**:165–173.

O'Neil, R. G., and Boulpaep, E. L., 1979, Affect of amiloride on the apical cell membrane cation channels of a sodium-absorbing, potassium-secreting renal epithelium, *J. Membr. Biol.* **50**:365–387.

O'Neil, R. G., and Helman, S. I., 1977, Transport characteristics of renal collecting tubules: Influence of DOCA and diet, *Am. J. Physiol.* **233**:F544–F558.

O'Neil, R. G., and Sansom, S. C., 1984, Characterization of apical cell membrane Na^+ and K^+ conductances of cortical collecting duct using microelectrode techniques, *Am. J. Physiol.* **247**:F14–F24.

Pallota, B. S., Magleby, K. L., and Barrett, J. N., 1981, Single channel recordings of Ca^{2+}-activated K^+ currents in rat muscle cell culture, *Nature* **293**:471–474.

Seaman, K. B., and Daly, J., 1981, Forskolin: A unique diterpene activator of cyclic-amp generating systems, *J. Cyclic Nucleotide Res.* **7**:201–224.

Stoner, L. C., Burg, M. B., and Orloff, J., 1974, Ion transport in cortical collecting tubule; effect of amiloride, *Am. J. Physiol.* **227**(2):453–459.

Wong, B. S., Lecar, H., and Adler, M., 1982, Single calcium-dependent potassium channels, *Biophys. J.* **39**:313–317.

Yellen, G., 1984, Ionic permeation and blockage in Ca^{2+}-activated K^+ channels of bovine chromaffin cells, *J. Gen. Physiol.* **84**:157–186.

Chapter 13

Channels in Photoreceptors

Juan Bacigalupo

1. INTRODUCTION

Photoreceptors are cells of the visual system specialized for responding to changes in light intensity. Their response consists of a change in membrane voltage.

The absorption of light is a role played by the photoreceptor integral membrane protein rhodopsin, which under the action of light undergoes an isomerization that enables this protein to trigger the transduction machinery of the cell. The final result is a change in membrane conductance [light-sensitive conductance, $G(L)$], and this has been the subject of study in a number of laboratories in the last few decades (for a review see Fain and Lisman, 1981). However, our knowledge of it is still confined to a few photoreceptor types, mainly because the small dimensions of most photoreceptors make it difficult to apply the electrophysiological methods necessary for the study of membrane conductances. I refer here only to studies done on the amphibian rods and the *Limulus* ventral photoreceptors, which have been most extensively studied.

Photoreceptors are remarkably sensitive to the sensory stimulus. It has been shown that they are able to respond to the absorption of single

---

JUAN BACIGALUPO • Departmento de Biología, Facultad de Ciencias, Universidad de Chile, Santiago, Chile.

photons of light by a change in membrane conductance in which a large number of channels (or carriers) participate (Fuortes and O'Bryan, 1972; Baylor et al., 1979a; Detwiler et al., 1982; Wong, 1978; Bacigalupo and Lisman, 1983). This fact indicates that transduction involves a considerable gain. This immediately suggests that the rhodopsin molecule activated by the absorption of one photon does not alter $G(L)$ directly but through the mediation of an internal messenger (Cone, 1973). The search for the "internal transmitter" has been one of the most active subjects of research in visual transduction in recent years and as yet remains unresolved. In vertebrate photoreceptors, several candidates are being considered: Ca^{2+} ions (Yoshikami and Hagins, 1971), protons Liebman et al., 1984), and cGMP (Bownds, 1980). These three undergo a change in concentration in the cytoplasm of the cell on illumination, but it is still a matter of debate whether their kinetics are compatible with the light response. It is also controversial whether the magnitude of the change in their concentration is sufficient to account for their putative role as the internal transmitter. In invertebrates, there are no good candidates at the present.

Another remarkable feature of the response of photoreceptors to light is that it is finely regulated. When the cell is stimulated by a very dim light, the gain is very high, which means that the cell has its maximal sensitivity to light. If the intensity of the stimulus is increased, the size of the response initially increases linearly, but beyond a certain intensity it undergoes an attenuation. This implies a decrease in the gain as the stimulus is made more intense, allowing the photoreceptor to have an enormous dynamic response range. This phenomenon is termed adaptation to the stimulus and is common in sensory cells. In *Limulus* photoreceptors, it has been clearly established that Ca^{2+} mediates light adaptation, although the mechanism by which this ion acts remains obscure (Lisman and Brown, 1971; Fain and Lisman, 1981).

Recent studies in vertebrate rods have provided interesting data about the molecular nature of the light-sensitive conductance (Detwiler et al., 1982). In *Limulus* photoreceptors, experimental evidence revealed that ionic channels are responsible for this conductance (Bacigalupo and Lisman, 1983). A detailed characterization of these conductance units could make possible the understanding of a number of important aspects of transduction. Most of this chapter is devoted to reviewing the recent advances in this topic.

There is a remarkable difference between the vertebrate and the invertebrate photoreceptors with regard to the effect of light on the light-dependent conductance. In vertebrate photoreceptors, light causes a decrease in conductance. On the other hand, in the invertebrate photoreceptors, the light-dependent conductance increases on illumination.

Therefore, it seems convenient to separate the discussion into two main topics, namely, vertebrate and invertebrate photoreceptors.

2. VERTEBRATE PHOTORECEPTORS

The electrophysiology of vertebrate photoreceptors, rods and cones, has been widely studied in recent years. As mentioned above, most of these studies have been done on the rods of lower vertebrates, such as toads and salamanders, which can be impaled with very fine microelectrodes. Because of their very small size, this technique cannot be applied to mammalian photoreceptors. Baylor *et al.* (1979) developed a technique for recording current from rods by using a suction pipette. Such a technique has been applied to amphibian rods and more recently to mammalian photoreceptors (Nunn *et al.*, 1984).

Rods are elongated cells subdivided into two morphologically distinguishable lobes, one is proximal to the synaptic terminal and termed the "inner segment," and the other is distal to the synaptic terminal and it is called the "outer segment." The inner segment contains the nucleus and the other organelles. The outer segment contains a tight pile of about a thousand closed membrane sacs commonly called "disks," which are separate from the plasma membrane and contain most of the rhodopsin molecules of the cell. Penn and Hagins (1969) first suggested that the light-dependent conductance is localized in the outer segment and is mainly permeant to Na^+ ions. This has been confirmed more recently by Baylor *et al.* (1979a,b). There is a steady Na^+ influx into the outer segment in the dark, and the effect of light is to transiently block the entrance of Na^+. The Na^+ ions do not accumulate in the cytoplasm because of the active extrusion of Na^+ by ATPases located in the inner segment.

The maximal dark current of a rod is about 30 pA, and its reversal potential is near 0 mV. The input resistance of this cell is about 1 GΩ. The small maximal light-sensitive current can result from (1) a small overall number of light-sensitive channels in the rod outer segment, or (2) a large number of channels of which only a small fraction are "functional" at a given time or a large number of channels (or carriers) with either a very small unitary conductance or a very short mean open time or both. A great deal of effort has been spent by several laboratories in trying to elucidate this problem and particularly in trying to decide which kind of molecular mechanism is responsible for the light-sensitive transport of ions across the plasma membrane of vertebrate photoreceptors.

Patch clamping on the outer segment of rods has failed to produce recordings of single-channel currents. Detwiler *et al.* (1982) studied the noise difference between dark and light on cell-attached membrane patches from rods of the amphibian *Gekko gekko*. They found that illumination produced a decrease in noise, which is consistent with the drop in membrane conductance that has been observed in similar conditions by means of macroscopic techniques (Fig. 1). The noise can be analyzed in the way Anderson and Stevens (1973) did for the acetylcholine-induced noise in single muscle cells. This noise analysis can give the value of the unitary conductance (channel or carrier), which in the case of the *Gekko* rods was 50 fS (Detwiler *et al.*, 1982). This value is an estimation based on the assumption that the only source of noise is the fluctuation of the light-sensitive channels (or carriers) between their different conductance states. The unitary conductance value obtained is two orders of magnitude or more below that measured for any other ionic channel studied so far. This suggests that this low-conductance channel may have either a large energy barrier for the passage of the conductive ions or a strong binding site for such ions within the conduction pathway. Alternatively, the channel may have a conductance value in the picosiemens range but a very short open time. In this case, the channel would present a smaller apparent conductance because of the limited time resolution of the recording technique.

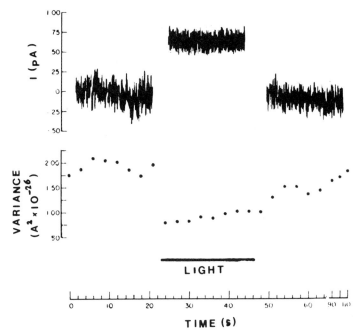

Figure 1. Light produces a reduction in the noise recorded from a *Gekko gekko* rod outer segment using a patch-clamp electrode. The lower trace shows the variance of the noise measured at regular intervals. (Modified from Detwiler *et al.*, 1982.)

Another possibility is that the ion conductive mechanism consists of a carrier rather than a channel. Useful information for making the distinction between a channel and a carrier would be provided by the power spectrum of the noise. Recently, Bodoia and Detwiler (1984) reported that the power spectrum from *Gekko* rods can be fitted by two Lorenzians. Assuming that such a power spectrum results from the fluctuations of the unitary components of the light-sensitive conductance, it could be taken as an indication that they are ionic channels rather than carriers, since the power spectrum for a carrier is not fitted by a Lorenzian (Hall, 1981).

3. INVERTEBRATE PHOTORECEPTORS

3.1. Morphology

I focus the discussion in this section on the photoreceptor of the ventral eye of *Limulus*, on which most of the relevant work has been done. *Limulus* ventral photoreceptors have the great advantage over most photoreceptor types of being large cells well over 100 µm long and about 80 µm in diameter, which makes them convenient for electrophysiological studies. Another advantage of photoreceptors in the ventral eye is that most of them are not electrically coupled. This characteristic is important for the study of the electrophysiology of the cell unencumbered by interactions with neighboring cells. Most other photoreceptors are electrically coupled. Furthermore, a method to free ventral photoreceptors of their surrounding glia and connective tissue has recently been developed (Stern *et al.*, 1982). By this method the photoreceptor plasma membrane is exposed, making it possible to do the kind of experiments I describe below.

As in the vertebrate photoreceptor, the ventral photoreceptor of *Limulus* also has the cell body subdivided into two kinds of lobes (Stern *et al.*, 1982). However, one major anatomic difference is that the plasma membrane of *Limulus* photoreceptor has microvilli on its surface. The microvilli, typical among invertebrate photoreceptors, contain the rhodopsin molecules. Stern *et al.* (1982) were able to show, using focal illumination, that only one of the lobes is light transducing, the one distal to the axon (termed the "R-lobe").

3.2. Electrophysiology: Macroscopic Studies

The electrophysiological properties of the ventral photoreceptor are relatively well understood at the macroscopic level. By means of the conventional voltage-clamp technique, various ionic conductances have been characterized (Millecchia and Mauro, 1969; Lisman and Brown, 1971; Pepose and Lisman, 1978; Lisman et al., 1982). The photoreceptor membrane has a conductance that is activated by light and several voltage-dependent conductances that are activated as a result of the depolarizing inward Na^+ current that passes through the light-sensitive conductance. The light-sensitive current that develops in response to a moderately bright light consists of an initial transient inward current that decays to a smaller and steady level when the stimulating light is maintained. The drop in the inward current is caused by the action of intracellular Ca^{2+}. During illumination, the free Ca^{2+} concentration increases and produces a decrease in the light-sensitive conductance (light adaptation).

From macroscopic studies much information has been obtained about the properties of the light-sensitive conductance (Fain and Lisman, 1981). Brown and Mote (1974) found that this conductance does not discriminate between Na^+ and Li^+ and is selective for Na^+ over K^+ by a factor of 2. We also know from macroscopic studies the characteristics of the activation of this conductance by light (for a review see Fain and Lisman, 1981). The effect of intracellular Ca^{2+}, which is to decrease the light-sensitive conductance, has been well established (Lisman and Brown, 1971; Brown and Lisman, 1975). Macroscopic studies have also revealed that the light-sensitive conductance is strongly voltage dependent, although it is not activated by voltage in the absence of light (Millecchia and Mauro, 1969; Bacigalupo et al., 1986).

3.3. Single-Channel Studies

In order to have a more profound understanding of these and other properties of the light-sensitive conductance, it would be desirable to study it at the single-channel level. Recently, John Lisman and I were able to apply the patch-clamp technique for single-channel current recording to the study of the light-sensitive conductance of the ventral photoreceptor (Bacigalupo and Lisman, 1983, 1984). Our first goal was to find out whether we could detect single-channel currents activated by light. The next step was to determine whether such channels displayed the properties of the light-sensitive conductance that had been characterized macroscopically. The first question can be answered if the patch-clamp current recordings reveal the discrete square jumps that result from the fluctuations of an individual channel between an open and a closed state. These current jumps cannot reflect carrier activity, since in principle a carrier molecule would be unable to transport enough ions per unit time to produce a current detectable by the patch-clamp method (Hall, 1981). I present evidence indicating that there are light-sensitive ionic channels in *Limulus*. As for the second question, the approach that we took was to compare the properties of the single channels with the macroscopic conductance.

In our patch-clamp experiments the photoreceptors were bathed in artificial sea water, and the patch pipette was filled with the same solution. In some experiments, the cell was impaled with a microelectrode in order to record the membrane potential. In order to apply the patch-clamp technique, one needs to remove the glia and connective tissue from the photoreceptor, which is done by teasing them out using a suction pipette after a pronase treatment (Stern et al., 1982). This procedure is not sufficient for obtaining gigaseals in the transducing lobe, presumably because of the densely packed microvilli of its surface. The diameter of a single microvillus is about 0.1 µm, a few times smaller than the opening of a small patch pipette (about 0.5 µm), and therefore, in no case could one patch clamp an individual microvillus. We used mild sonication as an additional treatment in an attempt to alter the cell surface so that the seals could form (Bacigalupo and Lisman, 1983). This treatment makes possible the formation of tight seals, although these still occur rather infrequently. This mild sonication does not impair the cell physiology in any obvious way. Sonication probably alters the morphology of small regions of the membrane, which become appropriate for patch electrodes to seal onto them.

Figure 2 shows the activation of single channels when the cell is stimulated with a steady light of moderate intensity. These channels remained silent in the dark. Careful examination of the current trace reveals

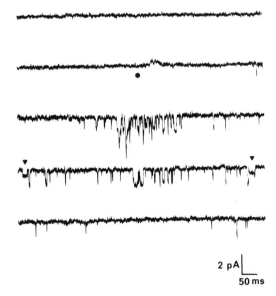

Figure 2. Light produces the activation of ionic channels in the ventral photoreceptor of *Limulus polyphemus*. This patch-clamp recording shows no channels activity on the dark. During the application of a maintained light (the onset of which is indicated by the dot), single-channel currents are observed. The arrowheads point to events of smaller amplitude also occurring during light. (Modified from Bacigalupo and Lisman, 1983.)

that there are two types of events in it, one about half the other in magnitude. The "large" and the "small" channels are both active during light. It is not clear yet whether the two types of single-channel events correspond to two different types of light-sensitive channels or to two different conductance states of the same channel. In the majority of our experiments we observed both types of events, although typically the "small" channel opened much more infrequently than the "large" channel. I refer mostly to the work done on the "large" channel, since the "small" channel has yet been characterized with the same detail.

Since light depolarizes the photoreceptor, we tested the possibility that the channels opened as a secondary result of the light-induced depolarization by depolarizing the patch in the dark. Such a possibility was ruled out because the channels were not activated under these conditions, although they were if a light stimulus was given while the patch was maintained depolarized (Bacigalupo and Lisman, 1983). Another possibility for the activation of the channels was that Ca^{2+}, whose intracellular free concentration has been shown to rise during illumination (Brown and Blinks, 1974; Brown et al., 1977; Levy, 1983), could produce the channel activation. We tested this possibility by injecting EGTA into the photoreceptor. If we were dealing with a Ca^{2+}-activated channel, EGTA should abolish its activation during light by preventing the rise in intracellular free Ca^{2+}. In experiments described in detail elsewhere (J. Bacigalupo et al., 1986), we found that the activation of the channels by light is actually enhanced after EGTA injection, indicating that the activation of the channel is not mediated by Ca^{2+}. Furthermore, this behavior is consistent with the effect of EGTA injection on the macroscopic light-sensitive conductance, which is to prevent the normal reduction in this conductance that follows the rise in Ca^{2+} (light adaptation). The evidence, therefore, supports the possibility that the channel is activated by light. This is a very important point, because it shows for the first time that a sensory stimulus can activate ionic channels in a sensory cell. Let us not forget, though, that the channel activation is not directly caused by light but rather is mediated by the internal transmitter.

Let us now compare the properties of the channels with those of the macroscopic light-sensitive conductance in order to determine whether these channels can account for the macroscopic conductance.

The magnitude of the light-sensitive current is graded with light intensity. We found that the probability of the light-sensitive channels being open is also graded with light intensity (Bacigalupo and Lisman, 1983). Furthermore, the probability of the channels being open is decreased when the photoreceptor is in the light-adapted state (Bacigalupo and Lisman, 1984), which is analogous to the reduction in the magnitude of the macroscopic current evoked by a test stimulus that occurs in a light-adapted cell.

The activity of the light-sensitive channels evoked by a maintained light of moderate intensity is preceded by a latency. The initial period of high channel activity rapidly decays to a lower sustained level. The macroscopic response develops in a qualitatively similar way in response to the same stimulus, as mentioned above (Fig. 2). However, we found in our experiments that the latency preceding the single-channel activation is consistently longer than that of the macroscopic response by a factor that varies considerably (Bacigalupo and Lisman, 1984; Fig. 2). The reason for this discrepancy remains obscure, although a plausible explanation for it is that the membrane regions where we were able to obtain seals are free of microvilli (presumably as a result of sonication) and therefore contain channels that are more distant from the microvilli than the average channel. If transduction is initiated at the microvilli (where rhodopsin is located), it would take longer for the activation of a more distant channel than for a channel closer to the microvilli. If most channels are in the microvillar membrane and only a small fraction of the channels are located in the microvilli-free areas of the R-lobe, the kinetics of the macroscopic response would show a normal latency.

The reversal potential of the light-sensitive current is about $+10$ mV (Millecchia and Mauro, 1969; Brown and Mote, 1974; Bacigalupo *et al.*, 1986). The reversal potential of the single-channel currents was found to be the same (Bacigalupo and Lisman, 1983, 1984). The current–voltage curve for the single channels is linear, indicating that the single-channel conductance (40 pS) is independent of membrane potential within the voltage range covered in our experiments (-70 to 60 mV). However, this is an apparent discrepancy with the marked outward rectification of the macroscopic conductance found by Millecchia and Mauro (1969; see also Fain and Lisman, 1981). This apparent discrepancy was resolved when we examined the behavior of the light-activated channels at different potentials. We found that the probability of the channel being open is strongly voltage dependent (Fig. 3), and the consequent rectification occurs at the same potential range as that of the macroscopic current (Bacigalupo *et al.*, 1986). This can account for the peculiar strong voltage dependence of the macroscopic conductance. Altogether, the evidence presented so far is consistent with the possibility that the channel that we have studied is the channel responsible for the macroscopic conductance.

As I mentioned above, it would be highly desirable to know the identity of the internal transmitter, since this would allow for a direct test of whether the channels under study are those responsible for the light-

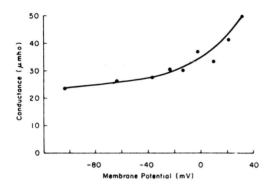

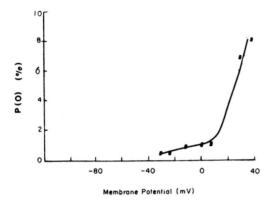

Figure 3. Voltage dependence of the light-sensitive channel. The upper plot shows how the macroscopic light-sensitive conductance varies with membrane potential. (Modified from Fain and Lisman, 1981.) Single light-sensitive channels present a similar behavior, as illustrated in the lower plot, which depicts the probability of the channel being in the open state times the number of channels, $N \times P(o)$, versus the membrane potential.

activated conductance. The experiment to test for this may involve obtaining excised membrane patches from the photoreceptor and adding the internal transmitter to the solution bathing the cytoplasmic side of the membrane patch. However, even without knowing the identity of the transmitter, one could learn something from the behavior of the channels in isolated patches, for example, the way in which the transmitter may act on the channels in excised-patch experiments. I would like to end this chapter by describing some experiments that John Lisman and I have done along these lines.

The experiments consisted of excising membrane patches containing light-sensitive channels in the inside-out configuration and exposing the cytoplasmic side of the membrane patch to a solution that mimicked the ionic composition of the cytoplasm. Therefore, the solution was high in K^+ and low in Na^+, and Ca^{2+} was maintained low (10^{-8} M) by the addition of 10 mM EGTA (see legend of Fig. 4 for details). Under these conditions, the excised patch exhibited a rather high channel activity (Fig. 4, upper trace) that was unaffected by light. The reversal potential of this

Channels in Photoreceptors

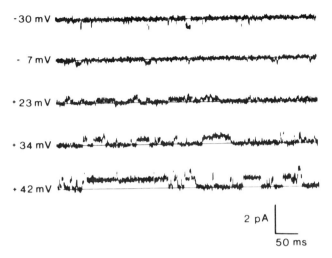

Figure 4. Membrane patches containing light-sensitive channels presenting a high, light-independent channel activity. The patches were exposed to a solution containing 300 mM K$^+$-aspartate, 20 mM NaCl, 2 mM MgCl$_2$, 10 mM EGTA, and 10^{-7} M Ca^{2+}, buffered at pH 7.0. The figure shows traces of current obtained at various patch potentials.

channel current (near +10 mV) and its average lifetime (in the millisecond range) are the same as those of the light-sensitive channel. The conductance of the channel in the excised membrane was 15 pS, which is the same as the low-conductance channel observed in cell-attached patches as mentioned above (J. Bacigalupo and J. E. Lisman, unpublished data).

It is tempting to suggest that the channel in the excised patches is the same as the "small" channel that we found in the cell-attached patches (Fig. 2, arrows). Since the light-sensitive channels in the cell-attached membrane patches were closed in the dark and opened during light, and the removal of the patch membrane from the cell caused the channel activity to rise, the action of the transmitter on the light-sensitive channels could be explained if this molecule were a blocker of the channels ("negative transmitter"). In such a case, the transmitter may be at a high concentration in the dark, and light would cause the concentration of the transmitter to drop, therefore increasing the probability of the channel to open. Another plausible explanation is that the removal of the patch membrane from the cell would cause the loss of some component of the channel complex that normally would prevent the channel from opening. In this case, the transmitter molecule, whose concentration would increase during illumination, would interact with the channel complex ("positive transmitter"), causing the channel to open.

Another problem that one may address by this technique is to examine the mechanism by which Ca^{2+} reduces the light-sensitive conductance causing light adaptation. Particularly, one could examine whether Ca^{2+} acts directly on the channel or whether it acts on some earlier step of the transduction process. For this purpose, one could manipulate the Ca^{2+} concentration in the bath solution and investigate if the channel activity of the excised patch is affected.

Within the next few years, these and other experiments of this sort may allow us to learn more about the last steps of the transduction process.

ACKNOWLEDGMENTS. I thank Drs. Enrico Nasi and John E. Lisman for thoughtful comments on the manuscript.

REFERENCES

Anderson, C. R., and Stevens, C. F., 1973, Voltage clamp analysis of acetylcholine produced end plate current fluctuations at frog neuromuscular junction, *J. Physiol. (Lond.)* **235**:655–691.

Bacigalupo, J., and Lisman, J. E., 1983, Single-channel currents activated by light in *Limulus* ventral photoreceptors, *Nature* **304**:268–270.

Bacigalupo, J., and Lisman, J. E., 1984, Light-activated channels in *Limulus* ventral photoreceptors, *Biophys. J.* **45**:3–5.

Bacigalupo, J., Chinn, K., and Lisman, J. E., 1986, Ion channels activated by light in *Limulus* ventral photoreceptors, *J. Gen. Physiol.* **87**:73–89.

Baylor, D. A., Lamb, T. D., and Yau, K. W., 1979a, Responses of retinal rods to single photons, *J. Physiol. (Lond.)* **288**:613–634.

Baylor, D. A., Lamb, T. D., and Yau, K. W., 1979b, The membrane current of single rod outer segments, *J. Physiol. (Lond.)* **288**:589–611.

Bodoia, R. D., and Detwiler, P. B., 1984, Patch clamp study of the light response of isolated frog retinal rods, *Biophys. J.* **45**:337a.

Bownds, M. D., 1980, Biochemical steps in visual transduction: Role of nucleotides and calcium ions, *Photochem. Photobiol.* **32**:487–490.

Brown, J. E., and Blinks, J. R., 1974, Changes in intracellular free calcium concentration during illumination of invertebrate photoreceptors. Detection with aequorin, *J. Gen. Physiol.* **64**:643–665.

Brown, J. E., and Lisman, J. E., 1975, Intracellular Ca modulates sensitivity and time scale in *Limulus* ventral photoreceptors, *Nature* **258**:252–254.

Brown, J. E., and Mote, M. I., 1974, Ionic dependence of reversal voltage of the light response in *Limulus* ventral photoreceptors, *J. Gen. Physiol.* **63**:337–350.

Brown, J. E., Brown, P. K., and Pinto, L. H., 1977, Detection of light-induced changes of intracellular ionized calcium concentration *Limulus* ventral photoreceptors using arsenazo III, *J. Physiol. (Lond.)* **267**:299–320.

Cone, R. A., 1973, The internal transmitter model for visual excitation: Some quantitative implications, in *Biochemistry and Physiology of Visual Pigments* (H. Langer, ed.), Springer-Verlag, Berlin, Heidelberg, New York, pp. 275–282.

Detwiler, P. B., Conner, J. D., and Bodoia, R. D., 1982, Gigaseal patch clamp recordings from outer segments of intact retinal rods, *Nature* **300**:59–61.

Fain, G. L., and Lisman, J. E., 1981, Membrane conductances of photoreceptors, *Prog. Biophys. Mol. Biol.* **37**:91–147.

Fuortes, M. G. F., and O'Bryan, P. M., 1972, Generator potentials in invertebrate photoreceptors, in *Physiology of Photoreceptor Organs. Handbook of Sensory Physiology*, Vol. II/2 (M. G. F. Fuortes, ed.), Springer-Verlag, New York, pp. 279–319.

Hall, J. E., 1981, Channels and carriers in lipid bilayers, in *Membrane Transport* (R. Bouting, and J. de Pout, eds.), North-Holland, Amsterdam, pp. 107–121.

Levy, S., 1983, Relationship between sensitivity and intracellular free Ca concentration in *Limulus* ventral photoreceptors. A quantitative study using Ca selective microelectrodes, Ph.D. Thesis, Boston University, Boston.

Liebman, P. A., Mueller, P., and Pugh, E. N., Jr., 1984, Protons supress the dark current of frog retinal rods, *J. Physiol. (Lond.)* **347**:85–110.

Lisman, J. E., and Brown, J. E., 1971, The effects of intracellular Ca^{2+} on the light response and on light adaptation in *Limulus* ventral photoreceptors, in *The Visual System: Neurophysiology, Biophysics and Their Clinical Applications* (G. B. Arden, ed.), Plenum Press, New York, pp. 23–33.

Lisman, J. E., Fain, G. L., and O'Day, P. M., 1982, Voltage dependent conductances in *Limulus* ventral photoreceptors, *J. Gen. Physiol.* **79**:187–209.

Millecchia, R., and Mauro, A., 1969, The ventral photoreceptor cells of *Limulus*. III. A voltage clamp study, *J. Gen. Physiol.* **54**:331–351.

Nunn, B. J., Schnapf, J. L., and Baylor, D. A., 1984, Spectral sensitivity of single cones in the retina of *Macaca fascicularis*, *Nature* **309**:264–266.

Penn, R. D., and Hagins, W. A., 1969, Signal transmission along retinal rods and the origin of the electroretinographic a-waves, *Nature* **223**:201–205.

Pepose, J. S., and Lisman, J. E., 1978, Voltage sensitive potassium channels in *Limulus* ventral photoreceptors, *J. Gen. Physiol.* **71**:101–120.

Stern, J., Chinn, K., Bacigalupo, J., and Lisman, J. E., 1982, Distinct lobes of *Limulus* ventral photoreceptors. I. Functional and anatomical properties of lobes revealed by removal of glial cells, *J. Gen. Physiol.* **80**:825–837.

Wong, F., 1978, Nature of light-induced conductance changes in ventral photoreceptors of *Limulus*, *Nature* **276**:76–79.

Yoshikami, S., and Hagins, W. A., 1971, Light, calcium and the photocurrent of rods and cones, in *Biochemistry and Physiology of Visual Pigments* (H. Langer, ed.), Springer-Verlag, Berlin, Heidelberg, New York, p. 245.

Chapter 14

Inactivation of Calcium Currents in Muscle Fibers from *Balanus*

M. Luxoro and V. Nassar-Gentina

1. INTRODUCTION

Our interest in calcium currents arose from our intention to approach an understanding of excitation–contraction coupling in muscle fibers from *Balanus* (Bacigalupo *et al.*, 1979; Hidalgo *et al.*, 1979; Luxoro and Nassar-Gentina, 1984). For a full account of recent developments on the subject, the student is referred to recent reviews (Reuter, 1979; Hagiwara and Byerly, 1981; Tsien, 1983; Kostyuk, 1984).

Several years ago these currents—as compared with the well-studied sodium currents—appeared to be curiosities of nature, but now they seem to be the rule rather than the exception. Indeed, they look omnipresent. In vertebrates, they have been described in heart (plateau of the action potential), in smooth muscle, in slow skeletal muscle, and even in twitch skeletal muscle fiber, whose action potential is almost a pure sodium spike. A strong Ca^{2+} component of the action potential of developing chick and rat muscle has also been detected, although it disappears in adults. Calcium spikes secm to be the rule in all types of invertebrate muscle. In protochordate, as in chick embryo, embryonic muscle cells depict a

---

M. LUXORO AND V. NASSAR-GENTINA • Laboratorio de Fisiología Celular, Facultades de Medicina y Ciencias, Universidad de Chile, Santiago, Chile.

Na^+–Ca^{2+} compound action potential, but tunicates preserve only the calcium component when they become adults, and amphioxus conserve both.

It appears that all nerve fibers (not only the nerve terminal where calcium currents are conspicuous and essential for neurotransmission) and nerve cell bodies, from both vertebrates and invertebrates, have calcium currents that contribute in different proportions to the generation of the spike. Moreover, receptors, different egg cells, epithelial cells, and paramecium have paid their tribute to the scientific literature related to calcium channels (see Hagiwara and Byerly, 1981).

An important difference appears to exist between Na^+ and Ca^{2+} channels: whereas the former present similar characteristics irrespective of origin, the latter can be classified into various subgroups. We discuss here what we have found regarding inactivation of calcium currents in isolated and internally perfused muscle fibers from *Megabalanus psittacus* (Darwin). Occasional references are made to significant similarities or differences in behavior of calcium channels from other preparations.

A more general comparison of the physiological properties of these fibers with those of twitch skeletal muscle of the frog is found elsewhere (Luxoro, 1982).

2. METHODOLOGICAL CONSIDERATIONS

We used voltage-clamped isolated muscle fibers that were internally and externally perfused with different solutions. The details of the dissection, the voltage clamping, and general setup have been described extensively (Bezanilla *et al.*, 1970; Keynes *et al.*, 1983; Hidalgo *et al.*, 1979).

The composition of the internal solution was (in mM) cesium acetate 200, sucrose 400, tetraethylammonium (TEA) 25, tris-acetate 5. Sometimes we used tris-HEPES instead of tris-acetate. The pH of the solution was adjusted to 7.3–7.4. This mixture was composed to prevent outward currents carried by potassium ions, which usually, and to different degrees, mask the inward currents transported by calcium, which are the object of these studies. The external solution (ASW) was (in mM) sodium chloride 440, calcium chloride 10, magnesium chloride 50, tris chloride 5. The pH was adjusted to 7.4–7.5. Occasionally calcium was replaced by barium or magnesium. If calcium concentration was increased above 10 mM, magnesium was reduced by an equivalent concentration.

The holding potential was kept between -35 and -45 mV, always 2 mV more negative than the resting value.

3. CHARACTERISTICS OF INWARD CURRENTS

If the membrane potential is pulsed from the holding potential to approximately 0 to 10 mV, one obtains maximal inward current (see Fig. 1). Under these conditions, this current is mostly, if not all, carried by Ca^{2+} ions, as evidenced by its disappearance in ASW in which Ca^{2+} was replaced by Mg^{2+}. A clear inactivation is seen as a decay in the current in spite of the fact that the membrane potential is held constant.

Both the total inactivation and its rate are dependent on the magnitude of the inward current. Thus, fibers that presented peak inward currents less than 1.3 mA/cm^2 were discarded.

3.1. Calcium Current Inactivation Is Current Dependent

The double-pulse technique introduced by Hodgkin and Huxley was used in order to study the dependence of inactivation on both potential and time. The prepulse was either depolarizing or hyperpolarizing and of various durations. The second (test) pulse was a depolarizing one of the order of 40 mV, i.e., one that would produce large inward currents.

Inactivation or suppression of inactivation was never induced with hyperpolarizing prepulses. On the other hand, if the prepulse was such that it produced significant inward current (30 to 60 mV depolarizing), it always induced inactivation (as detected by the current induced by the test pulse). If the prepulse was large enough to reach the reversal potential (90 to 110 mV), very little inactivation was induced (see Fig. 2). Thus, it seems that the inactivation is associated with the occurrence of large inward current rather than with a particular level of membrane potential. A dramatic illustration of this assertion was obtained in two muscle fibers in experiments in which the depolarizing prepulse had exactly the value

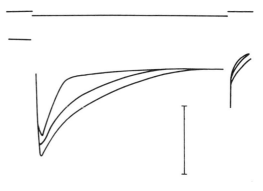

Figure 1. Upper trace is a 40-mV, 150-msec depolarizing pulse applied to the muscle fiber. The lower traces are the inward currents associated with the potential pulse. The smaller inward current was obtained under the standard conditions described in Section 2. Then, 10 mM EGTA was added to the internal perfusion solution. The following records were taken 6 and 12 min after the inclusion of EGTA. One sees that the inward currents increase and the rate of inactivation decreases. The vertical bar indicates 1 mA/cm^2. Time calibration is given by the 150-msec voltage pulse.

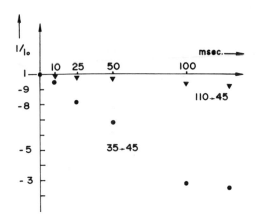

Figure 2. Inactivation (I/I_0) against duration of inactivating prepulse. The "test" pulse was 45 mV depolarizing (+5 mV absolute internal potential), and the prepulse was either 35 (black dots) or 110 mV depolarizing. The last prepulse (black triangles) produced little inward currents and, accordingly, a very small inactivation.

of the threshold for the production of inward current, that is, in one case 17 mV and in the other 23 mV. In such cases, the muscle fiber did not respond in a certain proportion of the applied pulses, but, when it did, it reacted with an inward current abruptly large, about two-thirds of the maximal. In spite of the fact that in all cases the prepulse had the same value, inactivation followed only when an inward current was associated with the threshold prepulse.

The current and not voltage dependence of calcium current inactivation is a property that is shared by almost all calcium currents described in the literature. Exceptions to this rule include component I of the calcium currents present in the egg cell membrane of a marine polychaeta (Fox, 1981) and the calcium currents of twitch skeletal muscle fibers from the frog (Cota et al., 1984).

3.2. Time Dependence

The time course of the development of inactivation depends strongly on the size of the inward currents. A detailed analysis of this dependence has not been made by us. Suffice it to say that it is at least one order of magnitude faster than previously thought (Figs. 1 and 2).

3.3. Recovery

As in the previous sections, the two-pulse technique was used, but with a variable interval (10 to 500 msec) inserted between them. The conditioning pulse was meant to produce a large inactivation (40 mV, 200 to 300 msec), and the second was, as usual, a test pulse (40 mV, 40 msec).

Within a 500-msec recovery interval, inward currents reach a plateau equivalent to 80–85% of their original value. The time constant of this process is of the order of 200 to 250 msec and is thus slower than the onset of inactivation. The plateau, nevertheless, is apparent, as evidenced by the response to the conditioning prepulse in the following pulse sequence, which is given every 90 sec. Indeed, after this period of time, the inactivation is fully absent, suggesting that there is more than one process involved in the mechanism of recovery.

4. MECHANISM OF INACTIVATION

The decay of the inward currents provoked by the currents themselves may be related to several factors:

1. Depletion of calcium in some restricted external compartment and/or accumulation of calcium at the inner side of the membrane, thus diminishing the Ca^{2+} chemical gradient across the channel.
2. Entering Ca^{2+} ions accumulating near the inner surface of the membrane may produce a screening of surface charges and modify the electrical potential gradient that ions crossing the channels effectively see.
3. Entering Ca^{2+} ions may activate potassium channels sensitive to Ca^{2+} ions, adding an outward current after the onset of the inward calcium current.
4. The existence of a deep well in the energy profile for ion permeation of calcium channels, which might bind the divalent cation with a net transient effect of blocking the channel.

A large effect of mechanism 1 does not seem to be compatible with our data. Indeed, the reversal potential for the inward currents obtained after inducing inactivation and in the presence of the prepulse is 8 to 10 mV smaller than the one obtained in the absence of the prepulse (66 to 72 mV). This shift implies a certain degree of Ca^{2+} depletion but at most accounts for 15% of the total current inactivation observed. This calculation assumes linearity between the current and $V_m - V_{reversal}$, which is a good assumption when the region of the I–V curve near the reversal potential is considered. A further support to our contention that Ca^{2+} depletion from external compartments plays only a small role in the process of inactivation is found in experiments using high external Ca^{2+} concentrations. When the muscle fiber is bathed in ASW with 60 mM Ca^{2+}, we found that inactivation is present but consistently slightly smaller than that obtained in normal-Ca^{2+} ASW. Peak calcium currents saturate when the Ca^{2+} concentration is raised beyond 25 mM. Therefore, it is expected that at 60 mM Ca^{2+} (well beyond the current saturation level) any calcium depletion effects would be absent.

Finally, one must consider that if one uses Ba^{2+}-ASW instead of Ca^{2+}-ASW in the external medium, the inward currents are larger (hence depletion or internal accumulation should also be larger), but the rate of inactivation is smaller.

Relaxations of calcium currents as a result of calcium depletion have been reported in component II of the calcium currents of the egg cell membrane of the polychaeta *Neanthes* (Fox and Krasne, 1984) and in muscle fibers of *Rana temporaria* (Almers et al., 1981). However, Cota et al. (1984) failed to detect this depletion in their experiments on twitch muscle fibers from the frog. Instead, they demonstrated a voltage dependence of calcium current inactivation. As they point out, one must expect a certain degree of Ca^{2+} tubular depletion from the rather long-lasting pulses used. Accordingly, two possibilities were considered to explain why tubular Ca^{2+} is not reduced during I_{Ca}: (1) a large tubular volume (as a consequence of working in hypertonic saline); and (2) the presence of a very active Ca^{2+} pump in the tubular membrane, which would transport calcium into the tubular lumen. It is likely that under the conditions of the experiments of Almers et al. (high intracellular EGTA), the pump could not replenish the lumen. Thus, one sees that by a slight alteration of the working conditions, the weight played by one mechanism may be completely changed.

If effect 2 were the most important, the inactivation of barium inward current should be dramatically larger: this should be so because the currents are greater and a larger accumulation should follow. Furthermore, one must consider that there is no internal mechanism as effective as the one represented by the removal of calcium by the sarcoplasmic reticulum that could remove barium. However, we found a smaller inactivation when Ba^{2+} was added to the external medium. Similar results have been found in paramecium (Eckert and Brehm, 1979) and in neuron bodies (Tillotson, 1979; Lux and Brown, 1984). Notwithstanding that, if 10 mM EGTA is added to the internal solution, the time course of calcium current inactivation is strongly reduced (Fig. 1). This figure shows that a rapidly inactivating muscle fiber (time constant of about 14 msec) is slowed down after 6 and then 12 min of intracellular perfusion with EGTA to time constants of 32 to 64 msec, respectively. The first effect of EGTA is an increase of the peak inward current, suggesting that the concentration of Ca^{2+} present inside the muscle fiber maintained a steady small level of inactivation.

Intracellular accumulation of Ca^{2+} is thought to be the main mechanism of inactivation in paramecium (Eckert and Brehm, 1979) and in neuron bodies (Tillotson, 1979), although the evidence claimed for that mechanism may be compatible with the mechanism of obstruction of the channel (4).

Effect 3 probably has to be discarded because most K^+ currents (not all of them) activated by Ca^{2+} are blocked by TEA. Also, Cs^+ was present instead of K^+ as intracellular cation. Finally, Ba^{2+} currents inactivated, though a little more slowly than Ca^{2+} currents. It is known that Ba^{2+} not only does not activate K^+ channels sensitive to Ca^{2+} but rather inhibits them.

Thus, by exclusion, we feel that the main mechanism causing calcium current inactivation in muscle fibers from *Balanus* is obstruction of the pore. We are planning to run experiments that might directly gather evidence in favor of this hypothesis. It is interesting to note that in a neuron body whose calcium currents behave quite similarly to those of *Balanus*, evidence has been presented that in our opinion is consistent with mechanism 4. Indeed, Brown *et al.* (1984) showed that at the molecular level, inactivation of calcium channels of neurons of *Helix pomatia* is manifested as a decrease in the probability of single channel opening, which develops with time. A theoretical model for a mechanism similar to 4 has been proposed by Standen and Stanfield (1982). This model describes rather well the inactivation of Ca^{2+} currents in insect skeletal muscle fibers.

Thus, our feeling is that in muscle fibers from *Balanus* there is more than one mechanism underlying the inactivation of calcium inward currents. The most important one seems to be an obstruction of channels as the result of a transient binding of the divalent charge carrier; this would be followed by depletion of Ca^{2+} from the transverse tubule system, though this last mechanism weighs very much less.

ACKNOWLEDGMENTS. This work was supported by the Departamento de Investigacion y Bibliotecas de la Universidad de Chile and by Proyecto N°1012/84 of FONDECYT, Chile.

REFERENCES

Almers, W., Fink, R., and Palade, P. T., 1981, Calcium depletion in frog muscle tubules: The decline of calcium current under maintained depolarization, *J. Physiol. (Lond.)* **312**:177–207.

Bacigalupo, J., Luxoro, M., Rissetti, S., and Vergara, C., 1979, Extracellular space and diffusion barriers in muscle fibres from *Megabalanus psittacus* (Darwin), *J. Physiol. (Lond.)* **288**:301–312.

Bezanilla, F., Rojas, E., and Taylor, R. E., 1970, Sodium and potassium conductance changes during a membrane action potential, *J. Physiol. (Lond.)* **211**:729–751.

Brown, A. M., Lux, H. D., and Wilson, D. L., 1984, Activation and inactivation of single channels in snail neurons, *J. Gen. Physiol.* **83**:751–769.

Cota, G., Nicola Siri, L., and Stefani, E., 1984, Calcium channel inactivation in frog (*Rana pipiens* and *Rana moctezuma*) skeletal muscle fibres, *J. Physiol. (Lond.)* **354**:99–108.

Eckert, R., and Brehm, P., 1979, Ionic mechanism of excitation in paramecium, *Annu. Rev. Biophys. Bioeng.* **8**:853–883.

Fox, A. P., 1981, Voltage-dependent inactivation of a calcium channel, *Proc. Natl. Acad. Sci. U.S.A.* **78**:953–956.

Fox, A. P., and Krasne, S., 1984, Two calcium currents in *Neanthes arenaceodentatus* egg cell membranes, *J. Physiol. (Lond.)* **356**:491–505.

Hagiwara, S., and Byerly, L., 1981, Calcium channel, *Annu. Rev. Neurosci.* **4**:69–125.

Hidalgo, J., Luxoro, M., and Rojas, E., 1979, On the role of extracellular calcium in triggering contraction in muscle fibres from barnacle under membrane potential control, *J. Physiol. (Lond.)* **288**:313–330.

Keynes, R. D., Rojas, E., Taylor, R. E., and Vergara, J., 1973, Calcium and potassium systems of a giant barnacle muscle fibre under membrane potential control, *J. Physiol. (Lond.)* **229**:409–455.

Kostyuk, P. G., 1984, Intracellular perfusion of nerve cells and its effects on membrane currents, *Physiol. Rev. (Lond.)* **64**:435–454.

Lux, H. D., and Brown, A. M., 1984, Patch and whole cell calcium currents recorded simultaneously in snail neurons, *J. Gen. Physiol.* **83**:725–750.

Luxoro, M., 1982, Propiedades fisiologicas de la fibra muscular estriada de *Megabalanus psittacus* (Darwin), *Arch. Biol. Med. Exp.* **15**:321–330.

Luxoro, M., and Nassar-Gentina, V., 1984, Potassium-induced depolarizations and generation of tension in barnacle muscle fibres: Effects of external calcium, strontium and barium, *Q. J. Exp. Physiol.* **69**:235–243.

Reuter, H., 1979, Properties of two inward membrane currents in the heart, *Annu. Rev. Physiol.* **41**:413–424.

Standen, N. B., and Stanfield, P. R., 1982, A binding-site model for calcium channel inactivation that depends on calcium entry, *Proc. R. Soc. Lond. [Biol.]* **217**:101–110.

Tillotson, D., 1979, Inactivation of Ca^{2+} conductance dependent on entry of Ca^{2+} ions in molluscan neurons, *Proc. Natl. Acad. Sci. U.S.A.* **76**:1497–1500.

Tsien, R. W., 1983, Calcium channel in excitable cell membrane, *Annu. Rev. Physiol.* **45**:341–358.

Chapter 15

Electrophysiological Studies in Endocrine Cells

Clay M. Armstrong

1. INTRODUCTION

It was not so many years ago that electrical activity was thought to be the exclusive property of relatively specialized cells found in nerve and muscle. During this era, electrophysiology and neurophysiology were virtually synonymous. In the last 10 or 15 years, however, the vistas of electrophysiologists have broadened greatly. All cells in the body have resting potentials, which must have developed early in evolution to solve the osmotic problem that arises when substances are sequestered in the cytoplasm (Tosteson, 1964). Membrane potentials also had an early use in energy production. According to the Mitchell hypothesis (Mitchell, 1967), the immediate source of energy for the synthesis of ATP is a transmembrane electrochemical gradient of protons. Volume control and energy production were problems confronted by cells very early on, and membrane potentials must thus have been around for a long time, and nature has devised many ways to use them.

In this chapter I want to talk about electrical activity as used in the functioning of endocrine cells. An early discovery was the finding that

CLAY M. ARMSTRONG • Department of Physiology, School of Medicine, University of Pennsylvania, Philadelphia, Pennsylvania 19104.

pancreatic β cells have membrane voltage changes that are closely related to their transduction function: the glucose concentration in the medium surrounding the cells affects their electrical activity, which in turn affects the secretion rate of insulin (Dean and Matthews, 1968; Meissner and Schmeltz, 1974). Electrical activity and action potentials were subsequently described in GH3 cells (a clonal cell line from a rat pituitary adenoma) (Kidokoro, 1975) and in chromaffin cells from the adrenal medulla. Parathyroid cells are reported to have a very unusual form of electrical activity (Bruce and Anderson, 1979, Lopez-Barneo and Armstrong, 1983). These cells respond to low serum calcium concentration by secreting parathyroid hormone, which acts on bone and other tissue to raise the serum calcium. The cells depolarize in response to raised external calcium concentration, and this depolarization seems closely correlated to the secretion of parathyroid hormone. This implies the existence in parathyroid cells of an unusual secretion mechanism, which does not follow the common rule that depolarization induces a calcium permeability increase, which causes a rise in intracellular calcium, which in turn triggers exocytosis.

The early work in this exciting area, which might be called electroendocrinology, was performed with great difficulty using high-resistance microelectrodes to penetrate the various endocrine cells and record their membrane voltages. Such recordings are quite difficult and require skill and patience, since the penetration almost invariably entails some degree of damage to the penetrated cell, which in general is quite small and sensitive to damage. The development of the patch-clamp technique (Sakmann and Neher, 1983) has made investigations of endocrine cells much easier in many respects and has provided a tool with very wide applicability to the study of small cells. I address exclusively the whole-cell variant of patch-clamp technique and discuss investigations performed by Dr. Rick Matteson and myself employing GH3 cells.

GH3 cells secrete prolactin and growth hormone, and secretion is enhanced by thyrotropin-releasing hormone (TRH), which chemically is the tripeptide L-pyroglutamyl-L-histidyl-L-prolineamide. Electrical studies have shown that GH3 cells fire action potentials spontaneously, and the rate of spontaneous firing is increased by TRH (Tareskevich and Douglas, 1980). Stated another way, TRH increases the pacemaking rate of GH3 cells. This is the simple behavioral repertoire that we would like to account for in detail. We want to know the permeabilities involved in the action potential, the pacemaking mechanism, the mode of action of TRH on pacemaking, the relationship of pacemaking to secretion, and the mechanism of secretion.

To one well versed in events at the neuromuscular junction, a hypothesis linking membrane electrical activity to secretion springs readily to mind: depolarization activates calcium channels, resulting in increased intracellular calcium, which stimulates exocytosis (Katz, 1969). The mechanism may well be more complicated, however. A number of possible paths have been implicated in the activation of secretion by TRH, including release of calcium from intracellular stores, adenylate cyclase activation, calcium-dependent phosphorylation, and increased turnover of phosphatidylinositol and its sequelae. The ability of TRH to enhance prolactin release in the absence of extracellular calcium seems well established (Gershengorn et al., 1983). Thus, if a neuromuscular-junction-type release mechanism is present in GH3 cells, it may be supplemented by other mechanisms.

Our approach to explaining GH3 cell behavior has so far focused on the electrical aspects and has had as its initial goal a characterization of the ionic channels present in the membrane. In the course of these studies we have discovered that GH3 cells have sodium channels, two types of calcium channels, and probably more than one type of potassium channel. Some of our work has been presented previously (Matteson and Armstrong, 1983, 1984; Armstrong and Matteson, 1985).

2. WHOLE-CELL PATCH-CLAMP METHODOLOGY

We have used the whole-cell variant of patch-clamp technique. In this method, a tight seal is formed between the membrane of the GH3 cell and the patch pipette, and the patch of membrane covering the pipette tip is broken by suction, bringing the lumen of the pipette in direct electrical contact with the interior of the cell. A Ag–AgCl wire in the pipette is then connected to the summing junction of a low-noise operational amplifier. The amplifier serves the dual functions of voltage-clamping the cell interior and measuring the membrane current. As shown in Fig. 1, the amplifier is wired in the configuration of a current-to-voltage transducer, and the amplifier output is proportional to the total membrane current plus the "command" voltage at the positive input of the amplifier. The command voltage is subtracted off in a subsequent stage to yield the membrane current signal alone. The voltage-clamp function results from the fact that in the steady state, the voltage difference between the positive and the negatve inputs of an operational amplifier is negligible. The negative input in the steady state thus follows the command voltage imposed on the positive input by the pulse generator. Since the negative input is directly connected to the cell interior, the membrane voltage is close to the voltage of the negative input if the resistance of the pipette is low.

To some degree the two functions assigned to the operational amplifier are incompatible. Good low-noise performance as a current-to-

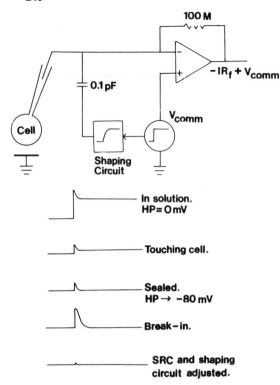

Figure 1. The core of a patch clamp is an operational amplifier in current-to-voltage transducer configuration but with positive input driven by a command voltage. Because of the high gain of the operational amplifier, the negative input follows the voltage applied to the positive input. This voltage is transmitted to the cell, thus clamping its voltage. During a voltage step, the current for charging the cell capacitance to the new voltage is provided by feeding the output of a pulse-shaping circuit through a small capacitance to the negative input. When the shaping circuit is ideally adjusted, all of the charging current is supplied through the capacitor, and the amplifier output supplies none. The procedure during seal formation is illustrated by the five current traces, which show the response to a voltage step at various stages during the process. When the electrode seals to the cell, the step component of the current becomes very small, leaving a small transient that is the charging current for the stray capacitance of the electrode walls. The transient increases when the electrode breaks into the cell, because the cell capacitance is added to the stray capacitance. After adjustment of the series resistance compensation and the shaping circuit, the transient is ideally zero. As discussed in the text, optimum frequency response requires careful adjustment of the shaping circuit.

voltage converter requires a high feedback resistor (Sigworth, 1983), whereas good performance as a clamp requires a low one. The latter is true because the current that charges the membrane capacitance during a voltage step is fed to the pipette from the amplifier output via the feedback resistor. A very high feedback resistor limits the current and slows the performance of the clamp. For this reason we use relatively low feedback resistors, of 100 MΩ or less.

Even so, clamp performance is slow and must be supplemented by a circuit that feeds in the charging current "ballistically." With this technique, the charging current is fed in through a small capacitor, one side of which is attached to the pipette lead and the other to a shaping circuit that rounds the voltage command step (Sigworth, 1983). When ideally adjusted, only the command voltage component appears at the output during the application of the voltage step, and the current component is zero because all of the charging current is supplied through the input capacitor and none is drawn through the feedback resistor.

Another step necessary for obtaining fast clamp performance is the use of very-low-resistance pipettes. Our pipettes usually have a resistance of 0.5 to 1.0 MΩ and, at the suggestion of Dr. Jim Rae, are drawn from Corning type 8161 glass. Finally, we use series resistance compensation to lessen the slowing imposed by the pipette resistance.

3. CELL CULTURE

GH3 cells were obtained from American Type Culture Collection. Maintenance of the cells is quite easy, but a rigorous schedule must be followed if they are to remain in good condition. The cells are grown in one or more horizontally placed flasks, and they adhere to the bottom of the flasks. About once a week, when the cells are almost confluent, they are "split"; i.e., the population in the flask is reduced by about 90%. During this procedure, some of the cells are plated on thin strips of coverglass for experimental use. In between splittings, the culture medium must be replaced every second day. Failure to do this in our experience makes the ionic currents from the cells small and in general makes them hard to work with. The cells must be treated as a biohazard.

4. IONIC CURRENTS IN GH3 CELLS

A family of ionic currents recorded from a GH3 cell is shown in Fig. 2. The cell was depolarized from a holding potential of −80 mV to the

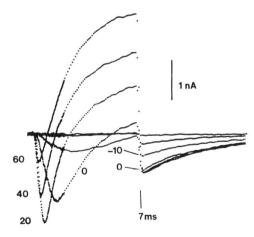

Figure 2. Family of currents recorded during steps to the indicated voltages in 130 Na, 10 Ca//140 K (outside//inside, concentrations in mM). Three major components of current are illustrated. At +20 mV, there is (1) a fast, transient inward current followed by (2) a large outward current and, after repolarization, (3) a slow tail of inward current. The following figures demonstrate that these currents are mediated by (1) Na^+ channels, (2) K^+ channels, and (3) Ca^{2+} channels, respectively.

values indicated in the figure. The initial current is inward at all voltages and is followed by a maintained outward current. These inward and outward currents are carried, respectively, by sodium and potassium ions mainly and resemble the currents of the same names recorded from other tissues, e.g., nerve fibers. The sodium current grows in magnitude as the step voltage is made more positive, and its kinetics become faster. On repolarization to -80 mV, there is an unusual tail of inward current that decays quite slowly. The tail is an unusual calcium current, as I show subsequently.

We have been mainly interested in inward currents in our investigations so far and have eliminated potassium current by replacing all the potassium in the pipette by cesium. The contents of the pipette equilibrate rapidly with the cell interior, and a few seconds after the pipette breaks into the cell, the potassium current is reduced to nil.

Both sodium and calcium carry inward current in GH3 cells, and their contributions can be separated kinetically by changing the ion content of the solutions or by using channel blockers, e.g., TTX. Figure 3 shows an experiment that gives kinetic information on the inward currents. The voltage was stepped from -80 mV to 20 mV just after the beginning of the trace and returned to -70 mV after various intervals, eliciting the tail currents, all of which have been plotted in a single figure. For a maintained depolarization, the current activates and reaches a peak in about 0.4 msec and then inactivates to a small but appreciable magnitude after a few milliseconds. This trace is labelled P in the figure, meaning current during the pulse.

The amplitude and time course of the tail currents depend on the duration of the preceding step. In all cases, the current jumps in magnitude when voltage is stepped to -70 mV because of the increased driving force

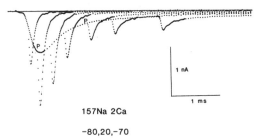

Figure 3. These superimposed traces show the current during a pulse from -80 to $+20$ mV, P, and the tail currents that resulted when the voltage was returned to -70 mV. Potassium current was eliminated by replacing internal potassium ion with cesium. The pulse current, P, has a transient phase that corresponds to activation and inactivation of the sodium channels and a small sustained phase that is primarily current though calcium channels. When the pulse is interrupted after a very short step (the two tails at the left), all of the current decays with the fast kinetics characteristic of sodium channels. After longer steps, there is an additional slow component of current through calcium channels. The experiment shows that the calcium channels that generate the slow tail current activate more slowly than the sodium channels.

on the ions. The maximum amplitude of the tail in each case is proportional to the magnitude of the pulse current just before the step back. After a short activating pulse, the tail current decays very rapidly, and current magnitude is near zero within 0.5 msec. This fast tail is absent when sodium is removed from the medium or when TTX is added and thus is sodium current carried through sodium channels with the customary properties. The fast tail inactivates within a few milliseconds. Concomitantly, a slow component appears in the tail. It is just evident in the third or fourth tail from the left and is progressively larger in tails from longer-duration pulses. The slow component is carried by one of two types of calcium channels that are found in GH3 cells. A calcium channel with rapid decay kinetics is demonstrated subsequently. In Fig. 3 the origin of the small fast component in the 3-msec tail is uncertain and could be either sodium current that has not inactivated or rapidly decaying calcium current.

From the technical point of view, Fig. 3 shows that good time resolution can be obtained even with these very small cells thanks to low-resistance pipettes, a relatively small feedback resistance, and careful adjustment of the ballistic charging circuit. Each sample point represents 10 μsec, and useful information can be obtained roughly 50 μsec after the step back to -70 mV.

5. CHARACTERISTICS OF CALCIUM CHANNELS

Current through calcium channels was investigated in several ways. Figure 4 shows an experiment in which sodium current was eliminated by adding TTX and potassium current was eliminated by replacing internal potassium with cesium. Barium is known to pass through calcium channels as well as or better than calcium itself, and the currents in Fig. 4 are Ba^{2+} moving through calcium channels. A small amount of calcium (0.9 mM) was left in the external medium to facilitate seal formation. Current during the pulse is just visible for a step to -10 mV and grows larger in magnitude up to about $+20$ mV because more and more of the channels open. Above 20 mV, amplitude begins to drop because of the smaller driving force. The current activates quite slowly at -30 mV and much more rapidly at 30 mV. The tails following steps to -10, 10, and 30 mV show two kinetic components, a fast component that is over in less than a millisecond and a slower component that is still visible at the ends of the traces. The slow component of the tail at 30 mV was fit with an exponential, which was extrapolated back to the time of the step. The fast component could then be obtained by subtraction. The time constants of fast and slow were, respectively, 121 μsec and 1.9 msec.

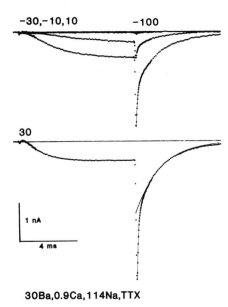

Figure 4. These traces show the activity of two types of calcium channels. They were recorded in barium-containing solution with tetrodotoxin to block the sodium current and cesium internally to block potassium current. The superimposed traces above are for steps to -30, -10, and 10 mV, with return to -100 mV, and the lower trace for a step to 30 mV. During the pulse, the inward current activates with a sigmoid time course and does not inactivate appreciably within 7 msec. The tail current has a fast and a slow component, both of which are particularly evident in the lower trace, where an exponential has been fitted to the slow phase. The two exponential components arise from the activity of two sets of calcium channels.

These two components can also be seen in solutions that contain only calcium and no barium, but the fast component is invariably smaller than that in Fig. 4, although the slow component is about the same size. That is, in 10 Ca^{2+} versus 10 Ba^{2+}, the slow component is about the same, and the fast component is much smaller in calcium. It should be pointed out that a good fraction of the fast component is lost because of limited time resolution. The major point seems clear, however: two components are clearly recognizable in the decay of the tail current.

The two components both become smaller as the calcium concentration is reduced (for technical reasons, calcium cannot be completely eliminated except on lucky occasions), proving that both are in fact calcium currents. The relative amplitudes of the fast and the slow component seem to vary capriciously from cell to cell, together with the amplitude of the sodium current. Figure 5 shows three examples of this, all recorded in identical conditions. In the upper trace, sodium current is large, and the tail current has small fast and slow components. In the middle trace, sodium current is relatively small, and the fast and slow components of the tail are large. In the bottom trace, sodium current is absent, and the tail is almost purely slow component. The cause of this variation is unknown.

In addition to their different rates of deactivation, the two calcium current components can be distinguished in several ways. A difference that is very useful experimentally is the fact that the slowly deactivating

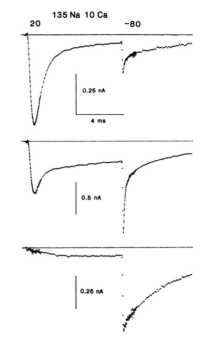

Figure 5. GH3 cells show considerable variation in the density of channels in their membranes. The three current traces shown were recorded under identical conditions from three different cells, using a pulse to 20 mV. The cell that generated the upper trace had many sodium channels and few calcium channels. The cell of the middle trace had some sodium channels and many fast deactivating (FD) and slowly deactivating (SD) calcium channels. The cell that generated the lower trace had mainly slowly deactivating calcium channels. The reason for the variability is not known.

(SD) channels inactivate, whereas the fast deactivating channels (FD) do not, as illustrated in Fig. 6. The figure shows currents recorded during steps to 20 mV with and without a prepulse to inactivate the SD channels. Both fast and slow components are evident in the tails when there is no prepulse. After the prepulse, the slow component is entirely absent from the tail, and the current during the pulse is reduced by about a third,

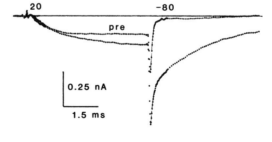

Figure 6. The SD channels can be inactivated by a prepulse. These traces show the currents during a pulse to 20 mV with and without a prepulse to inactivate the SD channels. The prepulse was to 20 mV for 100 msec and terminated 5 msec before the beginning of the illustrated traces. Inactivation of the SD channels reduces the current during the pulse by about a third and totally eliminates the slow component from the FD channels, which carry about two-thirds of the pulse current in this fiber and all of the fast tail current, are not affected by the prepulse. The traces were recorded in medium that contained 10 mM strontium ion, with TTX to eliminate sodium current and internal cesium to eliminate potassium current.

which is the contribution of the SD channels to the pulse current. We have also observed inactivation of the SD channels by observing the decay of current during long pulses. The time constant of inactivation is about 30 msec.

Other experiments have shown that both components are almost fully activated after a step of 10-msec duration. The slow component then deactivates with a time constant of about 30 msec (at 20°C), whereas the fast component remains almost unchanged. The slow component requires approximately a second to recover fully from inactivation.

There are several other differences in the properties of the two calcium channels that we have so far discovered. (1) The activation curve of the SD channels lies about 20 mV to the left of that for the FD channels. There is appreciable current through the SD channels at -30 to -40 mV, whereas the FD channel current is first apparent at about -10 to -20 mV. The SD channel activation curve is also to the left of that for sodium channels. Thus, the SD channels seem to be the most easily activatible of the inward current channels and the ones most likely to be open in the neighborhood of threshold. (2) Ion substitution experiments have shown that FD channels carry Ba^{2+} in preference to Ca^{2+}, whereas the SD channels carry the two ions about equally well. It is not clear which of the two channels carries more current in physiological calcium concentrations, since most of our experiments were performed in elevated calcium concentration, or in Ba^{2+}-containing solutions to make the currents easier to measure. (3) The FD channel usually disappears within a few minutes after the electrode breaks into the cell, whereas SD channel activity is maintained. The FD channels thus seem dependent on some constituent in the cytoplasm that is not contained in our artificial internal solution.

6. CONCLUSIONS

The invention of the patch clamp has opened many possibilities for studying the electrical properties of previously inaccessible cells, both at the level of the single channel and at the macroscopic level, in which currents are measured from the whole cell. Single-channel measurements have the advantage of studying the ultimate unit of electrical behavior, but macroscopic measurements are also of value. First, they give a quick average of behavior that is somewhat painful to reconstruct from single-channel data. Also, they make it possible to study channel behavior under ionic conditions in which single-channel currents are difficult or impossible to measure. Currents through single calcium channels, for example, are so far impossible to record in physiological calcium concentration, but macroscopic calcium currents are fairly easy to see.

Problems associated with combining the voltage-clamp and current measurement functions in a single amplifier were noted in Section 4. Good time resolution is not achieved through feedback, as in a conventional clamp, but instead requires careful adjustment of the ballistic charging circuit. At present our time resolution is adequate for some purposes, but we still lose about half of the rapidly decaying calcium tails. Further improvement clearly seems desirable for studying rapid events.

Calcium channels are thought to serve a number of purposes in cell functioning, and one is led to inquire whether all these purposes are served by a single type of channel. The first clear evidence that I am aware of for the existence of more than one type of calcium channel was provided by Fox and Krasne (1981), who studied the eggs of the polychaete *Neanthes*. They observed one calcium current that activated at about -45 mV and inactivated with a time constant of 10 to 50 msec. Both activation and inactivation were voltage dependent. The second calcium current had a more positive threshold and inactivated with a time constant of 300 to 600 msec. Inactivation of this current depended on calcium influx through the membrane.

Some of these features are shared by the two calcium channel types in GH3 cells (the present study; Matteson and Armstrong, 1984; Armstrong and Matteson, 1985) and in chick dorsal root ganglion cells (Carbone and Lux, 1984). In all cases, the two channel types differ in activation threshold, and the low-threshold channels inactivate by a voltage-dependent mechanism. Our studies show that the low-threshold channels (SD) have slightly slower activation kinetics and much slower deactivation kinetics. Carbone and Lux find that the low-threshold channels have a somewhat lower single-channel conductance than the high-threshold channels.

The physiological role of the two channel types is at present an interesting point for speculation with little supporting data. The SD channels in GH3 cells have a more negative activation range than either the sodium channels or the FD calcium channels and thus seem a logical choice for a role in events near threshold, such as pacemaking. Although we have no data on the question so far, it seems reasonable to predict that the SD channels have a lower single-channel conductance than the FD channels, thus providing a fine grading of behavior near threshold. It will be of great interest to see if the behavior of these channels is affected by TRH. The FD channels, on the other hand, seem well suited to the role of injecting calcium into the cell once the decision to fire has been made by the SD channels. Again, one looks forward eagerly to more data.

REFERENCES

Armstrong, C. M., and Matteson, D. R., 1985, Two distinct populations of calcium channels in a clonal line of pituitary cells, *Science* **4682**:65–67.
Bruce, B. R., and Anderson, N. C., 1979, Hyperpolarization in mouse parathyroid cells by low calcium, *Am. J. Physiol.* **236**(1):C15–C21.
Dean, P. M., and Matthews, E. K., 1968, Electrical activity in pancreatic islet cells, *Nature* **219**:389–390.
Gershengorn, M. C., Thaw, C., and Gerry, R. H., 1983, Calcium influx is not required for thyrotropin-releasing hormone stimulation of prolactin release from GH3 cells, *Cell Calcium* **4**:117.
Katz, B., 1969, *The Release of Neural Transmitter Substances*, Liverpool University Press, Liverpool.
Kidokoro, Y., 1975, Spontaneous calcium action potentials in a clonal pituitary cell line and their relationship to prolactin secretion, *Nature* **258**:741–742.
Lopez-Barneo, J., and Armstrong, C. M., 1983, Depolarizing response of rat parathyroid cells to divalent cations, *J. Gen. Physiol.* **82**:269–294.
Matteson, D. R., and Armstrong, C. M., 1983, Voltage clamp experiments on GH3 cells, *Biophys. J.* **41**:223a.
Matteson, D. R., and Armstrong, C. M., 1984, Evidence for two types of Ca channels in GH3 cells, *Biophys. J.* **45**:36a.
Meissner, H. P., and Schmeltz, H., 1974, Membrane potentials of beta-cells in pancreatic islets, *Pfluegers Arch.* **351**:195–206.
Mitchell, P., 1967, Proton-translocation phosphorylation in mitochondria, chloroplasts, and bacteria: Natural fuel cells and solar cells, *Fed. Proc.* **26**:1370–1379.
Sakmann, B., and Neher, E., 1983, *Single-Channel Recording*, Plenum Press, New York.
Sigworth, F., 1983, Electronic design of the patch clamp, in *Single-Channel Recording*, (B. Sakmann and E. Neher, eds.), Plenum Press, New York, pp. 3–35.
Tareskevich, P. S., and Douglas, W. W., 1980, Electrical behavior in a line of anterior pituitary cells (GH cells) and the influence of the hypothalamic peptide releasing factor, *Neuroscience* **5**:421–431.
Tosteson, D. C., 1964, Regulation of cell volume by sodium and potassium transport, in *The Cellular Functions of Membrane Transport*, (J. F. Hoffman, ed.), Prentice-Hall, Englewood Cliffs, New Jersey, pp. 3–22.

Part III

Ionic Channel Reconstitution

Chapter 16

Ion Channel Reconstitution
Why Bother?

Christopher Miller

1. INTRODUCTION AND BACKGROUND

Twenty years ago, membrane transport research was experiencing an exciting and, in some ways, intellectually chaotic time—a time when the extreme multiplicity of membrane transporters was being uncovered phenomenologically, a time when it seemed, at least to the bewildered student, that each different type of specialized cell carried its own unique constellation of dedicated solute "carriers" (as we then termed them). To that student, it was clear that the experimentalists often rubbing shoulders in physiology departments—the Na^+ pumpers, the sugar carriers, the oxidative phosphorylators, the renal electrolyte balancers, the bacterial energizers—that these people were working on totally different problems; and this list ignores the electrophysiologists.

One of the most exciting aspects of being a student at that time was the slow process of realizing that behind all of this membrane transport phenomenology, there is a fundamental unity. Membrane research of the 1960s led not only to the recognition of the three mechanistic classes of transporters—active pumps, coupled shuttles, and ion channels—but also

---

CHRISTOPHER MILLER • Graduate Department of Biochemistry, Brandeis University, Waltham, Massachusetts 02154.

to the realization that the ways that membranes of all types of cells capture and conserve free energy, though exhibiting marvellous variation in detail, operate on the same underlying principles—cells from the strangest archaebacterium to those of the most respectable department chairman.

At that time, only the first stirrings were being heard of what I consider turned out to be the next step in mechanistic membrane transport work: the solubilization, isolation, and purification of the membrane proteins responsible. Remember that in the 1960s, Singer and Nicholson (1972) had not yet drawn their famous cartoon of the fluid mosaic model, with its integral membrane battleships floating around in an amphipathic sea, and carrying the unspoken message that the interesting things done by membranes are done by membrane proteins. The main problem for the biochemists to overcome was not mere solubilization and purification but to insure that after these insults the proteins would continue to perform the same interesting functions that motivated the work in the first place. Slowly, over a period of a decade, biochemists learned enough physiology to realize that a closed membrane was required for most of the functions of these proteins and found ways to "reconstitute" a membrane from protein solubilized in detergent and exogenously added lipid (Racker, 1975).

What I want to discuss here is the application of this reconstitution approach to only one of the classes of membrane proteins: the ion channels. Compared with the great successes in attaining, through reconstitution, truly novel information about the other two classes of membrane transporters, pumps and shuttles (see Racker, 1975), the channels have been somewhat left out in the cold: reconstitution of ion channels remains a primitive field. This is really quite ironic, since mechanistically, the channels catalyze by far the simplest type of transport: simple diffusion towards electrochemical equilibrium. It would seem, then, that if the motivation of doing reconstitution is to gain mechanistic information about a system, it would be best to start with the simplest type of mechanism. This has clearly not been the case, and there are many reasons for this apparent lack of rationality on the part of membrane reconstitologists. But there is one reason that I think towers high above the rest in explaining this situation: a vicious combination of the surface/volume ratios of the various types of reconstitution systems with the high turnover rates of ion channels.

To see this, consider the two types of "artificial" membrane systems available to the budding reconstitologist (and I put that word in quotation marks because there really is nothing at all artificial about these membranes). There are liposomes—little round membranes—and planar bilayers—big flat ones. As Fig. 1. shows, the surface-to-volume ratios of the two systems differ by about 12 orders of magnitude. It is easy to

Ion Channel Reconstitution

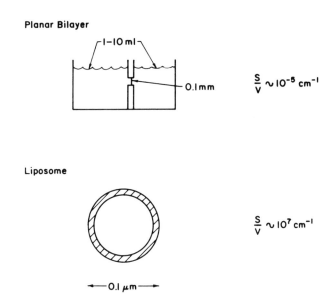

Figure 1. Liposomes and planar bilayers. A schematic diagram of these two types of model membrane systems with typical values of the ratio of membrane surface (S) to aqueous enclosed volume (V) for each system.

picture the planar bilayer system—a tiny hole separating two great big aqueous chambers, but also try to picture the liposome system. In a typical "reconstitution mix," we have, say, 1 ml of a 25 mM lipid suspension. It turns out that the total surface area in this mix is about 10 m², about the size of a slide-presentation screen. But this surface, crammed into the test tube, encloses a space of only about 10 µl, half the size of a drop of water.

The main consequence of the surface/volume difference in the two model membranes is the difference in time scales that one would have to work on in doing the conventional type of transport measurement: radioactive trace flux determination. Suppose we had a liposome system in which each liposome had a single pump or channel-type protein reconstituted into it. Then, we see in Table I that for a channel, the radioactive flux would reach equilibrium in about 1 msec, much too fast to follow using a fast filtration method, for instance. A pump, on the other hand, puts us in a very civilized time domain and even allows us a swallow of coffee or two between time points. Now one could in principle try to reconstitute these proteins into a planar bilayer at the same effective density as the above liposomes. But now, the time scale of the fluxes would be 3 years for a channel and 3 million years for a typical pump. The point is that there are fundamental constraints on measuring fluxes

Table I. Time Scales for Flux Equilibrium in Model Membranes[a]

Type of transporter	Time for 50% flux equilibrium		Unitary current
	Liposome	Planar bilayer	
Channel (10^7 sec^{-1})	1 msec	3 years	1 pA
Pump (10^2 sec^{-1})	2 min	10^6 years	1 aA

[a] Times for 50% equilibration of radioactive tracer flux are given for hypothetical experiments in which either a channel or a pump is reconstituted into either liposomes or planar bilayers at a density of one transporter per 10^5 molecules of lipid. This density represents one transporter per liposome (700 Å diameter) or 2×10^7 transporters per planar bilayer (100 μm diameter). The last column lists the unitary electric current associated with each type of transporter. These numbers are order-of-magnitude estimates.

in reconstituted systems now available, and this easily explains why pumps, not channels, have been furiously reconstituted in liposome systems.

There is a second irony in the mechanistically simple channels having been largely overlooked by the biochemists, and that is that they have been mechanistically persued by physiologists. From the high-time-resolution electrical methods of voltage clampers starting with Hodgkin and Huxley (1952) to the modern breakthrough of patch-recording techniques (Hamill *et al.*, 1981), which can observe individual channel molecules directly, electrophysiologists have been asking and answering questions of the catalytic mechanisms of these channel proteins, of the conformational changes involved in their opening and closing, and even of the gross structures of the protein complexes.

And they have been doing all this with the knowledge that their conclusions rest on the solid earth of cellular physiology; they need not face the perpetual nightmare ruining the slumbers of even the most fanatic reconstitologist: that possibly the channel protein has been twisted and functionally altered by the act of reconstitution. We all know the delicacy involved in the initial dissection steps required for electrophysiological experiments, the difficulty in internalizing these steps, and the requirements for and imperturbability in obtaining a usable specimen. So these nightmares are understandable when we contemplate the first step of a channel reconstitution experiment: violent tissue homogenization. Considering the brutality of the approach, it really is amazing that the channels observed in reconstituted systems do indeed come out behaving like those in cell membranes, at least in those cases (Na$^+$ channels and Ca^{2+}-activated K$^+$ channels) where comparison can be made, as is shown in the chapters by French *et al.* (Chapter 17) and by Wolff *et al.* (Chapter 19).

All this raises the question asked in my title about ion channel reconstitution: Why bother? I think that there are, in fact, some excellent reasons for beating one's head on the reconstitution of ion channels, even in this day of patch-recording technology. I illustrate this first not with reference to some of the odd channels of my own interest but rather by showing that these reconstitution tricks can now be accomplished with some old electrophysiological friends: the nicotinic AchR and the Na^+ channel.

Figure 2 shows the progress made in AchR reconstitution during the 1970s. In the early years of the decade, radioactive efflux from vesicles with AchRs in them followed qualitatively the correct behavior (Kasai and Changeux, 1971). But quantitatively there was a serious problem: the fluxes were three to four orders of magnitude too slow. By 1980, rapid flux methods had combined with an increased sophistication in handling solubilized membrane proteins to produce the flux records shown on the right, from Moore and Raftery's (1980) elegant fluorescence-quenching method. Moreover, when these reconstituted vesicles containing purified AchR are electrically probed with the patch-recording method, you can actually see good-looking single AchR channels (Tank et al., 1983). How do you do this? You certainly do not try to impale or patch these little liposomes, but you can transform them into a patchable preparation by enlarging them into the 10-μm range by freezing and thawing (Tank et al., 1982). When Dave Tank and Rick Huganir applied this method to reconstituted AchR, they easily observed single channels that by all cri-

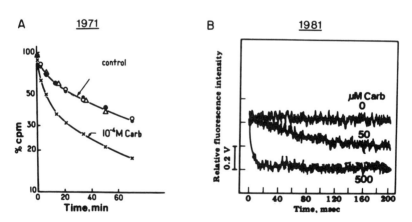

Figure 2. Acetylcholine receptor flux assays over a decade. A: Agonist-induced Na^+ fluxes in membrane vesicles prepared from eel electric organ. (From Kasai and Changeux, 1971, with permission.) B: Agonist-induced Tl^+ fluxes in liposomes containing purified AchR protein, measured by stopped-flow fluorescence-quenching method. (From Moore and Raftery, 1980, with permission.)

teria looked physiologically intact (Tank *et al.*, 1983). They could also adjust the average number of channels per patch by diluting the AchR protein into excess lipid. Likewise, AchR channels can be observed by the "tip-dip" method (Coronado and Latorre, 1983), which has been vigorously exploited by Montal's group to characterize in detail the behavior of the purified AchR protein (Montal *et al.*, 1983).

Work with Na^+ channels has also proceeded well during the past few years. Several groups are in the position, nearly unique in the history of membrane biochemistry, of being in basic agreement about something new, in this case the identification of the channel protein as a huge, 270-kD polypeptide, maybe with some smaller subunits purifying along with it (Agnew *et al.*, 1978; Hartshorne and Catterall, 1981; Barchi *et al.*, 1980). As to the question of function, this is not as far along as is the AchR work, but the Colombian poison arrow frog has made a great contribution to the work by synthesizing batrachotoxin (BTX) for these channel biochemists. This toxin acts by holding Na^+ channels open, and Barchi's group (Tanaka *et al.*, 1983) has shown in rapid mixing experiments that liposomes reconstituted with purified Na^+ channels have the capability of allowing the fluxes of channel-permeant cations. You can't see Na^+ going in—it's too fast, as it ought to be. But you can easily see the permeation of K^+, Rb^+, and even Cs^+, which is "impermeant" to channel electrophysiologists. In addition, recent exciting work from Agnew's and Montal's labs has shown that the purified Na^+ channel can be seen directly by single-channel recording methods, either in planar bilayers (Hartshorne *et al.*, 1985) or in liposome patches (Rosenberg *et al.*, 1984).

There are several lessons to be taken home from the AchR and the Na^+ channel work, lessons about bothering to reconstitute channels. First, it is clear that without this line of attack, of purifying the protein and putting it back into a reconstituted membrane, we would never have been able to identify the nicotinic AchR as the $\alpha_2\beta\gamma\delta$ pentamer that it is. There would always be someone suggesting that your purely cellular efforts might have overlooked some cellular cofactors needed for function. It is by now a truism, but no less true, that reconstitution allows the determination of the minimal functional unit. Once you have that, you can then start asking questions about modulation of this function. Tank and Huganir, for example, are presently using the liposome patch assay to ask whether the stoichiometric phosphorylation observed in the *Torpedo* AchR by an endogenous kinase (Huganir and Greengard, 1983) has any functional consequence. And they will be able to supply an answer to that question, one way or the other. Similarly, Montal and Lindstrom are beginning to test monoclonal antibodies on the reconstituted AchR single channels (Lindstrom *et al.*, 1983) to ask whether it is feasible to map the receptor protein's functional domains with this high-resolution method. Likewise, with the Na^+ channel, it is inevitable that the various groups will start falling out over whether the small subunits in the preparation are necessary for function, or which function they are necessary for, but with proper reconstitution methods, these questions will be answerable.

The previous examples are the real success stories of ion channel reconstitution, in that these channels are being reconstituted for all the right reasons. There are groups, however, that have been studying channels in model membranes, but on unpurified and biochemically unknown channel proteins. In this approach, native membrane fragments are by one trick or another transferred to a planar bilayer system. In one approach, they are fused into a planar bilayer (Miller, 1978), which may be made either by painting an alkane solution of phosholipid on a hole in a partition or by folding two air–water interface lipid monolayers over the hole. Alternatively, under some conditions, the crude membrane vesicles may be transferred to an air–water interface directly and thence to the end of a patch pipette (Coronado and Latorre, 1983).

This raises a serious question: if the protein isn't pure, if its purification has not even been attempted, then what is the point? Why bother? Why not just study these channels in the native membrane by the patch-recording method, using the living cell as the source of membrane? There are several answers to the question here, reasons for not giving up and leaving the channels to the electrophysiologists. The recognition of each of these answers was, to me at least, a big surprise. There are three areas in the use of "planar bilayer preconstitution," if you will, that justify their use: unexpected surprises, unconstrained variables, and unrealized hopes.

2. UNEXPECTED SURPRISES

The universal rule, almost unbroken so far in this preconstitution approach, is that whatever channel one chooses to study, one ends up finding something else. In searching for Ca^{2+} channels of sarcoplasmic reticulum, a K^+ channel was found (Miller, 1978); in attempted preconstitution of acetylcholine receptor, a voltage-dependent Cl^- channel announced itself (White and Miller, 1979); in a search for Ca^{2+} channels of paramecium ciliary membranes, Colombini found the mitochondrial porin channel (Schein et al., 1976); rabbit T-tubule vesicles surprised Latorre with a huge Ca^{2+}-activated K^+ channel (Latorre et al., 1982). Benos has discovered a jungle of channels in preconstitution work with apical membranes from a kidney cell line, and one of these may actually be the amiloride-sensitive Na^+ channel he seeks (Sariban-Sohraby et al., 1984).

None of these channels was "supposed" to be there when the research began, and we all worried a lot. But apparently there are more channels in heaven and earth than had been dreamed of, at least on the basis of squid axon and frog node work. In most of the examples above, the vesicles used for insertion into planar bilayers were derived from membranes not accessible to electrophysiological probes: the SR, T-tubule, and mitochondrial outer membrane because of small size, the noninnervated membrane of *Torpedo* electroplax because of its extremely high resting conductance. At this point, if one wants to study these channels, each of which has some interesting lessons to teach us, I think one is going to have to do it in a preconstitution experiment.

3. UNCONSTRAINED VARIABLES

The next reason for using this kind of approach is the wide latitude the experimenter has in varying the external conditions of the channel sitting in the planar bilayer. Some of these advantages are fundamental and some only technical, but even an advantageous technical detail can sometimes be more important in learning about the behavior of a system than a lot of deep thought. For example, we can study the variation in conductance of SR K^+ channels with salt concentration (Coronado *et al.*, 1980) or perform the same sort of measurement with the now-famous huge Ca^{2+}-activated K^+ channel. When we do this, we see that these channels appear to follow simple rectangular hyperbolas from 30 mM up to 1 M salt where KCl is the only ionic component of the medium (Latorre and Miller, 1983). We can go over a huge range of "substrate" concentration here. Latorre's group (Moczydlowski *et al.*, 1985), however, has shown that for the Ca^{2+}-activated K^+ channel, this simple picture is wrong, that there is a very strong deviation from Michaelis–Menten behavior below 30 mM; indeed, they have taken the channel down to 0.5 mM K^+, where it still has a conductance of 60 pS. Thus, it is clearly not a simple channel in its conduction behavior, but the constraints of physiological "patchability" would not have been able to uncover this fact. Likewise, we have measured the conductance–activity relationship for BTX-activated Na^+ channels (Moczydlowski *et al.*, 1984) from 1 mM up to 600 mM NaCl (Fig. 3); simple behavior appears to hold here, but the channel itself shows unexpectedly tight binding for Na^+, with a K_m of only 7 mM.

The ability to vary conducting ions and osmotic conditions widely, to make additions and perfusions on both sides of the membrane in an experiment, to control the number of channels in a bilayer, to change the lipid composition of the bilayer, these capabilities of the bilayer system

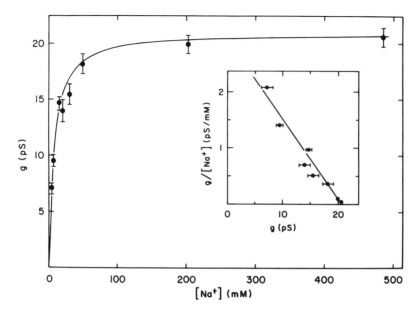

Figure 3. Sodium channel conductance–concentration curve. Batrachotoxin-activated Na$^+$ channels from rat skeletal muscle were inserted into planar bilayers, and single-channel conductance was measured in symmetrical solutions of NaCl at the indicated concentrations. Data are fit to a rectangular hyperbola with $K_m = 7$ mM and $\gamma^{max} = 22$ pS. Data are from Moczydlowski *et al.* (1984).

have enabled us to perform experiments with ease that have led to a physical picture of the diffusion pathway of one of the channels under study in my lab: a cation-selective channel from SR. Figure 4 shows the sort of cartoon we are no longer ashamed to draw after 5 years of looking at these single channels under a variety of cleanly defined conditions.

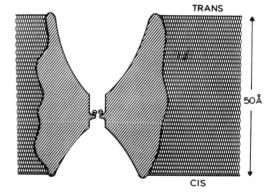

Figure 4. Cartoon of SR K$^+$ channel conduction pathway. Cartoon is drawn "to scale," illustrating the conclusion that this channel consists of three distinct regions: a "wide-mouth" region, a "short tunnel" of 10 Å length, and a small constriction within the tunnel. (From Miller, 1982, with permission.)

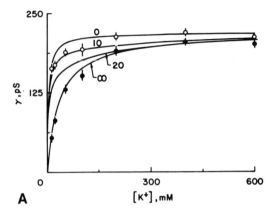

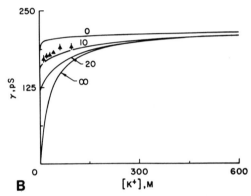

Figure 5. Effect of surface charge on SR K^+ channel conduction. Conductance–concentration relation for SR K^+ channels inserted into either neutral or charged phospholipid bilayers. A: Filled circles, neutral (PE/PC) bilayers; open circles, negatively charged bilayers containing 70% PS. Solid curves are predicted by Gouy–Chapman theory for a channel protruding a given distance (indicated in Angstroms) from bilayer surface. B: Filled triangles, data from a 25% PS bilayer. Solid curves as above.

What we have tried to do is to get a physical picture, admittedly vague, of the inside of this channel, to go feeling around in the dark (Miller, 1982b) using a variety of probes: organic cation selectivity, blocking, streaming potentials, lipid variations. In just one of these experiments, to illustrate the approach (Fig. 5), we wanted to know how far away from the lipid surface the channel senses the bulk aqueous concentration.

Joan Bell (Bell and Miller, 1984) attacked this question by following the conductance–concentration relationship in membranes of differing (and known) lipid surface charge densities—neutral, positive, and negative. She found that at low salt, where the Debye length is long, the surface charge makes a big difference in the channel conductance; at high salt, where the Debye length is short and the channel is near saturation anyway, the channel does not reflect the difference between the two membranes. By utilizing double-layer theory, which luckily has a lot of independent confirmation in the bilayers used here, we could place an estimate of about 10 Å on the distance of separation of the channel's "concentration sensor" and the lipid surface. Thus, this is not like the

AchR, which protrudes 50 Å out into solution on one side of the membrane. A similar set of experiments for the Torpedo Cl$^-$ channel shows that its "mouth" is removed from the bilayer surface by over 30 Å (White and Miller, 1981). Recently Latorre's group have performed similar experiments on Ca^{2+}-activated K$^+$ channels, finding that both the entryway to the conduction pathway and the Ca^{2+} regulatory site are located about 8 Å away from the bulk lipid surface (Moczydlowski *et al.*, 1985). These kinds of experiments would obviously be impossible to perform in a native membrane environment and can answer directly questions of lipid and charge interactions with channel proteins, questions that can be only very indirectly and tenuously approached with channels still embedded in their native membranes.

Likewise, the ability to vary conditions with abandon has allowed us to study the behavior of a very eccentric and unexpected channel, a Cl$^-$ channel from *Torpedo* electroplax (Miller, 1982a). This channel's bizarre unitary behavior has led us to the solid conclusion (Miller and White, 1984) that the conducting unit is actually a dimer of two "protochannels," each of which gates independently of the other on a rapid time scale, while being obligatorily coupled together on a slower time scale. Our basic picture of the channel is illustrated in Fig. 6, which also emphasizes that the "inactivated" state of the channel is thought to be the consequence of a change in quaternary interaction between the two monomers, leading to the shutting off of both of them. One recent result strongly supporting this picture is our ability to irreversibly inhibit one of the protochannels, using a covalent amino-group reagent, while leaving the other protochannel to gate normally (Miller and White, 1984).

4. UNREALIZED HOPES

This introduction to the Cl$^-$ channel leads me to the last reason for doing these channel-preconstitution experiments, the hopes for the system

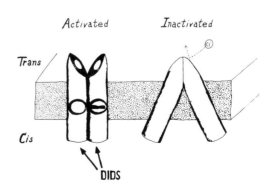

Figure 6. Cartoon of "double-barreled" Cl channel from *Torpedo* electroplax. (From Miller and White, 1984, with permission.)

as a way to purify the protein itself. These hopes are entirely unrealized as yet, but when you consider the problem, you will see why it is compelling enough to warrant much investment of time in difficult and refractory experiments. Here we have a Cl^- channel with some very interesting functional properties: a voltage-dependent gating mechanism that is activated by cytoplasmic proton activity; an "inactivation" gate, also voltage-dependent; and a hypothesis that says that the inactivation machinery is a direct functional consequence of the channel's dimeric structure. We also have a source, the *Torpedo* electroplax, which is, by all indications that we can gain, a very rich source of the channel protein. What is the next step? What should we do now with this channel? Obviously, we should try to purify the channel protein to allow much more detailed biochemical studies to be carried out simultaneously with biophysical planar bilayer studies. But there is good news and bad news here. The good news is that we have an excellent "reconstituted" membrane system appropriate for checking on the functional activity of our protein when it does get purified. The bad news is that there is no "probe" suitable for use as a purification assay: something like bungarotoxin which could be used to determine which of 50 column fractions the channel protein is present in.

So the unrealized hope, at this point, is to use the reconstitution itself as the assay of the channel protein. This has been done before, not with channels but with other membrane transport proteins (Racker, 1975), so it is possible both in principle and by precedent. But it will not be easy, and our recent efforts have all been geared just to set up a suitable reconstitution assay system in which we can detect the channel-mediated flux of Cl^- across some model membrane. Of course, once we have such a protein purified, that would be only the beginning of the real biochemistry on the system: the protein dissection, the reconstitution of the dimer from monomers, chemical and proteolytic modifications of function, and ultimately, on a very long time scale, the analysis of structure–function relations by a combination of site-directed mutagenesis with single-channel analysis.

The other unrealized hope is that by struggling with these reconstitution methods on this relatively "easy" Cl^- channel, we will learn enough to be able to go after more biochemically forbidding channels. We can easily detect, by planar bilayer preconstitution, a great host of K^+ channels, for instance, from many different types of sources: cardiac plasma membranes, epithelial membranes, neuronal preparations from *Helix* and even *Aplysia,* and from that supermarket of channels, the *Torpedo* electroplax. But the possibility of purifying any of these will require great luck, something that it would be nice not to have to rely on. By combining what we will learn from the Cl^- channel attempts, I would hope that we will be in a much better position to attack some of these very interesting and physiologically important K^+ channels: delayed rectifiers, inward rectifiers, and that most tantalizing, beautiful beast, the huge Ca^{2+}-activated K^+ channel.

What are the questions we would like to answer with all this? Like good enzymologists, we should be thinking about active-site mechanisms. Of course, for channels there are really two types of "active sites": the gating sites determining the protein's conformational changes leading to the conducting state, and the conduction sites, which select one ion over another while still allowing extremely fast ion transport. The latter of these will be easier to understand than the former, I think, since in thinking about it we can rest fairly comfortably on the assumption that the underlying mechanism will be a fundamentally simple one. Indeed, I find it delightful that the beautiful high-resolution structure work of AchR two-dimensional crystals (Kistler *et al.*, 1982) is basically confirmatory of what the electrophysiologists already knew on the basis of conduction and blocking studies: a wide mouth connected to a water-filled vestibule.

I feel that with a combination of the electrophysiological approaches in the native membrane, the hot pursuit of the biochemists and molecular geneticists, the unconstrained probing done by reconstitologists in model membranes, and, ultimately, the frozen crystal structures provided by the diffraction labs, each contributing its own special strength, we will begin to glimpse the workings of this most tantalizing and elusive of membrane protein types, the ion channels.

REFERENCES

Agnew, W. S., Levinson, S. R., Brabson, J. S., and Raftery, M. A., 1978, Purification of tetrodotoxin-binding components associated with the voltage-sensitive sodium channel from *Electrophorus electricus* electroplax membranes, *Proc. Natl. Acad. Sci. U.S.A.* **75**:2606–2610.

Barchi, R. L., Cohen, S. A., and Murphy, L. E., 1980, Purification from rat sarcolemma of the saxitoxin-binding component of the excitable membrane sodium channel, *Proc. Natl. Acad. Sci. U.S.A.* **77**:1306–1310.

Bell, J. E., and Miller, C., 1984, Effects of phospholipid surface charge on ion conduction in the K^+ channel of sarcoplasmic reticulum, *Biophys. J.* **45**:279–287.

Coronado, R., and Latorre, R., 1983, Phospholipid bilayers made from monolayers on patch-clamp pipettes, *Biophys. J.* **43**:231–236.

Coronado, R., Rosenberg, R. W., and Miller, C., 1980, Ionic selectivity saturation, and block in a K^+ selective channel from sarcoplasmic reticulum, *J. Gen. Physiol.* **76**:425–446.

Hamill, O. P., Marty, A., Neher, E., Sakmann, B., and Sigworth, F. J., 1981, Improved patch-clamp techniques for high resolution current recording from cells and cell free membrane patches, *Pfluegers Arch.* **391**:85–100.

Hartshorne, R. P., and Catterall, W. A., 1981, Purification of the saxitoxin receptor of the sodium channel from rat brain, *Proc. Natl. Acad. Sci. U.S.A.* **78**:4620–4624.

Hartshorne, R., Keller, B. U., Talvenheimo, J. A., Catterall, W. A., and Montal, M., 1985, Functional reconstitution of the purified sodium channel in planar lipid bilayers, *Proc. Natl. Acad. Sci. U.S.A.* **82**:240–244.

Hodgkin, A. L., and Huxley, A. F., 1952, A quantitative description of membrane current and its application to conduction and excitation in nerve, *J. Physiol. (Lond.)* **117**:500–544.

Huganir, R. L., and Greengard, P., 1983, cAMP-dependent protein kinase phosphorylates the nicotinic acetylcholine receptor, *Proc. Natl. Acad. Sci. U.S.A.* **80**:1130–1134.

Kasai, M., and Changeux, J. P., 1971, In vitro excitation of purified membrane fragments by cholinergic agonists. I. Pharmacological properties of the excitable membrane fragments, *J. Membr. Biol.* **6**:1–23.

Kistler, J., Stroud, R. M., Klymowsky, M. W., Lalancette, R. A., and Fairclough, R. H., 1982, Structure and function of an acetylcholine receptor, *Biophys. J.* **37**:371–383.

Latorre, R., and Miller, C., 1983, Conduction and selectivity in potassium channels, *J. Membr. Biol.* **71**:11–30.

Latorre, R., Vergara, C., and Hidalgo, C., 1982, Reconstitution in planar lipid bilayers of a Ca^{2+}-dependent K^+ channel from transverse tubule membranes isolated from rabbit skeletal muscle, *Proc. Natl. Acad. Sci. U.S.A.* **79**:805–809.

Lindstrom, J., Tzartos, S., Gullick, W., Hochschwender, S., Swanson, L., Sargent, P., Jacob, M., and Montal, M., 1983, Use of monoclonal antibodies to study acetylcholine receptors from electric organs, muscle and brain and the autoimmune response to receptor in myasthenia gravis, *Cold Spring Harbor Symp. Quant. Biol.* **48**:89–99.

Miller, C., 1978, Voltage-gated cation conductance channel from fragmented sarcoplasmic reticulum: Steady state electrical properties, *J. Membr. Biol.* **40**:1–23.

Miller, C., 1982a, Open-state substructure of single chloride channels from *Torpedo* electroplax, *Phil. Trans. R. Soc. Lond. [Biol.]* **299**:401–411.

Miller, C., 1982b, Feeling around inside a channel in the dark, in *Transport in Biological Membranes: Model Systems and Reconstitution* (R. Antolini, ed.), Raven Press, New York, pp. 99–108.

Miller, C., and White, M. M., 1984, Dimeric structure of single chloride channels from *Torpedo* electroplax, *Proc. Natl. Acad. Sci. U.S.A.* **81**:2772–2775.

Moczydlowski, E., Garber, S., and Miller, C., 1984, Batrachotoxin-activated Na^+ channels in planar lipid bilayers. Competition of tetrodotoxin block by Na^+, *J. Gen. Physiol.* **84**:665–686.

Moczydlowski, E., Alvarez, O., Vergara, C., and Latorre, R., 1985, Effect of phospholipid surface charge on the conductance and gating of a Ca^{2+}-activated K^+ channel in planar lipid bilayers, *J. Membr. Biol.* **83**:273–282.

Montal, M., Suarez-Isla, B., Wan, K., and Lindstrom, J., 1983, Single-channel recordings from purified acetylcholine receptors reconstituted in bilayers formed at the tip of patch pipettes, *Biochemistry* **22**:2319–2323.

Moore, H. P., and Raftery, M. A., 1980, Direct spectroscopic studies of cation translocation by *Torpedo* acetylcholine receptor on a time scale of physiological relevance, *Proc. Natl. Acad. Sci. U.S.A.* **77**:4509–4513.

Racker, E., 1975, Reconstitution of membrane pumps, in *Proceedings of 10th FEBS Meetings: Biological Membranes* (J. Montreuil and P. Mandel, eds.), North-Holland, Amsterdam, pp. 25–34.

Racker, E., 1976, *A New Look at Mechanisms in Bioenergetics*, Academic Press, New York.

Rosenberg, R. L., Tomiko, S. A., and Agnew, W. S., 1984, Single-channel properties of the reconstituted voltage-regulated Na channel isolated from the electroplax of *Electrophorus electricus*, *Proc. Natl. Acad. Sci. U.S.A.* **81**:5594–5598.

Sariban-Sohraby, S., Latorre, R., Burg, M., Olans, L., and Benos, D. J., 1984, Amiloride-

sensitive epithelial Na⁺ channels reconstituted into planar lipid bilayer membranes, *Nature* **308**:80–82.
Schein, S., Colombini, M., and Finkelstein, A., 1976, Reconstitution in planar lipid bilayers of a voltage-dependent anion-selective channel obtained from *Paramecium* mitochondria, *J. Membr. Biol.* **30**:99–120.
Singer, S. J., and Nicholson, G. L., 1972, The fluid mosaic model of the structure of cell membranes, *Science* **175**:720–731.
Tanaka, J. C., Eccleston, J. F., and Barchi, R. L., 1983, Cation selectivity characteristics of the reconstituted voltage-dependent sodium channel purified rat skeletal muscle sarcolemma, *J. Biol. Chem.* **258**:7519–7526.
Tank, D., Miller, C., and Webb, W. W., 1982, Isolated patch recording from liposomes containing functionally reconstituted chloride channels from *Torpedo* electroplax, *Proc. Natl. Acad. Sci. U.S.A.* **79**:7749–7753.
Tank, D. W., Huganir, R. L., Greengard, P., and Webb, W. W., 1983, Patch-recorded single-channel currents of the purified and reconstituted Torpedo acetylcholine receptor, *Proc. Natl. Acad. Sci. U.S.A.* **80**:5129–5133.
White, M. M., and Miller, C., 1979, A voltage-gated anion channel from electric organ of *Torpedo californica*, *J. Biol. Chem.* **254**:10160–10166.
White, M. M., and Miller, C., 1981, Probes of the conduction process of a voltage-gated chloride channel from *Torpedo* electroplax, *J. Gen. Physiol.* **78**:1–18.

Chapter 17

From Brain to Bilayer
Sodium Channels from Rat Neurons Incorporated into Planar Lipid Membranes

Robert J. French, Bruce K. Krueger, and Jennings F. Worley, III

1. PERSPECTIVES AND BACKGROUND

Contemporary studies on the molecular basis of membrane excitability spring from the pivotal conclusion reached by Hodgkin and Huxley (1952b) that the currents that generate an action potential must flow through a sparse array of voltage-sensitive conducting sites in the plasmalemma. They further showed that the squid axon membrane possessed two functionally distinct transmembrane pathways, which were selective for sodium and potassium ions, respectively (Hodgkin and Huxley, 1952a). These are the voltage-gated sodium and potassium "channels," and they are, indeed, discrete macromolecules that form hydrophilic pores through the membrane (Hille, 1984).

Crucially important to our ability to study the molecular properties of sodium channels are several potent and specific toxins that affect channel gating or flow of ions through the open channels. These toxins provide functional probes for recognition as well as ligands for isolation of the channels. From a biophysical standpoint, the recent development of techniques to measure current flow through single channels has enabled a

---

ROBERT J. FRENCH, BRUCE K. KRUEGER, AND JENNINGS F. WORLEY, III • University of Maryland School of Medicine, Baltimore, Maryland 21201.

direct approach to a variety of hitherto inaccessible questions about ion permeation and channel gating. In studying the channels after incorporation into an artificial lipid bilayer membrane, we place ourselves at the confluence of the biochemical and electrophysiological approaches and open the way to study channels either prior to or after various degrees of biochemical purification or modification.

1.1. Sodium Channel Biochemistry

By use of tritiated saxitoxin and tetrodotoxin as high-affinity labels, sodium channels from rat brain, rat muscle, and the electroplax of electric eel have been isolated and purified to near homogeneity. Several recent reviews describe these studies and provide extensive references to the original papers (Agnew, 1984; Barchi, 1982; Catterall, 1984).

1.1.1. Purification and Functional Reconstitution

In all preparations that have been studied, the sodium channel appears to be a glycoprotein with a large major chain of about 270,000 daltons. At present, it is not clear whether additional polypeptides that, in some tissues, copurify with the larger chain are necessary for channel function. For example, only the large peptide appears necessary for channel activity in the eel preparation; reconstitution into vesicles has allowed both tracer flux studies (Rosenberg *et al.*, 1984a) and electrical recordings (Rosenberg *et al.*, 1984b). The purified rat brain channel shows two smaller polypeptides of 37,000 and 39,000 daltons (Hartshorne and Catterall, 1984). Functional reconstitution of these components into vesicles (Tamkun *et al.*, 1984) and, more recently, planar bilayers (Hartshorne *et al.*, 1985) has been achieved. The major chain from the sarcolemmal channel shows a somewhat anomalous molecular weight (Barchi, 1983) with smaller subunit 38,000 (Kraner *et al.*, 1985). Veratridine and batrachotoxin activation of these channels has been demonstrated, and their selectivity has been studied by Tanaka *et al.* (1983). As techniques for handling the purified protein are refined, studies on reconstituted systems should enable different functions to be attributed to identified parts of the channel molecule.

1.1.2. Binding and Flux Studies Using Native Membranes

Demonstration that toxin binding and specific ion transport activity could be preserved in membrane vesicles isolated from living tissues was an important and direct prelude to our studies of channels incorporated into planar bilayers. Krueger *et al.* (1979) studied saxitoxin (STX) binding to synaptosomes, a membrane fraction (P3), and solubilized binding sites from rat brain (see also Weigele and Barchi, 1978; Catterall, 1984; Agnew, 1984). Fluxes stimulated by the alkaloid veratridine and by the steroidal activator batrachotoxin demonstrated the functional integrity of the sodium channels in the synaptosomes (Krueger *et al.*, 1980; Tamkun and Catterall, 1980). Such studies clearly identified these vesicle preparations as potential sources of sodium channels for incorporation experiments.

We chose the P3 membrane preparation for our initial attempts to incorporate functional sodium channels into planar bilayers. This fraction, obtained as the pellet resulting from a three-step centrifugation of a brain homogenate, provided a richer source of sodium channels (4–6 pmol STX bound/mg protein) than synaptosomes or whole-brain homogenate. Also, STX binding sites are almost 100% facing towards the outside of the vesicles, and there is minimal contamination with cellular organelles such as mitochondria. Freedom from mitochondrial membranes is important, since it avoids the complications posed by inadvertant incorporation of the large-conductance voltage-dependent anion channel (VDAC).

1.2. Looking at Single Molecules

Recording the electrical currents that pass through single channel molecules poses two problems. First, a preparation that contains one or a very few channels is needed. Second, the background electrical noise must be reduced to levels below the single-channel signal. These two requirements were first met when Bean *et al.* (1969) introduced very small amounts of the channel former EIM (excitability-inducing material) into a planar bilayer. Ehrenstein *et al.* (1970, 1974) demonstrated that the electrical excitability conferred on the bilayer by EIM could be entirely accounted for by the stochastic opening and closing of the individual EIM channels.

1.2.1. Patch Clamping

The high DC electrical resistance of a pure lipid bilayer keeps noise levels low enough to allow resolution of current fluctuations caused by single incorporated channels. In cell membranes, reduction of noise to the requisite low levels and recording from a very small number of channels are achieved by the single expedient of recording from a very small area or "patch" of membrane. Patch clamping, by pressing a polished pipette against a clean cell surface, allowed Neher and Sakmann (1976) to record from single acetylcholine receptor channels. A further advance, the attainment of spontaneous gigohm seals between cell membrane and pipette, enabled Sigworth and Neher (1980) to make the first recordings from single sodium channels. Isolation of a very small patch of membrane to increase the background resistance has the further benefit that the capacitance of the patch is very small. Thus, not only is the DC resistance high, but the AC impedance, at frequencies up to about 10 kHz, is high enough to allow single-channel signals to be resolved (see Table I). This ability to record very rapid fluctuations (open and closed times as short as a few tens of microseconds) with high resolution is essential if we are fully to understand fast channel gating and blocking processes. On the other hand, resolution of single channels at a lower bandwidth is sufficient for many studies of ion permeation and of the kinetics of tightly binding blockers such as STX.

1.2.2. Taming the Fluctuations

Regardless of the type of experimental preparation, tricks to slow down the unitary fluctuations can be useful. Even with the effective recording bandwidth offered by the patch clamp, good resolution of single channels can be difficult. Precise resolution requires fluctuations in current with reasonable dwell times at each current level. To resolve individual channel fluctuations of the anomalous rectifier channel, Ohmori et al. (1981) added a low concentration of barium to the external bathing solution. Barium, a blocker of the channel studied earlier by macroscopic current measurements, occluded the channels for periods of tens of milliseconds, allowing easy resolution of the small unitary current fluctuations. In our first recordings of single sodium-channel fluctuations from planar bilayers, a similar tactic was adopted, with STX being the blocker of choice (see below). Use of a potent, specific blocker to induce fluctuations has the additional advantage that it confirms the identity of the unitary events.

It is practical to look for blocker-induced fluctuations only if the channels would otherwise be open for relatively long periods. As we began our own studies, the results of Quandt and Narahashi (1982) indicated that, for BTX-treated sodium channels, resolution would be feasible in bilayers of relatively large area.

Table I. Impedance of Bilayers and Native Membranes

Preparation	Diameter (μm)	Specific resistance (Ω cm^2)	Specific capacitance (F cm^{-2})	Resistance (Ω)	Capacitative reactance at 1 kHz (Ω)
Patch	0.5	10^4	10^{-6}	5.1×10^{12}	8×10^{10}
Bilayer	50	10^8	10^{-6}	5.1×10^{12}	8×10^6
Bilayer	500	10^8	10^{-6}	5.1×10^{10}	8×10^4

2. ELECTROPHYSIOLOGY WITHOUT CELLS

Electrical recording from channels inserted into an artificial bilayer offers the obvious advantage, over studying the channels in the cell, that one can ask directly what components of the system are necessary for normal channel function and regulation. Composition of the membrane lipid and aqueous phases can be arbitrarily chosen, and only necessary components added. Conversely, studies in artificial systems are most meaningful when physiological properties of the channels are well known from experiments on the cell. Ideally, studies in the two systems should go forward in parallel.

2.1. Choosing a Membrane System

At the beginning of a channel incorporation/reconstitution study, a choice must be made of the technique for formation of the membranes and of the means of incorporation of the channels. If there is little experience in handling the particular channel protein in question, these choices are likely to require some trial and error before a workable procedure is found.

2.1.1. To Paint or to Fold?

For our own experiments, we chose to attempt incorporation of brain channels into painted decane–lipid membranes based on the successful experiments on a variety of channels in the laboratory of Christopher Miller (e.g., Miller, 1978; White and Miller, 1979; Labarca and Miller, 1981; and Miller, 1984). Since that time, folded membranes, formed by the apposition of two lipid monolayers (Montal and Mueller, 1972), have been used in a variety of channel studies (e.g., Nelson *et al.*, 1981; Schindler and Quast, 1981), though it seems generally to be more difficult to obtain incorporation by vesicle fusion in folded membranes (e.g., Bell and Miller, 1984; Moczydlowski *et al.*, 1984a). Incorporation into a preformed bilayer from vesicles avoids exposure of the protein at an air–water interface and is thus a gentler procedure, less likely to cause denaturation, than is the inclusion of the protein in the monolayers from which the bilayer is formed. This latter technique has been used in studies of the acetylcholine receptor channel (Nelson *et al.*, 1981; Schindler and Quast, 1981). It may well be that the sodium channel and other proteins of interest are not as resistant to this drastic treatment as is the acetylcholine receptor channel. However, folded membranes offer the advantages that they can be formed on very small holes, are virtually solvent-free, and can be made from monolayers of two different lipids. These features reward the effort required to develop methods for incorporation into these membranes.

2.1.2. Pipette or Partition?

Our membranes were painted on 0.1- to 0.5-mm holes in a plastic (polystyrene, Lexan, or Kel-F) partition separating two aqueous solutions. With these membranes, it is difficult to work with hole diameters below about 0.15 mm, and even at that size, the probability of incorporation seems to drop precipitously, out of proportion to the decrease in bilayer area. If smaller membranes are desired to increase bandwidth, patches of bilayer may be excised onto a glass pipette after incorporation of channels as introduced by Mueller (1975, Fig. 5), and described by Andersen (1983), at the expense of easy access to one of the aqueous solutions.

Folded membranes may be formed directly on either plastic partitions or pipettes and, having virtually no solvent annulus bounding them, may more easily be made on very small holes (see other chapters in this volume).

2.2. Installing and Seeing Channels: By Choice and Chance

When channels from a crudely fractionated native membrane vesicle preparation are incorporated, numerous channel types might be inserted into the bilayer. This rich variety may either be of great physiological interest or simply a nuisance. By a restrictive choice of the ionic species in solution, currents through channels selective for other ions may be reduced to negligible levels. For example, potassium-selective channels are not seen when sodium is the only monovalent cation species present. Then one hopes that the desired channel is incorporated before enough other proteins enter the membrane to make it too noisy and unstable for recording.

Incorporation of channels from our rat brain preparation does not seem to be particularly sensitive to conditions required for incorporation in other systems. Sodium channels were incorporated into neutral membranes containing phosphatidylethanolamine (PE) as the only lipid or into various mixtures of the negatively charged lipid phosphatidylserine (PS) with PE. Free Ca^{2+}, usually at a concentration of 100 μM, was included in the solutions. In general, we made no attempt to establish an ionic gradient across the vesicle membrane, since this was not necessary for incorporation. Orientation of incorporated sodium channels was consistent with incorporation by vesicle fusion, but we have no other direct evidence that this is the mechanism.

To emphasize that the membrane preparation is indeed heterogeneous, we give a brief description of three channel types that have been identified. Other potassium channels have been observed casually, and probably transmitter-activated channels are also present, though we have made no effort to study them to date.

2.2.1. Calcium-Activated K$^+$ Channels

A calcium-activated "maxi" K$^+$ channel can be observed by exposing bilayers bathed by KCl solutions to P3 vesicles (Krueger et al., 1982). Similar channels have been observed in a variety of other preparations (e.g., Marty, 1981; Pallotta et al., 1981; Latorre et al., 1982). They open only in the presence of free Ca^{2+}, show voltage-modulated gating, and have a conductance of more than 200 pS in 100 mM KCl (see Fig. 1). Generally, these channels incorporate with their intracellular (Ca^{2+} sensitive) ends facing the *cis* side. This is of particular interest since it implies that they are incorporated either by a different mechanism from the sodium channels or from a different population of vesicles. In contrast, sodium channels are inserted with their extracellular ends facing the *cis* solution.

2.2.2. Sodium Channels

Sodium channels may be studied following incorporation from P3 membrane vesicles into a bilayer bathed in NaCl solutions with BTX present (usually on the *trans* side). Adding vesicles to make the protein concentration about 1 µg/ml, with a hole size of 250 µm or less, gives a reasonable probability of incorporation of one or a very few sodium channels (see Fig. 2). The sodium channels consistently enter the bilayer with their STX binding sites facing the *cis* side and thus may all be blocked by addition of STX on that side. The *cis* side is thus identified as the extracellular side. Examples of current records from bilayers containing several sodium channels are shown in Fig. 3. The orientation of the channels is consistent with a fusion mechanism of incorporation from the vesicles, which have the STX sites facing outwards. Studies of STX binding to P3 vesicles have confirmed this orientation (Krueger et al., 1979).

2.2.3. Calcium Channels

Rat brain membranes have also proved a useful source of calcium channels for study in planar bilayers (Nelson et al., 1984). Even though unitary conductances are only about 2–8 pS, by using large driving forces and high divalent ion concentrations, single-channel current fluctuations were well resolved in 100–250 µm bilayers. For these studies, the membrane preparation was not the P3 fraction but rather was a synaptosomal preparation from the nerve-terminal-rich median eminence, although cal-

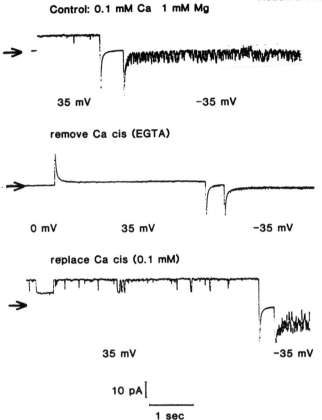

Figure 1. Calcium-dependent K$^+$ channels incorporated into a painted decane–lipid bilayer. Current records from a membrane first containing one channel (top) and subsequently two channels (bottom) are shown. The arrows indicate the level of current corresponding to no open channels. The concentrations of free (ionized) calcium and magnesium are indicated on the figure. Voltages (*cis* minus *trans*) correspond to the normal physiological convention (inside minus outside). Note that closing fluctuations occur more frequently at negative voltages because of the intrinsic voltage dependence of the channel's gating mechanism. Channels stay closed when free Ca^{2+} is removed from the *cis* side (inside) only by addition of EGTA.

cium channels have also been observed using P3 membranes. Judging by their voltage dependence, the orientation of the calcium channels was the same as that of the calcium-activated potassium channels and opposite to that of the sodium channels.

3. A CLOSER LOOK AT BATRACHOTOXIN-ACTIVATED SODIUM CHANNELS IN BILAYER MEMBRANES

Batrachotoxin-activated sodium channels in bilayers show a broad range of characteristics similar to those of sodium channels in cellular

Figure 2. Voltage-dependent gating of sodium channels. Current records at four different voltages, showing that closing fluctuations become more and more frequent as potential (inside minus outside) is made progressively more negative. The horizontal lines on the left indicate the current level when no channels are open. Only a single channel is present in the membrane at -70 and -80 mV and for the first record at -90 mV. A second channel inserted into the membrane near the beginning of the second trace at -90 mV, and two channels are present for the remainder of the recording at -90 and -100 mV. Records taken in the presence of 250 mM NaCl, 0 Ca^{2+} in both bathing solutions, and filtered (low-pass) at 200 Hz.

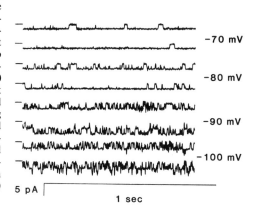

preparations. They are selectively permeable to sodium over other ion species; they show voltage-dependent gating; they are blocked by STX and TTX; they are blocked by external Ca^{2+}, with hyperpolarization enhancing the block; they have similar single-channel conductances under comparable conditions. In the following paragraphs we review some of these properties and present some new observations on the effect of ionic conditions on the single-channel conductance.

3.1. Selectivity

Krueger *et al.* (1983) showed by measurements of single-channel current reversal potentials that the channels were selective for sodium over potassium, cesium, and chloride. The channels were about 15-fold more permeable to sodium than to potassium, and there was no measurable permeability to cesium or chloride.

3.2. Voltage-Dependent Gating

Since BTX prevents inactivation of sodium channels, dependence of gating on voltage may be easily studied by recording single-channel current fluctuations at a series of maintained, different voltages (see Fig. 2; also Krueger *et al.*, 1983; French *et al.*, 1984; Moczydlowski *et al.*, 1984a; Hartshorne *et al.*, 1985). By fitting a Boltzmann function to the steady-state activation curve—a plot of fractional open time versus voltage—an apparent gating charge of about 4 e/channel can be determined. Both open and closed times are voltage dependent, open times increasing and closed times decreasing with depolarization. These studies are complicated by infrequent but abrupt changes in the probability of the channel being open;

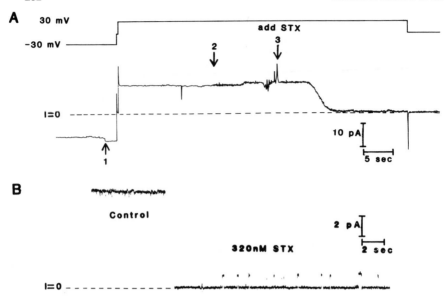

Figure 3. Saxitoxin (STX) block of sodium channels. The membranes were bathed in symmetrical 0.5 M NaCl solutions. A: A heavily filtered (2 Hz) record from a membrane containing more than 10 sodium channels. A steady current, indicating the presence of several open channels, is seen at either +30 or −30 mV. The jump in current at 1 probably indicates the insertion into the membrane of one or two additional channels. Stirring began at 2, Saxitoxin (300 nM) was added to the *cis* side only at 3, resulting in almost complete block of current flowing across the membrane. Block of the current by one-sided addition of the toxin demonstrates that all channels were oriented in the same direction in the membrane, with their extracellular ends facing the *cis* side. B: Segments of record (not continuous) from a membrane containing seven channels, taken at higher resolution. Here, E = −30mV and records were filtered at 20 Hz. Some downward fluctuations, representing individual channel closures, are seen in the control record, but all of the channels remain open almost all of the time. In the presence of 320 nM STX, all channels are blocked most of the time, but occasional upward fluctuations are visible. These represent the dissociation of an STX molecule from one of the channels, with the consequent unblocking of the channel. The rate constant for dissociation (0.05 sec^{-1}) of the STX from the channels was calculated from the times spent in the nonconducting (blocked) state, and that for association (1.1 × 10^{-7} sec^{-1}) from the times spent in the conducting (unblocked) state. These predict a dissociation constant of 4.5 nM, about the same value as was estimated from the steady-state block.

i.e., there are apparent nonstationarities in the gating (see also Weiss *et al.*, 1984).

3.3. Sensitivity to STX

As noted above, channels are blocked by application of STX on the *cis* side. Only after the breaking and reforming of a membrane in the presence of vesicles are channels likely to appear in the membrane with the opposite orientation. At small depolarizations, the apparent dissociation constant for STX block of the channels is 5–10 nM (Krueger *et al.*,

1983), very close to values seen in intact neurons. This also corresponds closely to values obtained in binding studies, indicating an intimate relationship between toxin binding and channel block. The fact that the toxin-induced current fluctuations are equal in magnitude to the spontaneous fluctuations caused by intrinsic channel gating was the first direct proof that the toxin produced an all-or-none block of the channels, as was long assumed on the basis of macroscopic measurements.

3.3.1. Voltage Dependence of STX and TTX Block

A surprising feature of the block by STX, noted at that time, was that it exhibited a striking voltage dependence, with depolarization reducing the potency of action of the toxin. The apparent dissociation constant changed by e-fold for about 35–40 mV (French *et al.*, 1984). This resulted from roughly symmetrical changes in the association rate constant and the dissociation rate constant with changes in voltage, as determined from a dwell-time analysis of the toxin-induced fluctuations. Depolarization slowed association and speeded dissociation.

The voltage dependence places a strong constraint on the possible molecular mechanism of toxin action, especially when considered alongside observations on TTX and other related toxins. Enhancement of block by hyperpolarization is an effect in the direction that would be expected if the cationic toxin—STX has a charge of +2 at physiological pH—were to penetrate part of the transmembrane voltage in reaching the blocking site. Because the two charges on the STX molecule are close together, it is unlikely that one charge would penetrate a substantial fraction of the transmembrane voltage without the other. With this in mind, we discover a paradox when we consider toxins with different net charge. TTX, which has a net charge of +1, has been studied in our laboratories as well as by Green *et al.* (1984), Hartshorne *et al.* (1984), and Moczydlowski *et al.* (1984a). Experiments by Moczydlowski *et al.* (1984b) have examined a variety of naturally occurring STX derivatives that have net charges of 0–2. All of these toxins act with identical dependence on voltage. The most reasonable conclusion at this time seems to be that some voltage-dependent conformational change within the channel protein is a necessary concomitant of binding and block (see comment by R. L. Barchi in discussion following French *et al.*, 1984; also Moczydlowski *et al.*, 1984a,b). Examination of the voltage dependence over a wider range and under varied conditions will allow this hypothesis to be tested further.

3.3.2. Toxin-Induced Fluctuations as an Aid to Permeation Studies

When a channel is activated by BTX or held open by some other drug like tetramethrin (Yamamoto *et al.*, 1984), nanomolar concentrations of STX and TTX cause long-lasting blocking events separated by long open periods. This allows accurate resolution of small single-channel currents even when one is recording from a relatively large bilayer and heavily filtering the records. We have used this approach to obtain the single-channel conductance down to low sodium concentrations (see Fig. 4), and it has also been used by Hartshorne *et al.* (1985) to identify and measure currents carried by potassium and rubidium through the sodium channel.

3.4. Conductance–Concentration Relationship

Single-channel conductance is a saturating function of sodium concentration as shown in Fig. 5 (see also Moczydlowski *et al.*, 1984a). We estimate a maximum conductance of 31 pS and an apparent dissociation constant of 37 mM for sodium with symmetrical solutions bathing neutral (PE) membranes. Lower maximal conductance, 21 pS, and apparent dissociation constant, 8 mM, were found for the sarcolemmal channel in PE/PC membranes by Moczydlowski and his co-workers. Saturating conductance–activity relationships are expected for a variety of models when the channel can accommodate only a limited number of ions. Much data on the sodium channel can be explained under the assumption of one ion, at most, being able to enter the channel at any instant (e.g., Hille, 1975). Perhaps the most surprising feature of these data is the relatively high affinity for sodium that is observed. Hille (1975) estimated an apparent dissociation constant of 368 mM for sodium in the frog node channel. The difference between this value and the one seen in the bilayer preparations probably results mostly from the different ionic conditions under which

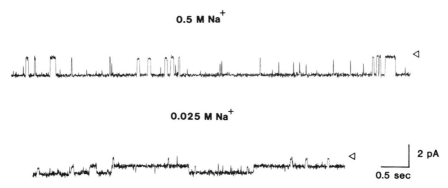

Figure 4. Current records obtained in the presence of symmetrical solutions containing 500 mM, and 25 mM Na$^+$ (<150 nM free divalent ions). Note the reduction in single-channel current at the lower Na$^+$ concentration. To easily resolve the small fluctuations seen with 25 mM Na$^+$, a low concentration of STX was added to induce slow blocking and unblocking fluctuations. Voltage, −60 mV; records filtered at 150 Hz; membrane lipid, PE.

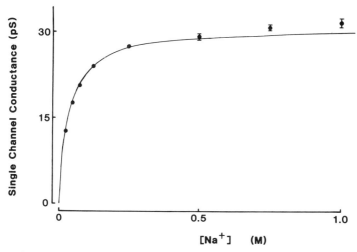

Figure 5. Single-channel conductance as a function of sodium concentration, obtained under the same conditions as the records in Fig. 4. Fitting a Michaelis–Menten type of saturation isotherm to these data required a concentration for half-maximal conductance of 37 mM and a maximum conductance of 31 pS. Here, the smooth curve is drawn from a four-barrier rate theory model describing ion permeation through the open channel.

the measurements were made, though some differences caused by the different species use of and by BTX modification are possible.

Another quantity of interest is the absolute value of the unit conductance. Even at 120–140 mM sodium, the levels in normal extracellular fluid, values we observe under symmetrical conditions are higher than those seen in such patch-clamp studies as those of Sigworth and Neher (1980) and Horn *et al.* (1984). These differences are again likely to be caused largely by the different ionic conditions and temperatures used. In particular, partial block of the channels by calcium can reduce the conductance and seems to be responsible for the low unit conductance of BTX-modified channels reported by Quandt and Narahashi (1982). Under physiological conditions, unit conductance is probably lowered somewhat by having high potassium at the intracellular surface rather than high sodium (Krueger *et al.*, 1983, Fig. 3). Our experiments with "physiological" monovalent ion gradients and 2 mM external Ca^{2+} give a slope conductance of 10 pS in the range 0 to -40mV (see Fig. 6 and Section 3.5) at 20–25°C. A Q_{10} of 1.3 can be used to estimate values at other temperatures. In the following paragraphs we discuss the effects of calcium in more detail.

3.5. Voltage-Dependent Block of the Channels by Calcium

Under stylized physiological conditions, calcium causes a voltage-dependent reduction of the single-channel currents (Fig. 6; see also Ya-

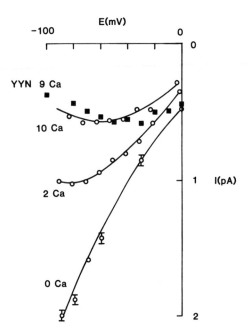

Figure 6. Voltage-dependent Ca^{2+} block of sodium channels in the presence of physiological monovalent ion gradients. Monovalent ion concentrations (mM), given as external//internal, were 125 Na^+, 5 K^+//5 Na^+, 125 K^+. Open symbols represent data obtained from sodium channels in bilayers. The closed symbols represent data taken from the patch-clamp experiments of Yamamoto et al. (1984) under similar, although not identical, ionic conditions. Concentrations of free Ca^{2+} ions (in mM) are shown on the figure. The lines were drawn by eye.

mamoto et al., 1984). This result is consistent with the macroscopic recordings of Woodhull (1973) and Taylor et al. (1976) and confirms the suggestion that channel blocking occurs independently of any effects on gating parameters. Taylor et al. suggested that the linear instantaneous I–E relationship seen for squid axon sodium channels resulted directly from the interference of calcium and magnesium with current flow through the channels. This is consistent with our own observations in which single-channel current in the absence of free calcium increases in a supralinear manner for hyperpolarizing voltages. Even at voltages near normal resting and threshold levels, 2 mM Ca^{2+} reduces currents by 10–30% from control values. Thus, one way in which interstitial calcium levels would modulate firing behavior of a nerve under physiological conditions is by a direct effect on the single-channel conductance.

The bilayer system is advantageous for studies of calcium action. Two points are significant. A true control measurement may be made since the bilayers are stable in the absence of free calcium. Many cell membranes, by contrast, become unstable if calcium is removed. Secondly, the surface charge on the membrane may be varied independently of the calcium concentration. This is an important variable because some calcium effects are thought to arise from its interactions with membrane surface charges.

3.6. Apparent Affinity of the Channels for Sodium in the Presence of Calcium

A dramatic illustration of the effect that competition between divalent ions and sodium can have on the apparent affinity of the channel for sodium can be obtained by examining the conductance–sodium concentration relationship in the presence of calcium. With symmetrical solutions, 10 mM Ca^{2+} increased the apparent dissociation constant for sodium by 3–40 times, depending on the voltage. Thus, for determinations of the parameters that describe the interaction between different ions and the channels, it is extremely important to be able to perform the measurements under conditions that are as simple and well controlled as possible.

4. LOOKING AHEAD

Incorporation and reconstitution are powerful new weapons in the arsenal of techniques for the study of channels. How can we use them to best advantage? One unique advantage lies in the ability to choose the lipid composition; a second is the potential to study channels after purification or modification. Still another advantage is access to membranes of organelles and tissues not approachable by other methods.

4.1. Further Studies on Sodium Channels

The use of fractionated membranes or purified channels offers the possibility of applying chemical modifiers either while the channels are being prepared or after they are inserted into a membrane. Charge-modifying reagents, studied in conjunction with changes in surface charge obtained by choice of different lipids, may enable us to build an electrostatic map of the functional channel and its surroundings, saying which processes—gating, conduction, and drug action, for example—are influenced by charges on protein and lipid membranes, respectively. Ultimately, modification or substitution of specific amino acids should help to illuminate the functional molecular architecture of the channel.

4.2. Wider Applications

The potential for study of poorly accessible membranes, has already been successfully exploited in studies of brain, sarcoplasmic reticulum, and T-tubular channels. Since incorporation from small tissue samples is possible, a new pathway of exploration into localized regions of the central nervous system is unfolding to challenge the ingenuity, and the patience, of adventurous investigators.

ACKNOWLEDGMENTS. The authors' work described in this paper was supported by NIH research grants NS 16285 and NS 20106, by contract DAMD17-82-C-2188 from the U. S. Army Medical Research and Development Command, and by the Bressler Research Fund of the University of Maryland School of Medicine. J. F. Worley was the recipient of a University of Maryland Graduate Fellowship.

REFERENCES

Agnew, W. S., 1984, Voltage-regulated sodium channel molecules, *Annu. Rev. Physiol.* **46**:517–530.

Andersen, O. S., 1983, Ion movement through gramicidin A channels. Single-channel measurements at very high potentials, *Biophys. J.* **41**:119–133.

Barchi, R. L., 1982, Biochemical Studies of the excitable membrane sodium channel, *Int. Rev. Neurobiol.* **23**:69–101.

Barchi, R. L., 1983, Protein components of the purified sodium channel from rat skeletal muscle sarcolemma, *J. Neurochem.* **40**:1377–1385.

Bean, R. C., Shepherd, W. C., Chan, H., and Eichner, J., 1969, Discrete conductance fluctuations in lipid bilayer protein membranes, *J. Gen. Physiol.* **53**:741–757.

Bell, J. E., and Miller, C., 1984, Effects of phospholipid surface charge on ion conduction in the K^+ channel of the sarcoplasmic reticulum, *Biophys. J.* **45**:279–287.

Catterall, W. A., 1984, The molecular basis of neuronal excitability, *Science* **223**:653–661.

Ehrenstein, G., Lecar, H., and Nossal, R., 1970, The nature of the negative resistance in bimolecular lipid membranes containing excitability-inducing material, *J. Gen. Physiol.* **55**:119–133.

Ehrenstein, G., Blumenthal, R., Latorre, R., and Lecar, H., 1974, Kinetics of opening and closing of individual excitability-inducing material channels in a lipid bilayer, *J. Gen. Physiol.* **63**:707–721.

French, R. J., Worley, J. F., III, and Krueger, B. K., 1984, Voltage-dependent block by saxitoxin of voltage-dependent sodium channels incorporated into planar lipid bilayers, *Biophys. J.* **45**:301–310.

Green, W. N., Weiss, L. B., and Andersen, O. S., 1984, Voltage and Na^+-dependent tetrodotoxin (TTX) block of batrachotoxin (BTX)-modified sodium channels, *Biophys. J.* **45**:68a.

Hartshorne, R. P., and Catterall, W. A., 1984, The sodium channel from rat brain. Purification and subunit composition, *J. Biol. Chem.* **259**:1667–1675.

Hartshorne, R. P., Keller, B. U., Talvenheimo, J. A., Catterall, W. A., and Montal, M.,

1985, Functional reconstitution of the purified brain sodium channel in planar lipid bilayers, *Proc. Natl. Acad. Sci. U.S.A.* **82**:240-244.

Hille, B., 1975, Ionic selectivity, saturation, and block in sodium channels. A four-barrier model, *J. Gen. Physiol.* **66**:535-560.

Hille, B., 1984, *Ionic Channels of Excitable Membranes*, Sinauer Associates, Sunderland, Massachusetts.

Hodgkin, A. L., and Huxley, A. F., 1952a, Currents carried by sodium and potassium ions through the membrane of the giant axon of *Loligo*, *J. Physiol. (Lond.)* **116**:449-472.

Hodgkin, A. L., and Huxley, A. F., 1952b, A quantitative description of membrane current and its application to conduction and excitation in nerve, *J. Physiol. (Lond.)* **117**:500-544.

Horn, R., Vandenberg, C. A., and Lange, K., 1984, Statistical analysis of single sodium channels. Effects of N-bromoacetamide, *Biophys. J.* **45**:323-335.

Kraner, S. D., Tanaka, J. C., and Barchi, R. L., 1985, Purification and functional reconstitution of the voltage-sensitive sodium channel from rabbit T-tubular membranes, *J. Biol. Chem.* **260**:6341-6347.

Krueger, B. K., and Blaustein, M. P., 1980, Sodium channels in presynaptic nerve terminals. Regulation by neurotoxins, *J. Gen. Physiol.* **76**:287-313.

Krueger, B. K., Ratzlaff, R. W., Strichartz, G. R., and Blaustein, M. P., 1979, Saxitoxin binding to synaptosomes, membranes, and solubilized binding sites from rat brain, *J. Membr. Biol.* **50**:287-310.

Krueger, B. K., French, R. J., Blaustein, M. B., and Worley, J. F., III, 1982, Incorporation of calcium-activated K^+-channels from rat brain into planar lipid bilayers, *Biophys. J.* **37**:170a.

Krueger, B. K., Worley, J. F., III, and French, R. J., 1983, Single sodium channels from rat brain incorporated into planar lipid bilayers, *Nature* **303**:172-175.

Labarca, P., and Miller, C., 1981, A K^+-selective three-state channel from fragmented sarcoplasmic reticulum of frog leg muscle, *J. Membr. Biol.* **61**:31-38.

Latorre, R., Vergara, C., and Hidalgo, C., 1982, Reconstitution in planar bilayers of a Ca^{2+}-dependent K^+ channel from transverse tubule membranes isolated from rabbit skeletal muscle, *Proc. Natl. Sci. U.S.A.* **79**:805-809.

Marty, A., 1981, Ca-dependent K channels with large unitary conductance in chromaffin cell membranes, *Nature* **291**:497-500.

Miller, C., 1978, Voltage-gated cation channel from fragmented sarcoplasmic reticulum: Steady-state electrical properties, *J. Membr. Biol.* **40**:1-23.

Miller, C., 1984, Integral membrane channels: Studies in model membranes, *Physiol. Rev.* **63**:1209-1242.

Moczydlowski, E., Garber, S., and Miller, C., 1984a, Batrachotoxin-activated Na^+ channels in planar lipid bilayers; competition of tetrodotoxin block by Na^+, *J. Gen. Physiol.* **84**:665-686.

Moczydlowski, E., Hall, S., Garber, S., Strichartz, G. R., and Miller, C., 1984b, Voltage dependent blockade of muscle Na^+ channels by guanidinium toxins: Effect of toxin charge, *J. Gen. Physiol.*, **84**:687-704.

Montal, M., and Mueller, P., 1972, Formation of bimolecular membranes from lipid monolayers and a study of their electrical properties, *Proc. Natl. Acad. Sci. U.S.A.* **69**:3561-3566.

Mueller, P., 1975, Membrane excitation through voltage-induced aggregation of channel precursors, *Ann. N.Y. Acad. Sci.* **264**:247-264.

Neher, E., and Sakmann, B., 1976, Single-channel currents recorded from the membrane of frog muscle fibers, *Nature* **260**:799-802.

Nelson, M. T., French, R. J., and Krueger, B. K., 1984, Voltage-dependent calcium channels from brain incorporated into planar lipid bilayers, *Nature* **308**:77-80.

Nelson, N., Anholt, R., Lindstrom, J., and Montal, M., 1981, Reconstitution of purified acetylcholine receptors with functional ion channels in planar lipid bilayers, *Proc. Natl. Acad. Sci. U.S.A.* **77**:3057–3061.

Pallotta, B. S., Magleby, K. L., and Barrett, J. N., 1981, Single channel recordings of Ca^{2+}-activated K^+ currents in rat muscle cell culture, *Nature* **293**:471–474.

Ohmori, H., Yoshida, S., and Hagiwara, S., 1981, Single K^+ channel currents of anomalous rectification in cultured rat myotubes, *Proc. Natl. Acad. Sci. U.S.A.* **78**:4960–4964.

Quandt, F. N., and Narahashi, T., 1982, Modification of single Na^+ channels by batrachotoxin, *Proc. Natl. Acad. Sci. U.S.A.* **79**:6732–6736.

Rosenberg, R. L., Tomiko, S., and Agnew, W. S., 1984a, Reconstitution of neurotoxin-modulated transport by the voltage-regulated sodium channel from the electroplax of *Electrophorus electricus*, *Proc. Natl. Acad. Sci. U.S.A.* **81**:1239–1243.

Rosenberg, R. L., Tomiko, S., and Agnew, W. S., 1984b, Single channel properties of voltage-regulated Na channel isolated from the electroplax of *Electrophorus electricus*, *Proc. Natl. Acad. Sci. U.S.A.* **81**:5594–5598.

Schindler, H., and Quast, U., 1981, Functional acetylcholine receptor from *Torpedo marmorata* in planar membranes, *Proc. Natl. Acad. Sci. U.S.A.* **77**:3052–3056.

Sigworth, F. J., and Neher, E., 1980, Single sodium channel currents observed in cultured rat muscle cells, *Nature* **287**:447–449.

Tamkun, M. M., and Catterall, W. A., 1980, Ion flux studies of voltage-sensitive sodium channels in synaptic nerve-ending particles, *Mol. Pharmacol.* **19**:78–86.

Tamkun, M. M., Talvenheimo, J. A., and Catterall, W. A., 1984, The sodium channel from rat brain. Reconstitution of neurotoxin-activated ion flux and scorpion toxin binding from purified components, *J. Biol. Chem.* **259**:1676–1688.

Tanaka, J. C., Eccleston, J. F., and Barchi, R. L., 1983, Cation selectivity characteristics of the reconstituted voltage-dependent sodium channel from rat skeletal muscle sarcolemma, *J. Biol. Chem.* **258**:7519–7526.

Taylor, R. E., Armstrong, C. M., and Bezanilla, F., 1976, Block of sodium channels by external calcium ions, *Biophys. J.* **16**:27a.

Weigele, J. B., and Barchi, R. L., 1978, Analysis of saxitoxin binding in isolated rat synaptosomes using a rapid filtration assay, *FEBS Lett.* **91**:310–314.

Weiss, L. B., Green, W. N., and Andersen, O. S., 1984, Single channel studies on the gating of batrachotoxin (BTX)-modified sodium channels in lipid bilayers, *Biophys. J.* **45**:67a.

White, M. M., and Miller, C., 1979, A voltage-gated anion channel from the electric organ of *Torpedo californica*, *J. Biol. Chem.* **254**:10161–10166.

Woodhull, A. M., 1973, Ionic block of sodium channels in nerve, *J. Gen. Physiol.* **61**:687–708.

Yamamoto, D., Yeh, J. Z., and Narahashi, T., 1984, Voltage-dependent calcium block of normal and tetramethrin-modified sodium channels, *Biophys. J.* **45**:337–344.

Chapter 18

Ionic Channels in the Plasma Membrane of Sea Urchin Sperm

A. Darszon, J. García-Soto, A. Liévano, J. A. Sánchez, and A. D. Islas-Trejo

1. INTRODUCTION

The exchange of information between a cell and its environment is frequently mediated by ionic movements through the plasma membrane. The physiology of sea urchin sperm is a representative case in which this phenomenon occurs. For instance, when these cells are spawned into sea water, their motility and respiration activate (Ohtake, 1976; Nishioka and Cross, 1978; Christen et al., 1982). This activation is mostly dependent on the transport of Na^+ and H^+ through the plasma membrane (Nishioka and Cross, 1978). The sperm acrosome reaction, a prerequisite for egg fertilization, is also modulated by ionic fluxes (Schakmann et al., 1978). This reaction in sea urchin sperm consists of the exocytosis of the acrosomal vesicle located at the anterior region of the sperm head (Dan, 1952; Summers et al., 1975) and leads to the exposure of a protein required for sperm–egg binding (Vacquier and Moy, 1977; Glabe and Lennarz, 1979) and of lytic enzymes that digest the coat of the egg (Levine and

---

A. DARSZON, J. GARCÍA-SOTO, A. LIÉVANO, AND A. D. ISLAS-TREJO • Departmento de Bioquímica, Centro de Investigación y de Estudios Avanzados del I. P. N., México, D.F. 07000, México.   J. A. SÁNCHEZ • Departmentos de Bioquímica y Farmacología, Centro de Investigación y de Estudios Avanzados del I.P.N., México, D.F. 07000, México.

Walsh, 1979; Green and Summers, 1980). Also during the process, the intracellular cyclic nucleotides increase (Garbers and Kopf, 1980), and an actin-containing acrosomal tubule is formed (Tilney et al., 1973).

In sea urchin sperm, the acrosome reaction is triggered by a high-molecular-weight component of the jelly coat that surrounds the egg (SeGall and Lennarz, 1979; Kopf and Garbers, 1980). This reaction requires Ca^{2+} and Na^+ in sea water at pH 8 (Dan, 1954; Collins and Epel, 1977; Schackmann and Shapiro, 1981) and is accompanied by an increase in the intracellular levels of Ca^{2+} and Na^+ and an efflux of H^+ and K^+ (Schackmann and Shapiro, 1981). These ionic movements and the acrosome reaction are inhibited by Ca^{2+} channel blockers such as verapamil and D600 (Schackmann et al., 1978). Ionophores such as A23187 and nigericin induce the acrosome reaction in the absence of jelly (Decker et al., 1976; Collins and Epel, 1977; Schackmann et al., 1978). Extracellular pH also modulates the acrosome reaction. Sperm suspended in sea water at pH 9 suffer the acrosome reaction in the absence of jelly, whereas in sea water at pH below 8, jelly is unable to trigger the acrosome reaction and the accompanying ionic fluxes (Collins and Epel, 1977; Schackmann et al., 1978).

Analysis of the partitioning of ions, weak acids, and bases in sperm appears to indicate a depolarization of the membrane potential and an internal pH increase during the acrosome reaction (Schackmann et al., 1981; Lee et al., 1983). Apparently sperm have a negative resting potential that is sensitive to external K^+. Raising the external K^+ concentration or adding tetraethylammonium (TEA) ions inhibits the acrosome reaction (Schackmann et al., 1981). These results suggest that the membrane potential may participate in this phenomenon and that the reaction requires, in addition to an increase in the intracellular Ca^{2+} activity, a Na^+-dependent increase in intracellular pH (reviewed by Epel, 1978).

It is clear that ionic fluxes through the plasma membrane modulate the behavior of sea urchin sperm and play an important role in the egg-jelly-induced acrosome reaction. However, the transport mechanisms involved and the sequence and relationship of the ionic fluxes that occur during the reaction have not been established. The participation of channels in these ion movements remains to be determined, but they may be deeply implicated as in other excitable cells. Nonetheless, direct electrophysiological evidence has been precluded by the small size of sperm (1–2 μm). We have therefore combined ion-flux studies in whole sperm with a reconstitution approach using isolated sperm plasma membranes in an attempt to characterize at the molecular level the ionic transport mechanisms that regulate the behavior of sea urchin sperm and participate in the jelly-induced acrosome reaction.

2. ARE THERE CHANNELS IN SEA URCHIN SPERM?

The inhibition of the sperm acrosome reaction and Ca^{2+} uptake caused by D600 has suggested the involvement of Ca^{2+} channels in this process (Schackmann *et al.*, 1978). This blocker is now considered poorly specific; for instance, it also blocks slow Na^+ channels in cardiac tissue (Kohlhardt and Fleckenstein, 1977). Since Na^+ uptake is involved in the acrosome reaction, we decided to test nisoldipine, a 1,4-dihydropyridine derivative belonging to the most powerful class of Ca^{2+} antagonists (Garcia-Soto and Darszon, 1984). All these drugs act with high affinity and specificity on voltage-controlled and/or receptor-operated Ca^{2+} channels of different tissues (Gould *et al.*, 1982; Ferry and Glossmann, 1982; Tan and Tashjian, 1984). As shown in Fig. 1, nisoldipine inhibits both the egg-jelly-induced Ca^{2+} uptake (80%) and the acrosome reaction (>75%) in *Strongylocentrotus purpuratus* sperm. The dihydropyridine derivatives are frequently used at nanomolar concentrations, although higher concentrations are required in some cases (Kohlhardt and Fleckenstein, 1977). The lower affinity exhibited by these blockers in some tissues indicates the possible existence of various types of Ca^{2+} channels (Plotek and Atlas, 1983). In sea urchin sperm, micromolar concentrations of nisoldipine are required to block Ca^{2+} entry, suggesting the presence of Ca^{2+} channels of the

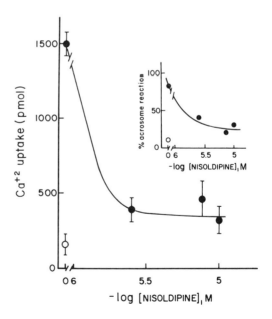

Figure 1. Effect of nisoldipine on the jelly-induced Ca^{2+} uptake by *S. purpuratus* sperm. Prior to the jelly addition, the sperm suspension (1% v/v in complete sea water, pH 8) was incubated with the drug for 1 min. Uptake was determined 2 min after the induction as described in Table I. Points represent the mean ± S.D. of three determinations. The insert shows the percent of acrosome reaction at each drug concentration tested for Ca^{2+} uptake. In the control (○), jelly and nisoldipine were not added. Similar results were obtained in three other sperm batches.

lower-affinity type; however, further studies are needed concerning the effect of this drug on sperm physiology.

The acrosome reaction can be triggered by pH 9 sea water in the absence of jelly (Collins and Epel, 1977). At this pH, the sperm plasma membrane permeability to Ca^{2+} increases (García-Soto and Darszon, 1984). As with egg jelly at pH 8, the high-pH-induced acrosome reaction requires the presence and uptake of Ca^{2+}. Table I shows that D600 and nisoldipine inhibit the high-pH-induced acrosome reaction and Ca^{2+} uptake. Thus, it is possible that the same transport system for Ca^{2+} uptake operates at both acrosome-reaction-triggering conditions, i.e., egg jelly or pH 9. These results suggest in addition that the Ca^{2+} channels in the sea urchin sperm plasma membrane might be regulated by the intracellular pH (García-Soto and Darszon, 1985).

Tetraethylammonium (TEA), a well-known blocker of K^+ channels (Stanfield, 1983), inhibits the sperm acrosome reaction (Schackmann et al., 1978). We have preliminary results that show that 4AP, another K^+ channel blocker (Ulbrich et al., 1982), inhibits the reaction. Both the acrosome reaction and the membrane potential are sensitive to external K^+ (Schackmann et al., 1978, 1981). These data taken together indicate the possible presence of K^+ channels in the plasma membrane of the sea urchin sperm and their participation in the acrosome reaction.

3. RECONSTITUTION STUDIES WITH ISOLATED SEA URCHIN SPERM PLASMA MEMBRANE

Two reasons have motivated us to isolate sperm plasma membranes and attempt to identify, characterize, and ultimately purify the membrane components responsible for regulating their permeability. (1) The sperm is a very tiny cell, and because of this it has been impossible to characterize its plasma membrane permeability properties using microelectrodes. (2) Permeability changes to several ions are involved in the acrosome reaction; thus, an attractive working hypothesis is that the triggering stimulus in the reaction is a jelly-induced permeability change to one or more of these ions. Having a sperm plasma membrane preparation has allowed us to use the reconstitution approach, which provides a variety of powerful tools to investigate the molecular basis of membrane function (Darszon, 1983).

We have isolated plasma membranes from sperm of two sea urchin species, *Strongylocentrotus purpuratus* and *Lytechinus pictus,* according to the method of Cross (1983). In brief, the method involves the separation of the sperm flagella by centrifugation after breaking them off from sperm by passing a sperm suspension through a hypodermic needle. Thereafter,

Table I. Inhibition of Acrosome Reaction and Ca^{2+} Uptake by D600 and Nisoldipine in Sperm Suspended in Complete Sea Water[a]

pH	Induction conditions	Acrosome reaction (%)		Ca^{2+} uptake (pmol/10^7 sperm)[b]		Acrosome reaction (%)		Ca^{2+} uptake (pmol/10^7 sperm)	
		−D600	+D600[c]	−D600	+D600[c]	−Nisoldipine	+Nisoldipine[f]	−Nisoldipine	+Nisoldipine
8	Control[d]	10	14	412 ± 88	479 ± 12	12	4	581 ± 78	812 ± 160
8	+Egg Jelly[e]	82	17	1227 ± 82	664 ± 80	76	34	1976 ± 306	1117 ± 246
9	−Egg Jelly	85	13	2019 ± 105	507 ± 148	70	18	2070 ± 150	1189 ± 174

[a] Sperm were collected and washed as in Darszon et al. (1984a). Jelly was prepared as in the same reference. For the Ca^{2+} uptake experiments, sperm were added to 300 μl (1.7 × 10^8 sperm/ml, final concentration) of sea water (SW = 486 mM NaCl, 10 mM KCl, 10 mM $CaCl_2$, 56 mM $MgCl_2$, and 2.4 mM $NaHCO_3$) at the indicated pH with or without jelly. Two minutes after sperm addition, three aliquots of 60 μl were removed from the incubation medium and filtered through GF/C filters (prewashed with the same medium). Each filter was washed under vacuum three times with 5 ml of ice-cold SW. Thereafter, the radioactivity of the filters was determined in 5 ml of Aquasol.® Blanks (medium + $^{45}Ca^{2+}$ without sperm) were subtracted from the uptake values. The acrosome reaction was determined in an aliquot from the same sample in which Ca^{2+} uptake was measured, observing sperm with a phase-contrast microscope as in Collins and Epel (1977).
[b] Uptake values represent the mean ± S.D. of three determinations.
[c] Sperm were exposed to 90 μM D600 for 1 min before the induction of acrosome reaction with jelly or high pH.
[d] No jelly addition.
[e] 5.4 μg of fucose equivalents of egg jelly in SW was added.
[f] Concentration of nisoldipine was 5 μM.

the flagella are subjected to a hypotonic shock, and the flagellar membranes are separated from the axoneme by floating the suspension on top of a sucrose solution. Two fractions are obtained, one that floats in the solution on top of the sucrose layer (top membranes) and one at the sucrose layer interphase (midmembranes) (Darszon et al., 1984a). There is evidence from ^{125}I surface labeling that these fractions are representative of the sperm plasma membrane as a whole (Cross, 1983).

We have shown that vesicles containing sperm membrane fractions respond to jelly with increases in Ca^{2+} and Na^+ uptake, whereas pure lipid vesicles do not. The Ca^{2+} uptake is increased in a species-specific manner; that is, *Arbacia punctulata* egg jelly, which does not induce the acrosome reaction in *S. purpuratus* sperm (SeGall and Lennarz, 1979, 1981), failed to induce significant Ca^{2+} uptake into vesicles made with *S. purpuratus* sperm membranes. Thus, the vesicles containing sperm membranes respond in a similar way to egg jelly as intact sperm in three respects, showing increased Ca^{2+} and Na^+ uptake and a species-specific response (Darszon et al., 1984a).

To further characterize and understand the ionic transport mechanisms present in the sea urchin sperm plasma membrane, we have formed planar bilayers at the tip of a patch-clamp pipette from monolayers derived from a mixture of isolated sea urchin plasma membranes and lipid vesicles (Darszon et al., 1984b). Monolayers spontaneously form at the air–water interface of vesicle suspensions; they are generated when either crude membranes mixed with liposomes or proteoliposomes formed with a purified protein interact with the air–water interface (Schindler and Rosenbush, 1978; Schindler, 1979). The lipid bilayers are formed at the tip of the patch-clamp pipette by successive removal of the pipette from the suspension and reimmersion through the monolayer-containing interface (Wilmsen et al., 1982; Hanke et al., 1983; Coronado and Latorre, 1983; Suárez-Isla et al., 1983). The membrane currents are recorded using patch-clamp instrumentation (Hamil et al., 1981). This strategy has allowed the functional reassembly of membrane proteins with the additional advantage over conventional planar bilayer formation of permitting measurements with faster time resolution and higher signal-to-noise ratio.

High-resistance membranes (1–30 GΩ) could be formed in about 40% of the attempts using *Strongylocentrotus purpuratus* sperm plasma membranes and soybean phospholipids, and from those around 25% showed channel activity. Bilayers formed of only lipid, in the absence of protein, did not display channel activity, although similar resistances were obtained. Gramicidin was added to some of these lipid bilayers, and after a short time, channels were observed, indicating the presence of a membrane at the tip of the microelectrode (Liévano et al., 1985).

The single-channel activity of a bilayer formed from sperm plasma membranes in symmetrical artificial sea water is illustrated in Fig. 2. Each deflection corresponds to an inward movement of thousands of ions through various types of ionic channels, as can be presumed from the different conductance levels. It was quite common to observe more than one type of channel in the same bilayer. We proceeded in the characterization of the ionic channels by using simpler solutions.

The results obtained in symmetrical 0.1 M KCl solutions are shown in Fig. 3. The original current records are displayed in part A of this figure. Current–voltage relationships of the data were created using only single-channel events lasting several digital sample intervals; some events were too short to be measured. The linear I–V relationships had conductances of 22, 46, and 86 pS (Fig. 3B). The permeant ions were identified in experiments in which the bilayers were subjected to a KCl concentration gradient; the results are illustrated in Fig. 4. Part A of this figure shows data from a bilayer that displayed two types of channels with linear I–V conductances as described in Fig. 3. Their zero-current potential was close to E_K, the K^+ equilibrium potential, indicating that K^+ was the main permeant ion. Figure 4B shows results from another bilayer in which three types of channels appeared, two of them with reversal potentials close to E_K and one close to E_{Cl}, the chloride equilibrium potential. In our experiments, chloride channels were only occasionally observed, whereas K^+ channel activity was recorded whenever single-channel events were observed.

The usual presence of several types of ionic channels in the same recording has complicated the kinetic analysis so far. We are now in the process of solving this problem with pharmacological agents that will also allow us to explore other features of the channels.

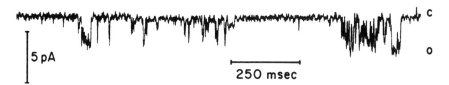

Figure 2. Single-channel events in a bilayer derived from sperm plasma membranes formed at the tip of a patch pipette. Monolayers were generated after 3 min at the air–water interface of a suspension (1 ml) of soybean phospholipids (1 mg/ml) and sperm plasma membranes (12.5 µg protein). Bilayers with high resistance (gigaohms) were formed as described in the text using pipettes pulled from VWR micropipettes or from Pyrex® glass. The holding potential (E_h) was 62 mV. The pipette contained 496 mM NaCl, 10 mM KCl, 2.4 mM NaHCO$_3$, pH 8.0, and the bath the same plus 5 mM CaCl$_2$. T = 20–22°C. Currents were low-pass filtered at 1 kHz (Par, Princeton New Jersey), stored on magnetic tape, and sampled every 200 µsec with the microcomputer for analysis and display.

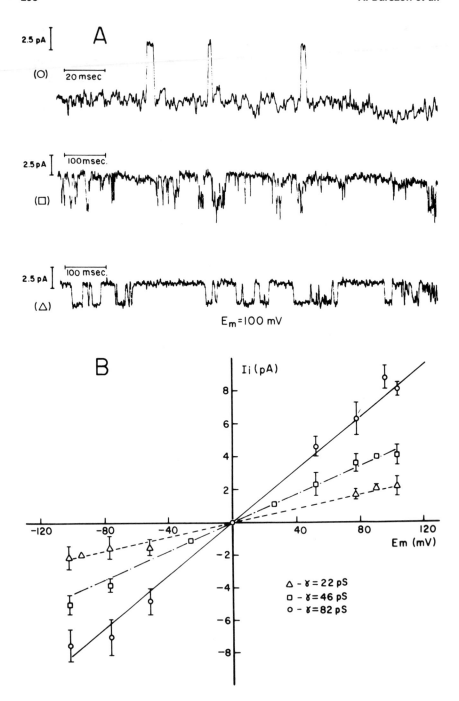

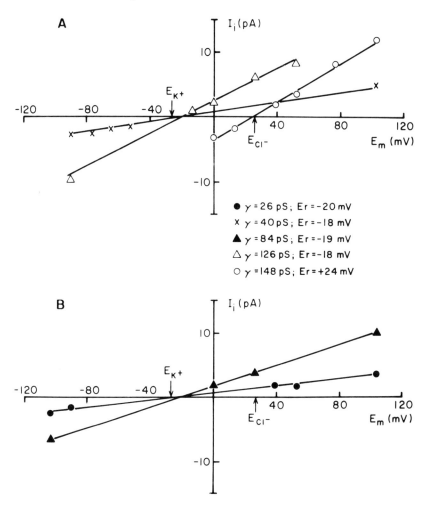

Figure 4. Current–voltage relationships in two different bilayers (A and B) in asymmetric KCl solutions. Bath contained 300 mM KCl, 1 mM $CaCl_2$, and 5 mM HEPES buffer, pH 8.0. Pipette solution was the same except KCl was 100 mM. $T = 20$–$22°C$.

Figure 3. A: Single-channel events in three different bilayers at 100 mV. Bilayers were formed as in Fig. 2. Currents were low-pass filtered at 1 kHz with a four-pole Bessel filter (Frequency Devices, Haverhill, MA). The bath and pipette solutions both contained 100 mM KCl, 74.4 μM $CaCl_2$, 80 μM EGTA (1 μM free Ca^{2+}), and 5 mH HEPES buffer, pH = 8.0. T = 20–22°C. B: Current–voltage relationships from experiments as in A. Each point is the average ± S.D. of data from up to seven different bilayers. The lines are the least-square fits to the data points giving single-channel conductance as indicated and a 0 mV reversal potential. $E_h = 0$ mV, and pulses lasting several seconds were applied in the range -110 mV to $+110$ mV.

4. CHANNELS IN THE PLASMA MEMBRANE OF SEA URCHIN SPERM: IMPLICATIONS FOR THE ACROSOME REACTION

The ionic requirements of the acrosome reaction mentioned previously indicate that the egg jelly regulates the sperm plasma membrane permeability. This point is further stressed by the inhibitory effects produced by blockers of ionic channels such as D600, nisoldipine, and TEA on this reaction. Thus, it is possible that as a result of egg jelly–sperm interaction, Ca^{2+} channels open, depolarizing the membrane. Subsequent activation of K^+ channels might conceivably be responsible for the K^+ efflux that accompanies the acrosome reaction. These K^+ channels could be gated either by voltage or by an increase in intracellular Ca^{2+}.

Until now, we have identified channels of relatively large unitary conductance. On the basis of that parameter alone, it is unlikely that the sperm channels we have observed are related to the delayed rectifier K^+ channels of classical preparations such as skeletal muscle and nerve (Schwarz et al., 1981) or the inward rectifier K^+ channels (Latorre and Miller, 1983), since they have smaller conductances. Other K^+ channels such as the Ca^{2+}-activated K^+ channel of several preparations and the sarcoplasmic reticulum K^+ channel (Latorre and Miller, 1983) do have conductances in the range that we have observed in our preparations.

Less frequently, we have also observed chloride channels with a large unitary conductance. Channels of this type are not uncommon either; they have been recorded, for instance, in rat myotubes with the patch-clamp technique (Blatz and Magleby, 1983). The identification of other ionic channels that we have observed in sea water solution remains to be settled.

5. ARE THERE RECEPTORS TO THE EGG JELLY IN THE SEA URCHIN SPERM PLASMA MEMBRANES?

We have found that sea urchin egg jelly fails to induce increases in Ca^{2+} and Na^+ uptake in pure lipid vesicles. Egg jelly was also unable to induce increases in electrogenic permeability to sea water ions in pure lipid planar bilayers, including those formed with a sperm lipid extract. In contrast, jelly induced Ca^{2+} uptake into vesicles containing sperm membranes in a species-specific manner (Darszon et al., 1984a). These results, together with the observations of species-specific egg jelly binding by sperm (SeGall and Lennarz, 1979) and induction of the acrosome reaction among some sea urchin species (SeGall and Lennarz, 1979) and the inhibition of the acrosome reaction by an antibody prepared against one of the main protein components of the sperm plasma membrane (Lopo and Vacquier, 1980), suggest that there may be a protein receptor for egg jelly in the sperm membrane (Darszon et al., 1984a).

We are attempting to isolate the putative sperm jelly receptor using affinity chromatography. The factor in sea urchin egg jelly responsible for inducing the sperm acrosome reaction was purified according to Kopf and Garbers (1980) and bound to CnBr-activated Sepharose. Triton extracts from sperm that had the external surface of their plasma membrane labeled with ^{125}I plus iodogen were added to the column in sea water. After 12

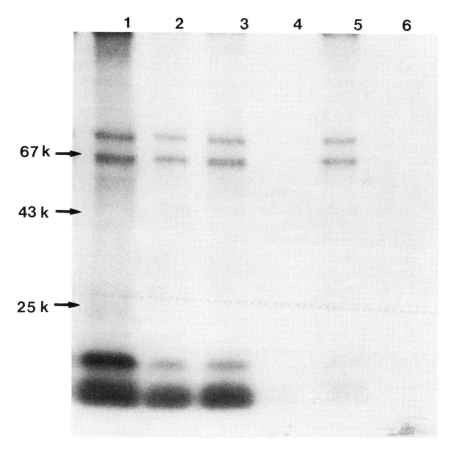

Figure 5. Electrophoretic distribution of ^{125}I-labeled sperm proteins on SDS/polyacrylamide gels (10%) that are retained by the affinity column. Triton X-100 (1% in SW) extracts of ^{125}I-labeled *Lytechinus pictus* sperm (as described in the text) were incubated in the affinity column (Sepharose-purified egg jelly factor, which induces acrosome reaction) for 12 hr. Thereafter the column was eluted with SW containing increasing concentrations of NaCl. The gels in the autoradiograph are as follows: (1) labeled sperm, (2) Triton extract of labeled sperm, (3) excess material that does not bind to the column, (4) elution with 1 M NaCl, (5) elution with 2 M NaCl, (6) elution with 3 M NaCl.

hr of incubation, the column was eluted with sea water that contained increasing NaCl concentrations. The fraction eluted with 2 M NaCl was enriched for three labeled proteins of 210, 80, and 60 kD (see Fig. 5). The molecular weights of these polypeptides coincide with those of the main sperm plasma membrane components that label with ^{125}I (Cross, 1983). Antibodies against two of those polypeptides inhibit the jelly-induced acrosome reaction. Thus, it is possible that the egg jelly receptor is composed of one or more of the polypeptides that bind to the affinity column.

6. PERSPECTIVES

The sperm plasma membrane fractions we have been using come preferentially from the flagella. Although there is evidence from surface ^{125}I labeling that these fractions are representative of the sperm plasma membrane as a whole (Cross, 1981), the acrosome reaction occurs in the sperm head. Thus, it is important to study the permeability properties of head plasma membrane and determine how they respond to jelly. We have isolated a crude fraction of sperm head membranes using polycationic beads (Jacobson, 1977). Vesicles formed by cosonicating sperm head membranes and soybean phospholipids respond to egg jelly in a species-specific manner with an increase in Ca^{2+} uptake (García-Soto et al., 1984). We are now in a position to compare the membranes isolated from the sperm head and flagella with respect to their composition, permeability properties, and response to egg jelly.

The identification and isolation of the components involved in the egg jelly response that are present in the membranes we have isolated require solubilization conditions that will preserve function (response to jelly). We have been able to show that vesicles reassembled from lipids and an octyl glucoside extract of flagellar sperm membranes are capable of responding to jelly with increased Ca^{2+} uptake (Darszon et al., 1984a). This indicates that we have the initial conditions to attempt the isolation and characterization of the sea urchin sperm plasma membrane components implicated in the egg-jelly-induced permeability changes. Now individual fractions can be tested in both planar and spherical bilayers.

ACKNOWLEDGMENTS. We are very grateful to Lucía de De la Torre and Irma Vargas for their technical support and to Agustín Guerrero for his comments. This work was supported by grants from NSF INT-80-03759, Fonda Ricardo Zevada, CONACYT and the Organization of American States.

REFERENCES

Blatz, A. L., and Magleby, K. L., 1983, Single voltage-dependent chloride-selective channels of large conductance in cultured rat muscle, *Biophys. J.* **43**:237–240.

Christen, R., Schackmann, R. W., and Shapiro, B. M., 1982, Elevation of intracellular pH activates respiration and motility of sperm of the sea urchin, *Strongylocentrotus purpuratus, J. Biol. Chem.* **257**:14881–14890.

Collins, F., and Epel, D., 1977, The role of calcium ions in the acrosome reaction of sea urchin sperm, *Exp. Cell Res.* **106**:211–222.

Coronado, R., and Latorre, R., 1983, Phospholipid bilayers made from monolayers on patch-clamp pipettes, *Biophys. J.* **43**:231–236.

Cross, N., 1983, Isolation and electrophoretic characterization of the plasma membrane of sea-urchin sperm, *J. Cell Sci.* **59**:13–25.

Dan, J., 1952, Studies on the acrosome. I. Reaction to egg water and other stimuli, *Biol. Bull.* **103**:54–66.

Dan, J., 1954, Studies on the acrosome. II. Effect of calcium deficiency, *Biol. Bull.* **107**:335–349.

Darszon, A., 1983, Strategies in the reassembly of membrane proteins into lipid bilayer systems and their functional assay, *J. Bioenerg. Biomembr.* **15**:321–334.

Darszon, A., Gould, M., de De la Torre, L., and Vargas, I., 1984a, Response of isolated sperm plasma membranes from sea urchins to egg jelly, *Eur. J. Biochem.* **144**:515–522.

Darszon, A., Liévano, A., and Sánchez, J., 1984b, Single channel recording from bilayer derived from isolated sea urchin plasma membranes, *Biophys. J.* **45**:308a.

Decker, G. L., Joseph, D. B., and Lennarz, W. J., 1976, A study of factors involved in induction of the acrosome reaction in sperm of the sea urchin *Arbacia punctulata, Dev. Biol.* **53**:115–125.

Epel, D., 1978, Mechanisms of activation of sperm and egg during fertilization of sea urchin gametes, *Curr. Top. Dev. Biol.* **12**:185–246.

Ferry, D. R., and Glossmann, H., 1982, Identification of putative calcium channels in skeletal muscle microsomes, *FEBS Lett.* **148**:331–337.

Garbers, D. L., and Kopf, G. S., 1980, The regulation of spermatozoa by calcium and cyclic nucleotides, *Adv. Cyclic Nucleotide Res.* **13**:251–306.

García-Soto, J., and Darszon, A., 1984, Effect of Ca^{2+}- antagonists on the Ca^{2+} uptake associated with the acrosome reaction of sea urchin sperm, *Biophys. J.* **45**:40a.

García-Soto, J., and Darszon, A., 1985, High pH-induced acrosome reaction and Ca^{2+} uptake in sea urchin sperm suspended in Na^+-free sea water, *Dev. Biol.* **110**:338–345.

García-Soto, J., de De la Torre, L., Vargas, I., and Darszon, A., 1984, Response of isolated sea urchin sperm head plasma membranes to egg jelly, in *8th International Biophysics Congress (Bristol, U.K.),* I.U.P.A.B. Abstracts. p. 289.

Glabe, C. G., and Lennarz, W. J., 1979, Species-specific sperm adhesion in sea urchins. A quantitative investigation of binding-mediated egg agglutination, *J. Cell Biol.* **83**:595–604.

Gould, R. J., Murphy, K. M. M., and Snyder, S. H., 1982, [^{3}H]Nitrendipine-labeled calcium channels discriminated inorganic calcium agonists and antagonists, *Proc. Natl. Acad. Sci. U.S.A.* **79**:3656–3660.

Green, J. D., and Summers, R. G., 1980, Ultrastructural demonstration of trypsin-like protease in acrosome of sea urchin sperm, *Science* **209**:398–400.

Hamill, O. P., Marty, A., Neher, E., Sackmann, B., and Sigworth, F., 1981, Improved patch-clamp techniques for high resolution recordings from cells and cell-free membrane patches, *Pfluegers Arch.* **391**:85–100.

Wilmsen, U., Methfessel, C., Hanke, W., and Boheim, G., 1983, Channel current fluctuations studies with solvent-free bilayers using Neher–Sackmann pipettes, in *Physical Chemistry of Transmembrane Ion Motions* (G. Spach, ed.), Elsevier/North-Holland, Amsterdam pp. 479–485.

Jacobson, B. S., 1977, Isolation of plasma membrane from eukaryotic cells on polylysine-coated polyacrylamide beads, *Biochim. Biophys. Acta* **471**:331–335.

Kohlhardt, M., and Fleckenstein, A., 1977, The inhibition of the slow sodium inward current by nifedipine in mammalian ventricular myocardium, *Naunyn Schmiedebergs Arch. Pharmacol.* **298**:267–272.

Kopf, G. S., and Garbers, D. L., 1980, Calcium and fucose-sulfate rich polymer regulate sperm cyclic nucleotide metabolism and the acrosome reaction, *Biol. Reprod.* **22**:1118–1126.

Lopo, A. C., and Vacquier, V. D., 1980, Antibody to sperm surface glycoprotein inhibits the egg jelly-induced acrosome reaction of sea urchin sperm, *Dev. Biol.* **79**:325–333.

Latorre, R., and Miller, C., 1983, Conduction and selectivity in potassium channels, *J. Membr. Biol.* **71**:11–30.

Lee, H. C., Johnson, C., and Epel, D., 1983, Changes in internal pH associated with initiation of motility and acrosome reaction of sea urchin sperm, *Dev. Biol.* **95**:31–45.

Levine, A. E., and Walsh, K. A., 1979, Involvement of an acrosin-like enzyme in the acrosome reaction of sea urchin sperm, *Dev. Biol.* **72**:126–137.

Liévano, A., Sánchez, J., and Darszon, A., 1985, Single channel activity of bilayers derived from sea urchin sperm plasma membranes formed at the tip of a patch electrode, *Dev. Biol.* **112**:253–257.

Nishioka, D., and Cross, N., 1978, The role of external Na^+ in sea urchin fertilization, in *Cell Reproduction* (E. R. Dirksen, D. Prescott, and C. F. Fox, eds.), Academic Press, New York, pp. 403–413.

Ohtake, H., 1976, Respiratory behaviour of sea-urchin spermatozoa. Effect of pH and egg water on the respiratory rate, *J. Exp. Zool.* **198**:303–312.

Plotek, Y., and Atlas, D., 1983, Characterization of benextramine as an irreversible α-adrenergic blocker and as a blocker of potassium-activated calcium channels, *Eur. J. Biochem.* **133**:539–544.

Schackmann, R. W., and Shapiro, B. M., 1981, A partial sequence of ionic changes associated with the acrosome reaction of *Strongylocentrotus purpuratus*, *Dev. Biol.* **81**:145–154.

Schackmann, R. W., Eddy, E. M., and Shapiro, B. M., 1978, The acrosome reaction of *Strongylocentrotus purpuratus* sperm. Ion requirements and movements, *Dev. Biol.* **65**:483–495.

Schackmann, R. W., Christen, R., and Shapiro, B. M., 1981, Membrane potential depolarization an increased intracellular pH accompanying the acrosome reaction of sea urchin sperm, *Proc. Natl. Acad. Sci. U.S.A.* **78**:6066–6070.

Schindler, H. G., 1979, Exchange and interactions between lipid layers at the surface of a liposome solutions, *Biochim. Biophys. Acta* **555**:316–336.

Schindler, H. G., and Rosenbush, J., 1978, Matrix protein from *Escherichia coli* outer membranes forms voltage-controlled channels in lipid bilayers, *Proc. Natl. Acad. Sci. U.S.A.* **75**:3751–3755.

Schwarz, W., Neumke, B., and Palade, P. T., 1981, K-current fluctuations in inward-rectifying channels of frog skeletal muscle, *J. Membr. Biol.* **63**:85–92.

SeGall, G. K., and Lennarz, W. J., 1979, Chemical characterization of the component of the egg jelly coat from sea urchin eggs responsible for induction of the acrosome reaction, *Dev. Biol.* **71**:33–48.

SeGall, G. K., and Lennarz, W. J., 1981, Jelly coat and induction of the acrosome reaction in echinoid sperm, *Dev. Biol.* **86**:87–93.

Stanfield, P. R., 1983, Tetraethylammonium ions and the potassium permeability of excitable cells, *Rev. Physiol. Biochem. Pharmacol.* **97**:1–67.

Suárez-Isla, B., Wan, K., Lindstrom, J., and Montal, M., 1983, Single-channel recordings from purified acetylcholine receptors reconstituted in bilayers formed at the tip of patch pippets, *Biochemistry* **22**:2319–2323.

Summers, R. G., Hylander, B. L., Colwin, L. H., and Colwin, A. L., 1975, The functional anatomy of the echinoderm spermatozoon and its interaction with the egg at fertilization, *Am. Zool.* **15**:523–551.

Tan, K. N., and Tashjian, A. J., Jr., 1984, Voltage-dependent calcium channels in pituitary cells in culture. I. Characterization by $^{45}Ca^{2+}$ fluxes, *J. Biol. Chem.* **259**:418–426.

Tilney, L. G., Hatano, S., Ishikawa, H., and Mooseker, M. S., 1973, The polymerization of actin: Its role in the generation of the acrosomal process of certain echinoderm sperm, *J. Cell Biol.* **59**:109–126.

Ulbricht, W., Wagner, H. H., and Schmidtmayer, J., 1982, Effects of aminopyridines on potassium currents of the model membrane, in *Aminopyridines and Similarly Acting Drugs: Effects on Nerves, Muscles and Synapses.* (Lechat, P. *et al.*, eds.) Pergamon, Oxford, pp. 29–41.

Vacquier, V. D., and Moy, G. W., 1977, Isolation of binding: The protein responsible for adhesion of sperm to sea urchin eggs, *Proc. Natl. Acad. Sci. U.S.A.* **74**:2456–2460.

Wilmsen, U., Methfessel, C., Hanke, W., and Boheim, G., 1982, in *Phisico Chimie des Mouvements Ioniques Transmembranaires,* Ecole Nationale Superieure de Chimie, Paris, p. 50.

Chapter 19

Characterization of Large-Unitary-Conductance Calcium-Activated Potassium Channels in Planar Lipid Bilayers

Daniel Wolff, Cecilia Vergara, Ximena Cecchi, and Ramon Latorre

1. INTRODUCTION

In this chapter we discuss the data obtained for a calcium-activated potassium channel of large unitary conductance. These channels are thought to be involved in the regulation of different cellular functions such as repetitive firing of neurons, hormonal secretion, potassium secretion in renal tubular cells, cyclic activity in smooth muscle, and others (Latorre et al., 1985).

2. CHANNEL GATING

In general, an ion channel can be in one of several possible states of conductance. Some of these states are electrically conducting (open states), and some are nonconducting (closed) states. Therefore, a transition between a closed and an open state can be observed directly as a fluctuation in the current flowing through the channel. This process is called channel gating. In the simplest case of a channel having a single open and a single

---

DANIEL WOLFF, CECILIA VERGARA, XIMENA CECCHI, AND RAMON LATORRE • Departmento de Bíología, Facultad de Ciencias, Universidad de Chile, Santiago, Chile; and Centro de Estudios Científicos de Santiago, Santiago, Chile.

closed state, it is expected that the distribution of lifetimes for these states would be exponentially distributed (see Alvarez, Chapter 1; Ehrenstein *et al.*, 1974). However, most of the channels studied so far present multiple conductance states. The existence of the different closed and open states can be made evident from the multiple-exponential distribution of open and closed times.

So far we are aware of three basically different modes of gating. Two of them have been known for quite a long time and are voltage-dependent gating and chemical- or neurotransmitter-mediated gating. In these cases, either the voltage applied across the membrane where the channel is inserted or the binding of an agonist modifies the equilibrium between the open and closed states of the channel. Recently, a third mode of altering the equilibrium between the open and closed states of a channel has been described. Guharay and Sachs (1984) have found a channel whose probability of being in the open state is a function of the stretch applied to the membrane (see Sachs, Chapter 10). There is also a class of channels in which gating properties depend both on the binding of an agonist and on voltage. This is the case for the calcium-activated K^+ channels that were first described in macroscopic terms by Meech and Standen (1975).

2.1. Kinetic Models

With respect to single-channel conductance, at least two different groups of Ca^{2+}-activated K^+ channels have been described: one of low conductance (10–20 pS), whose properties have not been studied in detail (Lux *et al.*, 1981), and one of high conductance (100–250 pS), which has been studied in some detail with respect to its conduction and gating properties and is the one we are referring to in this chapter. Calcium-activated K^+ channels of high conductance have been observed in very different cell types with the use of the patch-clamp technique and have also been incorporated into planar lipid bilayers (Krueger *et al.*, 1982; Latorre *et al.*, 1981; Marty, 1981; Methfessel and Boheim, 1982; Pallota *et al.*, 1981; Walsh and Singer, 1983; Wong *et al.*, 1982). Their activation kinetics are complicated. Single-channel current records from either patch-clamp or bilayer recordings show openings and closings with durations on the order of miliseconds and also very brief transitions (flickers) that last less than 1 msec. This activity is observed in bursts that are interrupted by silent periods that may last several hundred milliseconds. In all of the channels from this group studied so far, depolarizing membrane potentials and increases in the intracellular $[Ca^{2+}]$ increase the fraction of time that the channel is in the conducting state within a burst of activity. On the other hand, the periods of silence are more frequent at depolarizing potentials and rather high intracellular $[Ca^{2+}]$. Therefore, these channels can be treated as activated by an agonist, calcium, keeping in mind that voltage has a pronounced effect on the activation reaction.

How does one go from records like the ones described to a model that would propose possible molecular transitions underlying the observed behavior? The same strategy used for agonist-activated channels is used. This is basically to find the effect of the agonist, calcium in this case, on the distribution of open and closed times (Colquhoun and Hawkes, 1977, 1981). Different kinetic models will predict different relationships between agonist concentration and these two parameters. It is possible that more than one kinetic scheme will account for the observed behavior; in that case, the simplest scheme is normally the chosen one.

From the description of the activity of this channel one realizes that the kinetic scheme that would describe channel activation must include at least three closed and two open states to account for the different lifetimes observed. At this point, different authors have followed two different variations from the main strategy. On the one hand, Magleby and Pallota (1983a,b) have tried to find a kinetic scheme that will account for the effect of calcium on the distribution of all the open and shut intervals that they can resolve for this channel. Nevertheless, they have not included in their analysis the effect of voltage on channel activation. On the other hand, Methfessel and Boheim (1982) and Moczydlowski and Latorre (1983) have studied the effects of both voltage and calcium on channel activation, but they have excluded from their analysis fluctuations lasting less than 1 msec. Magleby and Pallota have found two and three exponentials for the distribution of open and closed time intervals, respectively. This finding by itself suggests the existence of at least two different open and three different closed states. Because in this analysis of the distribution of closed times the very long silent periods (>100 msec) were not included, they proposed the existence of a fourth closed or inactivated state.

A possible problem with this analysis comes from the fact that sodium ions block this channel (Vergara, 1983; D. Naranjo and D. Wolff, unpublished results), and these experiments were done in the presence of sodium. Therefore, it is possible that some of the shortest closed intervals they resolved correspond to brief blocking/unblocking transitions and are not part of the activation reaction. With respect to calcium, they found that it affects differently the distribution of short and long open and closed times. Based on these findings, they propose the following kinetic scheme:

$$C \rightleftharpoons Ca-C \rightleftharpoons Ca-C-Ca$$
$$\alpha' \updownarrow \beta' \qquad \alpha \updownarrow \beta \qquad \text{(R-1)}$$
$$Ca-O \qquad Ca-O-Ca$$

For a fixed calcium concentration and voltage, they were able to obtain a set of values for the rate constants that gave a very good description of the observed distribution of open and closed intervals. However, to account for the properties of the channel at different calcium concentrations, they must allow for changes in some of the rate constants. They must also include a third open state with three calciums bound coming from the doubly liganded open state.

A different approach to finding a basic kinetic scheme that can describe the channel behavior was followed by Methfessel and Boheim (1982) for patch-recorded channel activity and also by Moczydlowski and Latorre (1983) for bilayer data. After filtering the current at 1 kHz, both groups found that the mean open time increased and the mean closed time decreased linearly as the calcium concentration was increased. On analyzing the effect of voltage, Moczydlowski and Latorre found that both mean open and mean closed times are voltage dependent, but surprisingly, as calcium concentration decreases, the mean open time becomes voltage independent, whereas the main closed time becomes voltage independent at very high calcium concentrations. Considering these findings and that the activation curve as a function of calcium concentration suggests a minimum of two calcium ions bound for a complete opening, Moczydlowski and Latorre have proposed the following kinetic model:

$$C \underset{k_{-1}}{\overset{k_1[Ca]}{\rightleftarrows}} Ca-C \underset{\alpha}{\overset{\beta}{\rightleftarrows}} Ca-O \underset{k_{-2}}{\overset{k_2[Ca]}{\rightleftarrows}} Ca-O-Ca \quad (R-2)$$

In this scheme, the opening/closing transition rate constants are voltage independent. The voltage dependence of the overall reaction comes from the voltage-dependent binding of calcium ions. Methfessel and Boheim proposed a similar kinetic scheme. However, they assumed that the closed/open transition contained all the voltage dependence of the reaction. In the analysis of the data, the long closed intervals have also been excluded. Vergara and Latorre (1983) found that these long quiescent periods can be explained in terms of a slow calcium blockade of the channel (see below).

2.2. Surface Charge Effect on Channel Gating

Because cell membranes contain negatively charged lipids, the question of the possible effect of surface charge on channel gating and conduction has been raised for several different ion channels (Fohlmeister and Adelman, 1982; Bell and Miller, 1984; Wilson et al., 1983). In order to study the effect of lipid surface charge on the calcium-activated potassium channel, Moczydlowski *et al.* (1985) incorporated the channel into planar bilayers of well-defined composition.

A marked difference in the kinetics of activation of this channel was found when its calcium-dependent gating characteristics were studied in bilayers made of phosphatidylethanolamine (PE; neutral at pH 7) or phosphatidylserine (PS; negatively charged). Moczydlowski *et al.* (1985) found that at the same calcium concentration and voltage, the channel exhibits a greater probability of being in the open state in PS than in PE membranes. Channels incorporated into PS membranes are about tenfold more sensitive to calcium than channels incorporated into PE membranes. This difference can be ascribed to an increased calcium concentration in the neighborhood of the calcium activation sites because of the negatively charged PS.

In summary, although the kinetics of activation of calcium-activated potassium channels are complicated, some hints about possible mechanisms to explain its voltage and calcium dependence have emerged from the analysis of the effect of these parameters on the distribution of open and closed times. If we learn how to uncouple its voltage and calcium dependence, we may get more light on their mode of action.

3. CHANNEL CONDUCTANCE AND SELECTIVITY

The elucidation of the mechanisms of ion conduction and selectivity in ionic channels allows us to make some assumptions about the nature of the chemical groups of the conduction pathway, which can give some insight into the gross architecture of the channel. Recently a kinetic approach based on the absolute reaction rate theory has been developed to describe ion permeation through membrane channels. According to this view, the conduction pathway is visualized as a series of energy barriers that the permeant ion must cross in order to go through the channel. Using this approach, Lauger (1973) developed a description of the ion transport through a pore as a function of ion concentration and membrane potential. For a channel that can accept only one ion at a time, he showed that the conductance (g_0) is a hyperbolic function of the ion concentration as in an enzymatic reaction:

$$g_0 = g_{max} \cdot [S]/(K_s + [S]) \tag{1}$$

where g_{max} is the maximum conductance of the channel, K_s the dissociation constant, and [S] the concentration of the permeant ion. For multiply occupied channels, however, the conductance is a more complex function of ion concentration.

In his analysis, Lauger showed that g_{max} is a function of the energy of ions both at the binding sites and at the energy barriers. K_s, on the other hand, depends only on binding energies of ions. With this kind of transport model, the selectivity of a channel can be explained on the basis of the free energy of ions inside the channel, which depends on the type of ligands present in the protein.

4. CONDUCTANCE OF THE CALCIUM-ACTIVATED K⁺ CHANNELS

One striking characteristic of these channels is their high conductance: in 100 mM KCl, the conductance of the open channel is about 230 pS. The open single-channel current is a linear function of the applied voltage between -100 and $+100$ mV for concentrations ranging from 50 to 100 mM. This linearity might reflect a high degree of symmetry in the energy barrier profile of the channel. Figure 1 shows the conductance of the Ca-K channel from T-tubule and smooth muscle plasma membranes as a function of KCl concentration when incorporated to PE and PS bilayers. It is apparent that in the region of low ion activity the data deviate markedly from the rectangular hyperbola that would be expected for a single ion channel with fixed energy barriers. For the T-tubule channel incorporated in PE bilayers, the results can be fitted empirically as the sum of two Langmuir isotherms, and the following best-fit parameters are obtained: $g_{max1} = 206$ pS; $K_{s1} = 2.5$ mM; $g_{max2} = 404$ pS; $K_{s2} = 501$ mM. Similar conductance activity characteristics are observed in a Ca-K channel from smooth muscle (Cecchi *et al.*, 1984), although the conductance of this channel is slightly lower than the T-tubule channel conductance for all concentrations tested.

Blatz and Magleby (1984), however, have found a hyperbolic behavior for the Ca-K channel from cultured muscle cells. We have no explanation for the differences between these two channels, but a hyperbolic curve

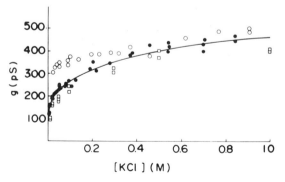

Figure 1. Conductance of the T-tubule and of the smooth muscle membrane Ca-K channels as a function of potassium concentration. The conductance values were determined from the slopes of current–voltage curves obtained between ±40 mV in symmetrical KCl concentrations (pH 7.0). ●, T-tubule channel in PE bilayer; ○, T-tubule channel in PE/PS bilayer; □, smooth muscle channel in PE bilayer.

is obtained for the T-tubule channel when the saline is buffered with tris-(hydroxymethyl)-aminomethane (tris), a cation that blocks the channel. This blockade originates an artifactual "Michaelis–Menten" kinetics (Latorre and Miller, 1983). Blatz and Magleby used N-methyl-D-glucosamine as buffer in their solutions, and it is possible that this cation also blocks the channel.

Several mechanisms could explain the observed behavior of the conductance in the low-activity range. (1) Coulombic interactions between K^+ and its binding site might produce a slow conformational change of the site (Lauger, 1980). (2) Fixed negative charges may be present in the protein or lipids surrounding the ion-binding site (Apell et al., 1979). (3) A multisite K^+ occupancy mechanism could produce the observed conductance–concentration curve in a pore that can contain two K^+ ions simultaneously if the two sites are close enough to allow for electrostatic repulsion.

At low potassium concentrations, T-tubule and smooth muscle Ca-K channels display a higher conductance when incorporated into PS bilayers as compared to PE bilayers (see Fig. 1). However, this difference disappears at concentrations above 400 mM. The differences observed in these two lipids can be explained by the increased potassium concentration in the vicinity of the channel induced by the fixed negative charge of PS membranes. Model calculations of the expected K^+ concentrations at various distances from a PS membrane surface, using the Guy–Chapman–Stern theory, suggest that the K^+ conduction sites sense a fraction of the surface potential equivalent to the local electrostatic potential at a distance of 0.9 nm from the surface (Moczydlowski et al., 1985).

5. SELECTIVITY OF THE Ca-K CHANNELS

The zero-current potential measured when a test ion X is added to one side of a symmetrical KCl solution or under biionic conditions has been used to determine the relative permeability of X to K^+. The relationship between the permeability ratio and this potential is obtained from the Goldman–Hodgkin–Katz equation.

5.1. Selectivity to Anions

The selectivity of K^+ over Cl^- can be measured in the presence of a KCl gradient across the bilayer. Under these conditions, the zero-current potential found corresponds to that expected for a perfectly cation-selective channel. Therefore, this channel allows K^+ to permeate and completely excludes Cl^- ions.

5.2. Selectivity to Monovalent Cations

The Ca-K channels from different cells show striking similarities in their selectivity characteristics (Vergara, 1983; Blatz and Magleby, 1984; Wolff et al., 1985). The following selectivity sequence has been found for the monovalent cations as determined from the reversal potential shifts of single-ion currents: $Tl^+ > K^+ > Rb^+ > NH_4$; Cs^+, Li^+, and Na^+ are not measurably permeant. Table I illustrates the permeability ratios relative to K^+ reported for different Ca-K channels.

The high selectivity displayed by these Ca-K channels may be attributed to a high degree of rigidity of the selectivity filter that can accomodate K^+ but not other ions. Both Tl^+ and NH_4^+ have been used as probes to obtain information about the chemical structure of the groups determining the selectivity. The selectivity sequence of these "fingerprint" ions with respect to K^+ suggests that the selectivity filter of Ca-K channels possesses groups with sixfold coordinate oxygens of etherlike characteristics. The permeability pattern of the Ca-K channel is similar to that found in other K^+-selective channels; the same selectivity sequence has been found for the delayed rectifier of squid axon, node of Ranvier, and frog skeletal muscle. These similarities suggest that these channels share some basic structural similarities in the region where the selectivity filter is located.

6. BLOCKADE OF THE Ca-K CHANNELS

6.1. General

If a second ion species is added to the solution of permeant ions bathing the channel, the two species may compete for sites in the pore.

Table I. Permeability Ratios (P_X/P_K) Determined from Reversal Potential Shifts in Different Ca-K Channels

Ion (X)	T-tubule[a]	Smooth muscle[b]	Chromaffin cell[c]	Cultured muscle cell[d]
Tl^+	1.5	—	—	1.2
K^+	1.0	1.0	1.0	1.0
Rb^+	0.7	0.66	0.83	0.67
NH_4^+	*[e]	—	—	0.11
Cs^+	*	*	*	<0.05
Li^+	*	*	—	<0.02
Na^+	*	*	<0.03	<0.01

[a] Vergara, 1983; D. Naranjo and R. Latorre, unpublished results.
[b] Wolff et al., 1985.
[c] Yellen, 1984a.
[d] Blatz and Magleby, 1984.
[e] Asterisk indicates value was unmeasurable.

If the second species can enter the channel but cannot permeate, this ion behaves as a blocker.

A simple model for blocking is described in the following reaction scheme for a channel with two conductance states:

$$C \underset{\beta}{\overset{\alpha}{\rightleftarrows}} O \underset{k_{-1}}{\overset{k_1[B]}{\rightleftarrows}} B \tag{R-3}$$

The model assumes that the blocking ion can only interact with the open channel and that when the blocker is bound to a site in the conduction pathway, the permeant ion cannot pass through. So, in the presence of the blocker, the channel can be closed (C), open (O), or blocked (B). If the binding site for the blocker is within the electric field, the dissociation constant for the blocking reaction, K_b, is voltage dependent (Woodhull, 1973):

$$K_b = K_b(0) \cdot \exp(-z\delta FV/RT) \tag{2}$$

where $K_b(0)$ is the dissociation constant at zero voltage, z the valence of the blocker, δ the fractional electrical distance at which the blocking site is located, V the electrical potential; F, R, and T have their usual meanings.

If the blocking reaction is much faster than the gating reaction and the time resolution of the recording system, the blockade is detected as an apparent reduction in the open-state conductance of the channel. In the presence of a fast blocker, and assuming a simple competition with the permeant species, the conductance of the channel (g) is described by the expression:

$$g = g_{max}[S]/\{K_s(1 + [B]/K_b) + [S]\} \tag{3}$$

In the absence of the blocker, g/g_0 is obtained from equations 1 and 3. This expression can be linearized:

$$\ln(g_0/g - 1) = \ln[B]/K_B(0) + z\delta V/RT \tag{4}$$

where $K_B(0) = K_b(0)\{(K_s + [S])/K_s\}$ is the apparent dissociation constant at zero voltage. Then, from a plot of $\ln(g_0/g - 1)$ versus voltage the values of δ and $K_B(0)$ can be obtained.

Blocking ions compete with permeant species for sites in the conduction pathway, so studies of blockade may give information on the location of the binding sites and also about the occupancy of the channel. For channels that can be simultaneously occupied by more than one ion, it is possible that the permeant ion increases the exit rate of the blocker through electrostatic repulsion.

6.2. Blockade by Alkali Cations

The addition of Cs^+, Li^+, and Na^+ to the *cis* side of a chamber containing symmetrical KCl solutions causes an apparent reduction of the single-channel conductance that is voltage dependent. This finding is consistent with a fast blockade, as shown in scheme R-1. From the voltage dependence of the blockade, information can be obtained about the location of the binding sites inside the channel. Values of $\delta = 0.7$ for Cs^+, 0.75 for Na^+, and 0.38 for Li^+ have been obtained. At 50 mM KCl, the dissociation constants for the blockers are 55 mM for Cs^+, 185 mM for Na^+, and 77 mM for Li^+.

Cesium ions added to the *trans* side also block the T-tubule and smooth muscle channels, but the affinity constant is in the micromolar range. Figure 2 shows the voltage dependence of the channel conductance at different Cs^+ concentrations. The values for the fractional electrical distance obtained by linearizing these curves is always higher than one indicating multiple occupancy of the channel. However, two observations appear to be inconsistent with this hypothesis. (1) When the channel conductance is plotted against the log of Cs^+ concentration at a constant voltage, a single-site adsorption isotherm is obtained. (2) In competition experiments, the symmetrical increase of K^+ concentration on both sides of the membrane relieves Cs^+ blockade as if both ions were competing for a single binding site (Fig. 3).

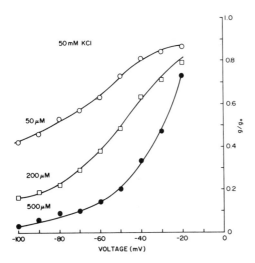

Figure 2. Voltage dependence of the *trans* Cs^+ blockade in the smooth muscle Ca-K channel. Experimental points are the ratio between single-channel conductance in symmetrical 50 mM KCl in the presence (g) and in the absence (g_0) of Cs^+ added to the *trans* side only at concentrations of (○) 50, (□) 200, and (●) 500 μM. Solid lines were drawn by eye.

Figure 3. *Trans* Cs^+ blockade as a function of Cs^+ concentration in the smooth muscle Ca-K channel. Experimental points are the ratios between the channel conductance in the absence and in the presence of the blocker, at different symmetrical KCl concentrations (mM): (●) 100, (▲) 300, and (■) 500. Membrane potential was -50 mV. The lines are the regression lines for a competitive inhibitor according to the following equation:

$$g_0/g = 1 + (K_s/K_b) [B]/(K_s + [S])$$

where g_0 is the channel conductance in the absence of the blocker, g is the channel conductance in the presence of the blocker, K_s is the binding constant of potassium, K_b is the binding constant of the blocker, and [S] and [B] are the concentrations of potassium and blocker, respectively.

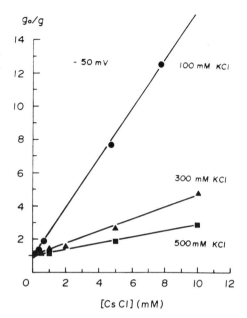

The release of the blockade observed in these competition experiments could be explained by assuming that K^+ ions on the side opposite Cs^+ increase the rate of exit of Cs^+ ions. In support of this explanation are the results obtained by Yellen (1984b), who found a relief of Na^+ blockade in the Ca-K channel of chromaffin cells by external ions. In that case, he showed that K^+ relieves the block by increasing the rate of exit of Na^+, suggesting that K^+ can enter the channel from the outside and expel the Na^+ by a knock-off mechanism. In this picture, two ions can be accommodated in the channel, but the sites are close enough to allow electrostatic repulsion between them. This model predicts that in the case of an increase in K^+ concentration on the side the Cs^+ is on, a reinforcement of the blockade should be observed, since Cs^+ could not get off because of the presence of K^+ in the more external site. Preliminary studies carried out with the Ca-K channel from smooth muscle show a small increase in the *trans* Cs^+ blockade when K^+ concentration is raised in the same side. These results, together with the shape of the conductance–concentration curve, suggest that the channel can accommodate at least two ions at a time. A barrier model considering this possibility could explain both the Cs^+ blockade characteristics and the concentration dependence of the channel conductance.

Cesium blockade has also been observed in Ca-K channels from chromaffin cells (Yellen, 1984a) and cultured muscle cells (Blatz and Magleby, 1984). In the chromaffin cell channel, Cs^+ blocks from either side but binds to sites with lower affinity than in the smooth muscle and T-tubule channels. The electrical distances found are approximately 1 for the external (*trans*) and 0.3 for the internal site.

6.3. Blocking by Quaternary Ammonium Ions

Large cations such as tetraethylammonium (TEA) and nonyltriethylammonium (C_9) are strong blockers of the Ca-K channels from either the *cis* (internal) or *trans* (external) side (Blatz and Magleby, 1984; Vergara *et al.*, 1984; Yellen, 1984b). Table II shows the values of K_B and δ reported for these channels. *Cis* TEA blockade is observed as an apparent reduction in the open-channel current, indicating that block and unblock events are too fast to be detected. The block increases for positive potentials and is relieved at negative membrane potentials, indicating that the blocking site senses the electric field. Similar results have been found in the T-tubule, chromaffin cell, and cultured muscle cell Ca-K channels (Vergara, 1983; Yellen, 1984a; Blatz and Magleby, 1984). *Trans* TEA is effective in lower concentrations, indicating that the affinity for the blocker is approximately 120-fold higher than from the *cis* side. In the chromaffin cell channel, Yellen found that *trans* blockade kinetics are slower than the *cis* process and can be detected as rapid flickers in the channel current. He also reported that the TEA blockade is voltage dependent (δ = 0.2). On the other hand, in the T-tubule channel the blockade is voltage independent.

The striking differences in the characteristics of the TEA blockade from *cis* and *trans* sides suggest the existence of two different binding sites for TEA in the channel.

Nonyltriethylammonium (C_9) is also an effective blocker of the Ca-

Table II. Values of δ and $K_B(0)$ for TEA Blockade in Different Ca-K Channels[a]

	T-tubule	Chromaffin cells	Cultured muscle cell
TEA *cis* (internal)			
δ	0.34	0.1	0.26
$K_B(0)$ (mM)	35[b]	24[c]	60[d]
TEA *trans* (external)			
δ	0.0	0.2	—
$K_B(0)$ (mM)	0.3[b]	0.2[c]	0.3[d]

[a] References are the same as in Table I.
[b] In symmetrical 100 mM KCl.
[c] In 160 mM internal KCl/160 mM external NaCl.
[d] In symmetrical 140 mM KCl.

K channels from chromaffin cells and T-tubule. When added to the *cis* (internal) side, C_9 produces a clearly resolvable block of the current through the channel. As in the case of TEA, the blockade is voltage dependent, but this ion presents a much higher affinity for the binding site. On the other hand, from the *trans* side, C_9 produces a slight decrease in the channel conductance. When the data are fitted to equation 4, a $K_B(0)$ of 3.5 μM and a δ of 0.34 are obtained. These values indicate that C_9 binds to a site located approximately at the same electrical distance as the site for TEA, but with much higher affinity.

Yellen (1984a) could measure both block and unblock rates for C_9 directly from the distributions of the open and closed times during an opening burst in the Ca-K channel from chromaffin cells. These time duration distributions were well fitted by single exponentials. At +60 mV, the average blocked time was 0.25 msec, and the average open time (between blocking events) was 0.49 msec. These results suggest that the properties of the TEA and C_9 binding sites are very similar to the internal receptor for these organic cations in the delayed rectifier. The characteristics of the TEA blockade indicate that the Ca-K channels possess *cis* and *trans* wide mouths that allow the entrance of these cations and that the binding sites are very different. The *cis* side is located in the electric field and appears to be more hydrophobic than the *trans* site in view of the higher affinity for C_9, a less polar organic cation.

6.4 Blocking by Calcium and Barium Ions

Calcium ions, in addition to activating the channels from the *cis* side, also produce a slow block. This blockade appears as long periods during which the channel dwells in a nonconductive state. Vergara and Latorre (1983) have studied in detail the slow blockade of the T-tubule channel, characterizing the mean burst time (t_b) and the mean time for the slow closings (t_s) as functions of Ca^{2+} concentration and voltage. Their data are consistent with a reaction scheme that considers that the blocker interacts only with the open channel and that blocking and unblocking events are much slower than the gating process.

$$C \underset{\beta}{\overset{\alpha}{\rightleftarrows}} O \underset{k_{-1}}{\overset{k_1[Ca^{2+}]}{\rightleftarrows}} B \qquad \text{(R-4)}$$

$$\text{fast} \qquad \text{slow}$$

At high probability of opening, they found, as predicted by scheme R-4, that the mean burst time was a linear function of $1/[Ca^{2+}]$ and that the mean block time was independent of $[Ca^{2+}]$. From the voltage dependence of the block they concluded that the fractional electrical distance of the binding site is 0.65 and the dissociation constant, $K_B(0)$, is 0.29 M.

Barium ions added to the *cis* side also produce slow closings in the Ca-K channel from T-tubule. Nevertheless, Ba^{2+} is 1000-fold more effective than Ca^{2+} as a blocker and does not activate the channel. It is assumed that the Ba^{2+} blockade is similar to that of Ca^{2+}: the cation blocks the open channel, and the blocking events are slow as compared with the gating ones. Barium blockade is voltage dependent: the binding site senses about 80% of the applied voltage, and the dissociation constant, $K_B(0)$, is 36 μM. Potassium ions compete with both divalent blockers, and an increase in the *trans* K^+ concentration decreases the entry probability of the blocker. This finding suggests that the binding site for the blocker is located in the potassium pathway. From the study of calcium and barium blockade, it can be concluded that the apparent slow gating kinetics displayed by the Ca-K channel are a consequence of the slow blockade by these ions. The blocking by both ions presents the same voltage dependence, suggesting that they bind to the same site.

In addition to the slow blockade, *cis* calcium ions produce a rapid blockade of Ca-K channels from T-tubule and smooth muscle, which is shown as an apparent decrease in the channel current. This blockade is also voltage dependent, and the site senses approximately 35% of the field; the dissociation constant $K_B(0)$ is 2.8 mM.

7. CONCLUSIONS

All the high-conductance calcium-activated potassium channels analyzed present similar characteristics of conduction and blockade, which makes it possible to think that they share a similar conduction mechanism. Any attempt to postulate a model for this mechanism has to take into account the following features:

1. The high conductance and strong selectivity to potassium, which rule out the possibility of an aqueous pore.
2. The saturation of the current with the ion concentration, which suggests the presence of binding sites inside the channel. However, the function is not a hyperbola, which may indicate multiple occupancy of the channel.
3. The blocking by other ions that can enter the channel but cannot cross it in a way that indicates the presence of saturable binding sites located someplace in the electric field.

The conductance, selectivity, and blocking data showed here represent a good starting point to postulate a conduction mechanism that can be modeled as a series of binding sites and energy barriers, accepting at least two ions at a time.

REFERENCES

Apell, H. J., Bamberg, E., and Lauger, P., 1979, Effects of surface charge on the conductance of the gramicidin channel, *Biochim. Biophys. Acta* **552**:369–378.

Bell, J. E., and Miller, C., 1984, Effect of phospholipid surface charge on ion conduction in the K^+ channel of sarcoplasmic reticulum, *Biophys. J.* **45**:279–287.

Blatz, A. L., and Magleby, K. L., 1984, Ion conductance and selectivity of single calcium-activated potassium channels in cultured rat muscle, *J. Gen. Physiol.* **84**:1–22.

Cecchi, X., Wolff, D., Alvarez, O., and Latorre, R., 1985, Incorporation of Ca^{2+}-activated K^+ channels from rabbit intestinal smooth muscle sarcolemma into planar bilayers, *Biophys. J.* **45**:38a.

Colquhoun, D., and Hawkes, A. G., 1977, Relaxations and fluctuations of membrane currents that flow through drug operated channels, *Proc. R. Soc. Lond. [Biol.]* **199**:231–262.

Colquhoun, D., and Hawkes, G., 1981, On the stochastic properties of single ion channels, *Proc. R. Soc. Lond. [Biol.]* **211**:205–235.

Ehrenstein, G., Blumenthal, R., Latorre, R., and Lecar, H., 1974, Kinetics of the opening and closing of individual excitability-inducing material channels in a lipid bilayer, *J. Gen. Physiol.* **63**:707–721.

Fohlmeister, J. F., and Adelman, W. J., 1982, Periaxonal surface calcium binding and distribution of charges on the faces of squid axon potassium channels molecules, *J. Membr. Biol.* **70**:115–123.

Guharay, F., and Sachs, F., 1984, Stretch-activated single ion channel current in tissue-cultured embryonic chick skeletal muscle, *J. Physiol. (Lond.)* **352**:685–701.

Krueger, B. K., French, R. J., Blaustein, M. B., and Worley, J. F., 1982, Incorporation of Ca-activated K channels from rat brain into lipid bilayers, *Biophys. J.* **37**:170a.

Latorre, R., and Miller, C., 1983, Conduction and selectivity in potassium channels, *J. Membr. Biol.* **71**:11–30.

Latorre, R., Vergara, C., and Hidalgo, C., 1982, Reconstitution in planar lipid bilayers of a Ca dependent K channel from transverse tubule membranes isolated from rabbit skeletal muscle, *Proc. Natl. Acad. Sci. U.S.A.* **79**:805–809.

Latorre, R., Coronado, R., and Vergara, C., 1984, K^+ channels gated by voltage and ions, *Annu. Rev. Physiol.* **46**:485–495.

Latorre, R., Alvarez, O., Cecchi, X., and Vergara, C., 1985, Properties of reconstituted ion channels, *Annu. Rev. Biophys. Biophys. Chem.* **14**:79–111.

Lauger, P., 1973, Ion transport through pores: A rate theory analysis, *Biochim. Biophys. Acta* **311**:423–441.

Lauger, P., Stephan, W., and Frehland, E., 1980, Fluctuations of barrier structure in ionic channels, *Biochim. Biophys. Acta* **602**:167–180.

Lux, H. D., Neher, E., and Marty, A., 1981, Single channel activity associated with the Ca-dependent outward current in *H. pomatia*, *Pfluegers Arch.* **389**:293–295.

Magleby, K. L., and Pallota, B. S., 1983a, Calcium dependence of open and shut interval distributions from calcium-activated potassium channels in cultured rat muscle, *J. Physiol. (Lond.)* **344**:585–604.

Magleby, K. L., and Pallota, B. S., 1983b, Burst kinetics of single calcium-activated potassium channels in cultured rat muscle, *J. Physiol. (Lond.)* **344**:605–623.

Marty, A., 1981, Ca dependent K channels with large unitary conductance in chromaffin cell membranes, *Nature* **291**:497–500.

Meech, R. W., and Standen, N. B., 1975, Potassium activation in *Helix aspersa* neurones under voltage clamp: A component mediated by calcium influx, *J. Physiol. (Lond.)* **249**:211–234.

Methfessel, C., and Boheim, G., 1982, The gating of single calcium-dependent potassium channels is described by an activation–blockade mechanism, *Biophys. Struct. Mech.* **9**:35–60.

Moczydlowski, E., and Latorre, R., 1983, Gating kinetics of Ca-activated channels from rat muscle incorporated into planar lipid bilayers, *J. Gen. Physiol.* **82**:511–543.

Moczydlowski, E., Alvarez, O., Vergara, C., and Latorre, R., 1985, Effect of phospholipid surface charge on the conductance and gating of a Ca-activated K channel in planar lipid bilayers, *J. Membr. Biol.* **83**:273–282.

Pallota, B. S., Magleby, K. L., and Barret, J. N., 1981, Single channel recordings of Ca-activated K currents in rat muscle cell culture, *Nature* **293**:471–474.

Vergara, C., 1983, *Characterization of a Ca-Activated K Channel from Skeletal Muscle Membranes in Artificial Bilayers*, Ph.D. disertation, Harvard University, Cambridge.

Vergara, C., and Latorre, R., 1983, Kinetics of Ca-activated K channels from rabbit muscle incorporated into planar bilayers, *J. Gen. Physiol.* **82**:543–563.

Vergara, C., Moczydlowski, E., and Latorre, R., 1984, Conduction, blockade and gating in a Ca-activated K channel incorporated into planar lipid bilayers, *Biophys. J.* **45**:73–76.

Walsh, A. V., and Singer, J. J., 1983, Identification and characterization of a Ca-activated K channel in freshly dissociated vertebrate smooth muscle cells using the patch clamp technique, *Biophys. J.* **41**:56a.

Wilson, D. L., Morimoto, K., Tsuda, Y., and Brown, A. M., 1983, Interaction between calcium ions and surface charge as it relates to calcium currents, *J. Membr. Biol.* **72**:117–130.

Wolff, D., Cecchi, X., Naranjo, D., Alvarez, O., and Latorre, R., 1985, Cation selectivity and Cs^+ blockade in a Ca-activated K channel from rabbit intestinal smooth muscle, *Biophys. J.* **47**:386a.

Wong, B. S., Lecar, H., and Adler, M., 1982, Single Ca-dependent K channels in clonal anterior pituitary cells, *Biophys. J.* **39**:313–317.

Woodhull, A. M., 1973, Ionic blockade of sodium channel in nerve, *J. Gen. Physiol.* **61**:687–798.

Yellen, G., 1984a, Ionic permeation and blockade in Ca-activated K channels of bovine chromaffin cells, *J. Gen. Physiol.* **84**:157–186.

Yellen, G., 1984b, Relief of sodium block of Ca-activated K channels by external cations, *J. Gen. Physiol.* **84**:187–199.

Part IV

Ionic Channel Modulation

Chapter 20

Metabolic Regulation of Ion Channels

Irwin B. Levitan

1. INTRODUCTION

During the last 30 years a great deal of progress has been made in our understanding of how neurotransmitters can regulate the activity of excitable cells. Much of our knowledge has come from studies on the vertebrate neuromuscular junction, in part because the preparation is easily accessible for experimental manipulation, but perhaps more importantly because a series of brilliant investigators have made it the focus of their highly imaginative studies. The resulting body of work has given us a detailed molecular picture of the nicotinic acetylcholine receptor/channel as a single macromolecular complex that can both bind acetylcholine and mediate the transport of ions across the plasma membrane. The opening of the channel (and transport of ions) is rapid in onset after the binding of acetylcholine to the receptor and is rapidly reversible when the receptor is no longer occupied by the agonist. Figure 1A summarizes schematically this way of thinking of rapidly reversible changes in the activity of an ion channel as being dependent on the continued occupation of a closely associated receptor by an agonist.

IRWIN B. LEVITAN • Graduate Department of Biochemistry, Brandeis University, Waltham, Massachusetts 02254.

A

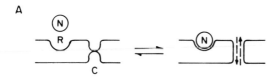

B

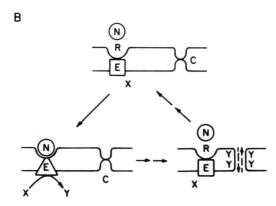

Figure 1. Short-term and long-term changes in neuronal membrane properties induced by neurotransmitters. A: Short term. According to this scheme, the neurotransmitter receptor (R) and the ion channel (C) are intimately associated in a macromolecular complex. Occupation of R by a transmitter (N) leads to a conformational change in C, allowing ions to pass through. This change is rapidly reversible and is dependent on the continued occupation of R by N. The most thoroughly studied directly coupled system of this type is the nicotinic acetylcholine receptor/channel at the vertebrate neuromuscular junction. B: Long term. According to this scheme, R and C are not necessarily in close physical proximity but are coupled indirectly by a second messenger (Y). When R is occupied by N, there is a conformational change in the closely associated enzyme E, which activates E and allows it to convert substrate X to product Y. Y then causes, perhaps via a series of steps, a change in C, which allows ions to pass through. The change can long outlast the initial stimulus, the occupation of R by N, and persists as long as Y exerts its effects.

Longer-lasting modulation of ion channels by neurotransmitters is difficult to explain in terms of the scheme described above. Although it is conceivable that in some instances occupancy of receptor by agonist might persist for prolonged periods, there are examples of physiological responses that outlast the initial stimulus by many seconds, minutes, or even hours. It has been convenient to think of such long-lasting changes in terms of a scheme such as that described in Fig. 1B: here the changes in ion channel properties are not dependent on the continued occupation of the receptor by the transmitter but rather result from some long-lasting metabolic modification of the channel. This scheme suggests that the receptor and channel need not necessarily be intimately associated in a single macromolecular complex but may communicate via some intracellular second messenger, which is produced on occupancy of receptor by agonist. The second messenger sets in motion a series of steps, culminating in some covalent modification of the channel that alters its activity, and the functional change persists until the covalent modification has been reversed (again, this may require a series of steps).

What second messengers might mediate long-lasting modulation of ion channel activity in excitable cells? Cyclic AMP (cAMP) and calcium ions are well established as second messengers in the regulation of carbohydrate metabolism in such tissues as liver and muscle, where they modulate the activity of certain enzymes by causing them to be phosphorylated (Cohen, 1982). It has long been suspected that they might play a similar role in the regulation of ion channels, and recent evidence from several laboratories has confirmed that this is indeed the case. The remainder of this chapter focuses on some of the experimental strategies that have been and are being used to investigate the modulation of ion channels in excitable cells by calcium, cAMP, and cAMP-dependent protein phosphorylation.

2. SECOND MESSENGERS

2.1. Regulation of Neuronal Activity

If one wants to determine whether a particular physiological response is mediated by cAMP, one obvious question to ask is whether cAMP can mimic the response. With this in mind, a large number of investigators have applied cAMP to cells while monitoring their membrane properties (Brunelli *et al.*, 1976; Deterre *et al.*, 1981; Drummond *et al.*, 1980; Kaczmarek *et al.*, 1978; Klein and Kandel, 1978; Levitan and Norman, 1980; Pellmar, 1981; Siegelbaum *et al.*, 1982; Treistman, 1981; Treistman and Levitan, 1976a,b; Tsien, 1973). In some cases, membrane-permeable derivatives of cAMP have been applied in the extracellular medium, whereas in others cAMP itself or derivatives that are resistant to hydrolysis by phosphodiesterases have been injected intracellularly via microelectrodes. In other experiments intracellular cAMP levels have been elevated by application of phosphodiesterase inhibitors, which inhibit its breakdown, or activators of adenylate cyclase, which increase its synthesis. Results from all of these approaches indicate that cAMP can indeed modulate membrane properties. In some of these studies voltage-clamp analysis has allowed the identification of particular ionic currents that can be affected by cAMP. There appears to be no single ion current that is a universal target for cAMP, but rather the current that is modulated is different from cell to cell.

A demonstration of a pharmacological effect of cAMP on membrane properties is, of course, not in itself evidence that cAMP plays a physiological role. However, in several of the studies referred to above, it has been shown that cAMP mimics the action of a particular physiological agonist and that the physiological agonist can cause the stimulation of adenylate cyclase and the accumulation of cAMP in the target cell. These and other pharmacological, biochemical, and electrophysiological experiments have indeed provided strong circumstantial evidence for a physiological role for cAMP in some cells. Among the physiological responses that appear to be mediated by cAMP are the activation of the slow inward calcium current by β-adrenergic stimulation of cardiac cells (Tsien, 1973, 1977) and the inactivation (Deterre et al., 1981; Klein and Kandel, 1978) or activation (Drummond et al., 1980) of several distinct potassium currents by serotonin in several different molluscan neurons. One example taken from work in our laboratory is illustrated in Fig. 2. Serotonin causes the identified *Aplysia* neuron R15 to hyperpolarize (Drummond et al., 1980) and, as shown in Fig. 2A, this is accompanied by an increase in the slope of the steady-state current–voltage curve at hyperpolarized potentials. A series of ion substitution and pharmacological blocking experiments have established that this results from the activation of an inwardly rectifying potassium channel. Furthermore cAMP mimics this response (Fig. 2B), and serotonin activates adenylate cyclase and causes cAMP to accumulate within R15. These and other combined biochemical and physiological experiments have provided strong evidence (Drummond et al., 1980) that the serotonin-evoked increase in the activity of this potassium channel is mediated by cAMP.

2.2. Regulation of Ion Channels

Convincing evidence for second messenger involvement in transmitter-evoked physiological responses in neurons has been surprisingly difficult to come by. As pointed out above, there are now several cases for which the data are becoming quite compelling, but much of it is indirect. One difficulty is that no general criteria have been available to identify

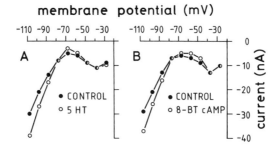

Figure 2. Steady-state current–voltage curves in voltage-clamped neuron R15. Membrane currents under control (●) and experimental (○) conditions for extracellular application of 1 μM serotonin (5 HT, A) or 0.7 mM 8-benzylthio-cAMP (8-BT cAMP, B). In both cases the control and experimental curves reverse sign between -75 and -80 mV, close to the potassium equilibrium potential in R15. (Modified from Drummond et al., 1980.)

second messenger coupling independent of what the messenger might be. However, appropriate use of recently developed patch-recording techniques does provide such a general test.

Since a gigohm seal between a patch-recording electrode and a membrane forms a lateral diffusion barrier, it follows that neurotransmitter applied outside the electrode cannot activate receptors in the patch of membrane inside the electrode. Thus, if channels inside a patch are affected by transmitter applied outside, this must be through occupancy of receptors located outside, and these receptors can only communicate with the channels inside the patch by means of a diffusible intracellular messenger. In contrast, if a receptor is coupled directly to an ion channel (e.g., the nicotinic acetylcholine receptor/channel), it should be possible to affect a channel in the electrode only by applying transmitter inside (Neher and Sakmann, 1976).

This general test has been used in at least two laboratories to identify second-messenger-mediated responses. Siegelbaum *et al.* (1982), working with sensory neurons from the abdominal ganglion of *Aplysia,* have demonstrated that serotonin applied in the extracellular medium can cause the closure of individual potassium channels inside a patch electrode. Other experiments have shown that the second messenger that mediates this response is cAMP (but this knowledge was not necessary for the demonstration that some second messenger must be involved). Maruyama and Petersen (1982) have used a similar approach to show that cholecystokinin and acetylcholine activation of single calcium-dependent cation channels in pancreatic acinar cells (which may be considered honorary neurons for our purposes) is mediated by intracellular calcium ions. These elegant experiments have introduced a new level of sophistication to the way we think about and study second-messenger mediated physiological responses in excitable cells.

3. PROTEIN PHOSPHORYLATION

3.1. Regulation of Neuronal Activity

It is thought (Kuo and Greengard, 1969; Glass and Krebs, 1980) that all actions of cAMP in eucaryotes are mediated by the activation of cAMP-dependent protein kinases, and several laboratories have investigated the possibility that cAMP-mediated physiological responses involve protein phosphorylation. The active catalytic subunit (CS) of cAMP-dependent protein kinase has been purified to homogeneity from several mammalian sources (Peters *et al.,* 1977). Its active site has been well conserved during evolution, and thus it is possible to use the mammalian enzyme as a pharmacological tool to examine membrane excitability in mammals and molluscs. The CS has been applied intracellularly in cardiac myocytes (Osterrieder *et al.*, 1982) and several different molluscan neurons (Alkon *et al.*, 1983; Catellucci *et al.*, 1980; De Peyer *et al.*, 1982; Kaczmarek *et al.*, 1980), and it has been found that it can affect membrane properties in a way predictable from the known actions of cAMP and neurotransmitters on these same cells. A complementary approach to the pharmacological use of exogenous CS is to ask whether inhibition of *endogenous* CS can affect a physiological response. A molecular probe that has been particularly useful in this regard is a naturally occurring specific protein inhibitor (Walsh and Ashby, 1973) of cAMP-dependent protein kinase. This 10,000-dalton inhibitor can also be purified from mammalian muscle (Demaille *et al.*, 1977), and it has been found that its intracellular injection can block the cAMP-mediated actions of serotonin in several different molluscan neurons (Adams and Levitan, 1982; Castellucci *et al.*, 1982). Again, an example from our work is illustrated in Fig. 3, which shows that the serotonin-evoked increase in the inwardly rectifying potassium channel in neuron R15 is blocked by the injection of the protein kinase inhibitor.

3.2. Regulation of Ion Channels

The approaches described above have provided strong evidence that cAMP-dependent protein phosphorylation is a necessary step in the actions of some neurotransmitters on membrane properties. However, they

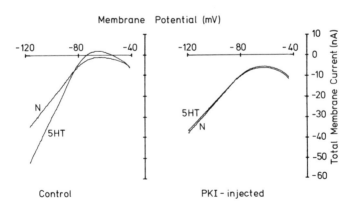

Figure 3. Effect of intracellular injection of protein kinase inhibitor (PKI) on the serotonin (5HT)-evoked increase in the inwardly rectifying potassium channel. The increase in slope of the steady-state current–voltage curve normally evoked by serotonin (left) is completely inhibited after PKI injection (right). (Modified from Adams and Levitan, 1982.)

do not make clear whether it is the ion channel itself that is phosphorylated or whether phosphorylation is simply an early step in some cascade of events that leads ultimately to modulation of channel activity. Again, single-channel recording techniques have provided a way to examine this question. The gigohm seal between an electrode and a membrane patch is not only electrically tight, it is also mechanically tight, and thus the patch can be pulled free from the cell under conditions that result in the cytoplasmic face of the membrane being exposed to the bathing medium (Hamill *et al.*, 1981). When such isolated patches from *Aplysia* (Shuster *et al.*, 1985) or *Helix* (Ewald *et al.*, 1985) neurons are bathed in CS and ATP, there is a change in the activity of individual ion channels in the patch, indicating that the target for phosphorylation must be some protein that remains associated with the patch of membrane when it is pulled away from the cell. This target might be the channel itself, or it might be some cytoskeletal component or other protein that comes away with the patch.

A related approach involves the examination of ion channel activity in a reconstituted system. It is possible to fuse membrane vesicles containing functional ion channels with artificial lipid bilayers under conditions that allow the current passing through individual channels to be measured (Miller, 1983). In fact, the bilayer can even be formed on the tip of a patch-recording electrode (Coronado and Latorre, 1983); this marriage between the patch and reconstitution methods provides single-channel recordings with particularly favorable signal-to-noise characteristics because of the small size of the lipid bilayer. When calcium-dependent potassium channels from *Helix* neurons are examined in these reconstituted systems, addition of CS and ATP to one side of the bilayer results in an increase in the activity of the individual channels (Fig. 4). Since proteins are essentially at infinite dilution in such planar bilayers (at least in the large ones), it can be concluded that the phosphorylation target must be either the channel itself or something intimately associated with the channel.

4. SUMMARY

The second messenger hypothesis has provided a conceptual framework for our thinking about mechanisms of long-term neuromodulation. The concepts far outstripped the experimental data for a long time, and only in the last few years has it become evident that cAMP and calcium can indeed mediate certain physiological responses in excitable cells. Although it seems likely that both of these second messengers exert their effects via protein phosphorylation, there is no reason to believe *a priori*

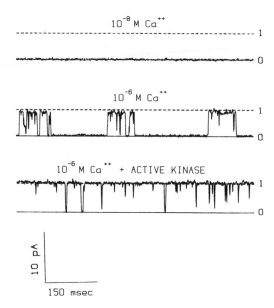

Figure 4. Effect of the catalytic subunit of cAMP-dependent protein kinase on the activity of a single calcium-dependent potassium channel reconstituted in an artificial bilayer on the tip of a patch electrode. This channel virtually never opened when the calcium concentration was 0.01 μM (top trace) but was open about 20% of the time at 1 μM calcium (middle trace). The channel activity was not changed by treatment with DTNB-inactivated catalytic subunit but was increased dramatically (open 90% of the time) following treatment with active catalytic subunit (bottom trace). (Modified from Ewald et al., 1985.)

that phosphorylation is the only covalent modification that can alter membrane excitability. Among the other possibilities that exist, methylation is a prime candidate, but this has not yet been studied in detail in anything other than unicellular organisms.

It is no accident that many of the major advances in this field have come from studies on molluscan neurons; their large size and ready identifiability allow combined biochemical, pharmacological, and electrophysiological experiments to be carried out on individual neurons, and such an interdisciplinary approach is certainly necessary for a complete understanding of long-term neuromodulation. The recent development of powerful and flexible single-channel recording techniques and their application to this problem should allow the extension of second messenger studies to cells that, for reasons of small size or complex geometry, have not previously been amenable to experimental manipulation.

REFERENCES

Adams, W. B., and Levitan, I. B., 1982, Intracellular injection of protein kinase inhibitor blocks the serotonin-induced increase in K$^+$ conductance in *Aplysia* neuron R15, *Proc. Natl. Acad. Sci. U.S.A.* **79**:3877–3880.

Alkon, D. L., Acosta-Urquidi, J., Olds, J., Kuzma, G., and Neary, J. T., 1983, Protein kinase injection reduces voltage-dependent potassium currents, *Science* **219**:303–306.

Brunelli, M., Castellucci, V., and Kandel, E., 1976, Synaptic facilitation and behavioral sensitization in *Aplysia:* Possible role of serotonin and cyclic AMP, *Science* **194**:1178–1181.

Camardo, J., Shuster, M., Siegelbaum, S., Eppler, C., and Kandel, E., 1983, Purified catalytic subunit of cyclic AMP-dependent protein kinase closes the serotonin-sensitive potassium channel of *Aplysia* sensory neurons in cell-free membrane patches, *Soc. Neurosci. Abstr.* **9**:22.

Castellucci, V. F., Kandel, E. R., Schwartz, J. H., Wilson, F. D., Nairn, A. C., and Greengard, P., 1980, Intracellular injection of the catalytic subunit of cyclic AMP-dependent protein kinase simulates facilitation of transmitter release underlying behavioral sensitization in *Aplysia*, *Proc. Natl. Acad. Sci. U.S.A.* **77**:7492–7496.

Castellucci, V. F., Nairn, A., Greengard, P., Schwartz, J. H., and Kandel, E. R., 1982, Inhibitor of adenosine 3':5'-monophosphate-dependent protein kinase blocks presynaptic facilitation in *Aplysia*, *J. Neurosci.* **2**:1673–1681.

Cohen, P., 1982, The role of protein phosphorylation in neural and hormonal control of cellular activity, *Nature* **296**:613–620.

Coronado, R., and Latorre, R., 1983, Phospholipid bilayers made from monolayers on patch-clamp pipettes, *Biophys. J.* **43**:231–236.

Demaille, J., Peters, K., and Fischer, E., 1977, Isolation and properties of the rabbit skeletal muscle protein inhibitor of adenosine 3',5'-monophosphate dependent protein kinases, *Biochemistry* **16**:3080–3086.

DePeyer, J. E., Cachelin, A. B., Levitan, I. B., and Reuter, H., 1982, Ca^{2+}-Activated K^+ conductance in internally perfused snail neurons is enhanced by protein phosphorylation, *Proc. Natl. Acad. Sci. U.S.A.* **79**:4207–4211.

Deterre, P., Paupardin-Tritsch, D., Bockaert, J., and Gerschenfeld, H. M., 1981, Role of cyclic AMP in a serotonin-evoked slow inward current in snail neurones, *Nature* **290**:783–785.

Drummond, A., Benson, J., and Levitan, I. B., 1980, Serotonin-induced hyperpolarization of an identified *Aplysia* neuron is mediated by cyclic AMP, *Proc. Natl. Acad. Sci. U.S.A.* **77**:5013–5017.

Ewald, D. A., Williams, A., Levitan, I. B., 1985, Modulation of single calcium-dependent potassium channel activity by protein phosphorylation, *Nature* **315**:503–506.

Glass, D., and Krebs, E., 1980, Protein phosphorylation catalyzed by cyclic AMP-dependent and cyclic GMP-dependent protein kinases, *Annu. Rev. Pharmacol. Toxicol.* **20**:363–388.

Hamill, O. P., Marty, A., Neher, E., Sakmann, B., and Sigworth, F. J., 1981, Improved patch-clamp techniques for high-resolution current recording from cells and cell-free membrane patches, *Pfluegers Arch.* **391**:85–100.

Kaczmarek, L., Jennings, K., and Strumwasser, F., 1978, Neurotransmitter modulation, phosphodiesterase inhibitor effects, and cyclic AMP correlates of afterdischarge in peptidergic neurites, *Proc. Natl. Acad. Sci. U.S.A.* **75**:5200–5204.

Kaczmarek, L. K., Jennings, K. R., Strumwasser, F., Nairn, A. C., Walter, U., Wilson, F. D., and Greengard, P., 1980, Microinjection of catalytic subunit of cyclic AMP-dependent protein kinase enhances calcium action potentials of bag cell neurons in cell culture, *Proc. Natl. Acad. Sci. U.S.A.* **77**:7487–7491.

Klein, M., and Kandel, E., 1978, Presynaptic modulation of voltage-dependent Ca^{2+} current: Mechanism for behavioral sensitization in *Aplysia californica*, *Proc. Natl. Acad. Sci. U.S.A.* **75**:3512–3516.

Kuo, J., and Greengard, P., 1969, Cyclic nucleotide-dependent protein-kinases, IV. Widespread occurrence of adenosine 3',5'-monophosphate-dependent protein kinase in various tissues and phyla of the animal kingdom, *Proc. Natl. Acad. Sci. U.S.A.* **64**:1349–1355.

Levitan, I. B., and Norman, J., 1980, Different effects of cAMP and cGMP derivatives on the activity of an identified neuron: Biochemical and electrophysiological analysis, *Brain Res.* **187**:415–429.

Maruyama, Y., and Petersen, O. H., 1982, Cholecystokinin activation of single-channel currents is mediated by internal messenger in pancreatic acinar cells, *Nature* **300**:61–63.

Miller, C., 1983, Integral membrane channels: Studies in model membranes, *Physiol. Rev.* **63**:1209–1242.

Neher, E., and Sakmann, B., 1976, Single-channel currents recorded from membrane of denervated frog muscle fibers, *Nature* **260**:799–802.

Osterrieder, W., Brum, G., Hescheler, J., Trautwein, W., Flockerzi, V., and Hofmann, F., 1982, Injection of subunits of cyclic AMP-dependent protein kinase into cardiac myocytes modulates Ca^{2+} current, *Nature* **298**:576–578.

Pellmar, T. C., 1981, Ionic mechanism of a voltage-dependent current elicited by cyclic AMP, *Cell Mol. Neurobiol.* **1**:87–97.

Peters, K., Demaille, J., and Fischer, E., 1977, Adenosine 3′:5′-monophosphate dependent protein kinase from bovine heart. Characterization of the catalytic subunit, *Biochemistry* **16**:5691–5697.

Shuster, M., Camardo, J., Siegelbaum, S., Kandel, E. R., 1985, Cyclic AMP-dependent protein kinase closes the serotonin-sensitive potassium channels of *Aplysia* sensory neurons in cell-free membrane patches, *Nature* **313**:392–395.

Siegelbaum, S. A., Camardo, J. S., and Kandel, E. R., 1982, Serotonin and cyclic AMP close single K^+ channels in *Aplysia* sensory neurones, *Nature* **299**:413–417.

Treistman, S. N., 1981, Effect of adenosine 3′,5′-monophosphate on neuronal pacemaker activity: A voltage clamp analysis, *Science* **211**:59–61.

Treistman, S. N., and Levitan, I. B., 1976a, Alteration of electrical activity in molluscan neurones by cyclic nucleotides and peptide factors, *Nature* **261**:62–64.

Treistman, S. N., and Levitan, I. B., 1976b, Intraneuronal guanylyl-imidodiphosphate injection mimics long-term synaptic hyperpolarization in *Aplysia*, *Proc. Natl. Acad. Sci. U.S.A.* **73**:4689–4692.

Tsien, R. W., 1973, Adrenaline-like effects of intracellular iontophoresis of cyclic AMP in cardiac purkinje fibers, *Nature (New Biol.)* **245**:120–122.

Tsien, R. W., 1977, Cyclic AMP and contractile activity in heart, *Adv. Cyclic Nucleotide Res.* **8**:363–420.

Walsh, D., and Ashby, C., 1973, Protein kinases: Aspects of their regulation and diversity, *Recent Prog. Horm. Res.* **29**:329–359.

Chapter 21

The Cell-to-Cell Membrane Channel
Its Regulation by Cellular Phosphorylation

Werner R. Loewenstein

1. INTRODUCTION

I shall deal with a channel that spans two cell membranes. This channel was a solution to an ancient evolutionary problem: how to interconnect the cyberneting loops of information of different cells and to do so without mixing the information sources. The seeds for this problem were sown by evolution herself as she grew, during the first one and a half billion years: a hydrophobic membrane for the cells, a barrier to keep the hard-won informational molecules from drifting apart and to keep nonsense molecules from drifting in. The barrier was instrumental in the selection of cybernetically apt units in an aqueous environment and was absolutely necessary for Darwinian evolution of cells. But, over the next two billion years, the organismic times, it became an obstacle to multicellular organization. The problem was then how to get signal molecules across the hydrophobic barrier to interconnect the units without abolishing its primitive insulating function.

The problem eventually met with three elegant solutions—three kinds of membrane bypasses that managed to keep the cell membrane leakproof

---

WERNER R. LOEWENSTEIN • Department of Physiology and Biophysics, University of Miami School of Medicine, Miami, Florida 33101.

by hydrophobic lipid–protein or lipid–lipid interactions (Fig. 1). One solution was to imbed the signal molecules into the cell membrane and to let them interact with each other on cell contact. The signals here complete a molecular bridge between the interiors of two cells. A second solution uses a vehicle with a membrane skin capable of fusing with the cell membrane to transport the intracellular hydrophilic signals across—a "wolf in the sheepskin" ruse. The continuity between the cells here is given by the extracellular medium, and the ruse may be used twice, on the way out of the cells and on the way in, assisted by a receptor molecule; or alternatively, a receptor associated with a membrane channel or a transmembrane molecular bridge might complete the way in. A third solution was to tunnel through the two membranes of contiguous cells.

The last was the most dangerous of the three evolutionary strategies. It risked losing the individuality of the information sources. But it was a vital risk because it afforded the most direct way of communication. The solution came—as it generally does—through the selection of the right protein for the job: a protein with a hydrophilic bore of 16–20 Å (mammalian cells), wide enough to let through the cytoplasmic inorganic ions, the metabolites, the building blocks of biomolecular synthesis, and the high-energy phosphates but not the macromolecules. The channel is made of two protein halves—one contributed by each membrane—tightly joined, forming a rather leakproof aqueous tunnel between cells (Loewenstein, 1974) (Fig. 2).

We find this cell-to-cell channel in nearly all organized tissues throughout the phylogenetic scale, constituting one of the most basic forms of cellular

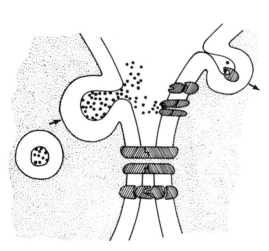

Figure 1. Membrane bypasses for intercellular communication. (1) Bridge of interacting molecules spanning the two cell membranes. (2) Intercellular tunnel: a channel spanning the two cell membranes. (3) Bypasses for the less direct humoral forms of communication (via the extracellular liquid): (a) membrane vesicles that fuse with the cell membrane. This bypass serves hydrophilic signals on the way out of cells (exocytosis) and on the way in assisted by a receptor molecule (receptor-mediated endocytosis). In relayed modalities of the humoral communication form, (b) a receptor associated with a membrane channel or (c) a transmembrane molecular bridge is the bypass completing the entry route into the target cell.

Cell-to-Cell Channels and Phosphorylation

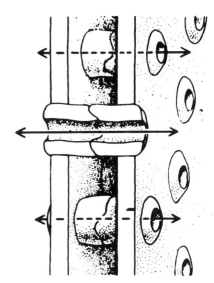

Figure 2. The cell-to-cell channel. The channel is made of two tightly joined transmembrane proteins, one from each apposing cell membrane. The aqueous permeation-limiting bore of the channel is 16–20 Å in mammalian cells. (From Loewenstein, 1974.)

communication. As a rule, a cell in a tissue is thus connected with several neighbors, and so whole organs or large organ parts are continuous from within (Loewenstein, 1966).

The channels tend to cluster in the junctional membrane regions, where they are recognizable electron microscopically as clusters of membrane particles, called a "gap junction." The particles are hexameric, six slanted subunits forming a half-channel (Unwin and Zampighi, 1980).

The channels form spontaneously when competent cells are brought into contact. They are then detectable as quantal increments of cell-to-cell conductance (Loewenstein *et al.*, 1978). Typically, the total conductance of such a nascent junction increases over a few minutes to a plateau (Ito *et al.*, 1974). This has led to the notion that the communicating junction forms by a progressive accretion of individual cell-to-cell channels (Ito *et al.*, 1974). Electron microscopically, the number of particles in the forming gap junction increases with time (Johnson *et al.*, 1974).

One physiological regulator of the channel is Ca^{2+}. The channel closes within fractions of a second when the concentration of this ion rises above 10^{-5} M in the junctional locale (Rose and Loewenstein, 1976). The six subunits of the gap junction particle then appear to slide radially towards one another, obliterating the bore—a shutter mechanism of the iris diaphragm sort (Unwin and Ennis, 1983). This closure mechanism is activated when a cell springs a leak (Oliveira-Castro and Loewenstein, 1971) or its ion pumps stand still (Rose and Loewenstein, 1976). It is a basic survival mechanism allowing the cell community to disconnect an unhealthy member and to seal itself off (Loewenstein, 1966).

Another physiologically interesting channel regulator has recently come to light: a cAMP-dependent protein kinase. This phosphorylating enzyme appears to control the formation process of the channel and, thus, to determine the number of (open) channels in the cell junction. This regulation is the subject of this chapter.

Jean Flagg-Newton and I accidentally came upon this mechanism in 1978. We were probing the channels of mammalian cells in cultures, trying to characterize their permeabilities. One week we had forgotten to feed the cultures and noticed, to our surprise, that these starved cells had unusually high junctional permeabilities. It confounds one that something should improve by hunger, but this discovery soon set us on the track of cAMP, for it was known that serum-starved cells have increased levels of cAMP. So it came that we joined the league of fans of this mighty substance.

2. THE CELL-TO-CELL CHANNELS ARE UP-REGULATED BY cAMP-DEPENDENT PHOSPHORYLATION

We did three kinds of experiments exploring the role of cAMP in cell-to-cell communication. In one kind of experiment we supplied the cells with cAMP. In another we elevated their endogenous cAMP level by treatments with phosphodiesterase inhibitors or adenylate cyclase activators. In a third kind we supplied them with the phosphorylating enzyme subunit. All had the same outcome: an increase in junctional permeability (Flagg-Newton, 1979; Flagg-Newton and Loewenstein, 1981; Flagg-Newton et al., 1981; Radu et al., 1982; Wiener and Loewenstein, 1983).

2.1. Experiments with cAMP and Phosphodiesterase Inhibitors

Figure 3 shows an example in which we supplied dibutyryl cAMP plus the phosphodiesterase inhibitor caffeine to cells of the rat fibroblast B line. We probed the cell junctions with fluorescent-labeled polyamino acids—di- and triglutamic acid labeled with lissamine rhodamine B (LRB-Glu-Glu, LRB-Glu-Glu-Glu). These probes were microinjected into one member of a cell pair, and their rates of transit cell to cell (junctional transit rate) and cell to exterior (loss rate) were measured. The junctional transit rates of both probes increased within 4 hr of the treatment, about tripling within 24 hr. The loss rates stayed unchanged, clearly showing that the junctional permeability had increased.

Increases of junctional permeability of this sort were typical of cells that make channels with one another readily, as many mammalian cell types do in culture (Flagg-Newton et al., 1981; Radu et al., 1982). The most dramatic effects were found in cells that make channels sparingly:

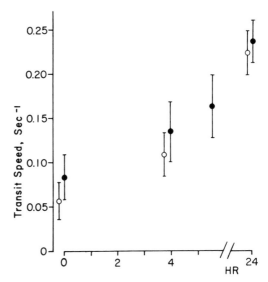

Figure 3. The rate of junctional molecular transfer is increased by treatment with dibutyryl cAMP (1 mM) and caffeine (1 mM). The junctions of isolated cell pairs of the rat B fibroblast line are probed with microinjected LRB-Glu$_2$ (●) and LRB-Glu$_3$ (○), and the junctional transit rate of these fluorescent molecules is measured (S.E. bars). Treatment starts at time zero. (From Flagg-Newton et al., 1981.)

in a cell type that displayed zero junctional permeability in basal conditions, a junctional permeability was "induced" by the cyclic nucleotide treatment (Azarnia et al., 1981).

The increases in junctional permeability went hand in hand with increases in the number of membrane particles of gap junction. In cells with basal expression of junctional permeability, both the number of particles in the gap-junctional clusters and the frequency of the clusters rose on treatment with cyclic nucleotides (Fig. 4). In the channel-deficient cells, gap-junctional particle clusters emerged where, before the treatment, there

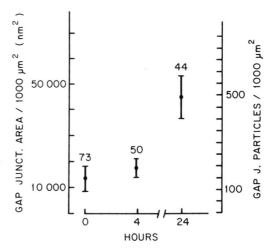

Figure 4. The number of gap-junctional membrane particles increases by treatment with dibutyryl cAMP (1 mM) and caffeine (1 mM). Mouse 3T3-BalbC cells. Abscissae: right, the number of particles per unit membrane area; left, the area of the particle clusters (gap junction area) per unit membrane area. Treatment starts at time zero. (From Flagg-Newton et al., 1981.)

were none. Treatment with the protein synthesis inhibitors cycloheximide or puromycin, but not treatment with the cytokinesis inhibitor cytochalasin B, blocked that effect (Azarnia *et al.*, 1981; Flagg-Newton *et al.*, 1981).

These results lead us to believe that the increase in junctional permeability reflects an increase in the number of open cell-to-cell channels.

2.2. The Up-Regulator Scheme

In summary of the results so far, we may write the following regulator scheme:

$$\text{cAMP} \uparrow \longrightarrow P_J \uparrow \tag{1}$$

where the concentration of cAMP in the cells determines the steady-state junctional permeability (P_J). A plausible mechanism is an up-regulation of the number of open channels, channels newly formed, or the opening of preexisting ones:

$$\text{cAMP} \uparrow \longrightarrow \text{number of channels} \uparrow \tag{2}$$

2.3. Hormones Drive the Up-Regulation

As one would expect, the mechanism can be driven by hormones (Radu *et al.*, 1982). Figure 5 illustrates this for a catecholamine. The cells chosen for this experiment, rat neuroglioma C-6 cells, had β-adrenergic receptors, and a few micromoles of isoproterenol set the channel up-regulation into motion. We were probing here with LRB-Glu-Glu, a relatively large molecule with three negative charges. In the basal condition, there were not enough channels to give a detectable transfer to any of the neighbors of the cell injected with that probe (Fig. 5a). About 7 hr after the hormone treatment (and the consequent rise in the endogenous cAMP concentration), the probe was typically transferred to several (first-order) neighbors (six out of the seven in Fig. 5b), and the number of gap-junctional particles about tripled.

Such enhancement of junctional permeability was shown for other cAMP-elevating hormones too: for gonadotropin in frog oocyte/follicle junction (Browne and Wiley, 1979) and for prostaglandin E_1 in human lung WI-38 cells (Radu *et al.*, 1982).

2.4. Protein Kinase I Is Instrumental in Regulation

What protein kinase mediates this channel up-regulation? The regulatory subunit of the cAMP-dependent protein kinase, the receptor for cAMP, occurs in two isoforms, I and II. In the holoenzyme, one or the

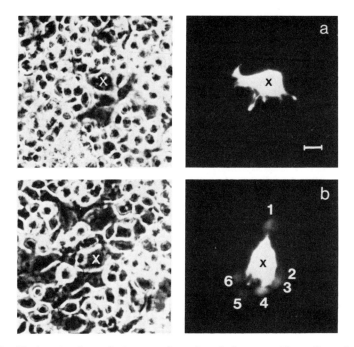

Figure 5. The junctional transfer increases by action of a hormone. Neuroglioma C-6 cells, which have β-adrenergic receptors. Fluorescent marker LRB-Glu$_2$ is microinjected into a cell (x) in control conditions (a) and 7.1 hr after application of 50 μM isoproterenol (b). The injected cell has seven contiguous neighbors in each case. None of these neighbors exhibits the fluorescent tracer in the control condition, but six do so in the hormone-treated condition. Left, phase-contrast micrograph; right, dark field. Calibration, 15 μm. (From Radu et al., 1982.)

other form is associated with the catalytic subunit (invariant), which splits off to catalyze the phosphorylation of the protein target:

$$2\ \text{cAMP} + 2\ \text{R-C} \longrightarrow 2\ \text{R-cAMP} + 2\ \text{C}$$

$$T + n\text{NTP} \xrightarrow{C} \text{T-P}_n + n\text{NDP}$$

where R and C are the regulatory and catalytic subunits, respectively, T the target protein, and NTP, a high-energy phosphate, commonly ATP (Krebs, 1972; Rosen and Krebs, 1981; see also Kuo and Greengard, 1969).

To learn which of the two isoenzymes operates in this channel regulation, we resorted to mutant cells deficient in one or the other enzyme form, namely, one-step mutants of the Chinese hamster ovary cell line derived by Michael Gottesman and his colleagues (1981). One of the mutants (I$^-$) lacked protein kinase type I and had little of type II, and another mutant (II$^-$) lacked protein kinase II and had type I. Mutant I$^-$

turned out to be clearly deficient in junctional permeability (Fig. 6). Mutant II^-, on the other hand, was only slightly less permeable than the wild-type cells (despite a lower cAMP sensitivity of its kinase) (Wiener and Loewenstein, 1983).

Thus, at least, enzyme type I is involved in the channel regulation process and seems sufficient for it; this was the only isoenzyme present in the communication-competent mutant II^-.

This correlation between enzyme phenotype and communication phenotype also held for backshifts of mutation. After carrying mutant I^- in culture for about half a year, clones appeared that had reverted to the normal communication phenotype. This reversion invariably went hand in hand with reversion to the normal enzyme phenotype (Table I).

Next we asked whether incorporation of the catalytic subunit into the mutant I^- would remedy its communication deficiency. Using a chemical permeabilization procedure (Kerrick and Krasner, 1975), we loaded the cells with a purified preparation of the enzyme subunit from bovine skeletal muscle (Wiener and Loewenstein, 1983). On loading, the junctional permeability of the mutant cells rose significantly (Fig. 7, parts 1 and 2) and to a level comparable to that of the wild-type cells (compare experiment series 2 and 4). Control loadings of the mutant cells with inactivated enzyme (3) or of wild-type cells with active enzyme (4) gave no change in junctional permeability.

2.5. The Regulator Scheme Enlarged: Heuristics of Interaction

Based on these results, we write the up-regulator scheme in the form

$$cAMP + R^I C \longrightarrow C \longrightarrow T\text{-}P \longrightarrow \text{channels} \uparrow \qquad (3)$$

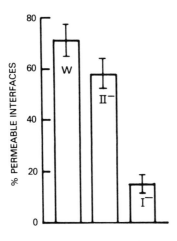

Figure 6. Communication deficiency is associated with deficiency of protein kinase type I. The incidence of LRB-Glu-permeable cell interfaces (% ± S.E.)—a sensitive index of junctional permeability—in Chinese hamster ovary cells, wild type (W) and mutants I^- and II^-. The statistical confidence level of the difference between I^- and W is $P < 0.0005$ (no significant difference between II^- and W). (From Wiener and Loewenstein, 1983.)

Table I. Cyclic-AMP-Dependent Protein Kinase and Communication Phenotypes of Clones from a Reverting I⁻ Culture

Clone	Kinase I[a] (units/ml)	Permeable interfaces[b] (%)
1	392	71 ± 6
2	360	69 ± 5
3	326	72 ± 7
4	403	78 ± 6
5	381	73 ± 6
6	—	7 ± 2
7	—	13 ± 4
8	—	8 ± 3
9	—	11 ± 4
10	—	6 ± 2
11	—	7 ± 2
W	380	72 ± 6
I⁻	—	12 ± 4

[a] One activity unit = 1 pmole of P transferred/min per mg protein. — denotes undetectable activity (<25 units/ml). Kinase II ranged 180–260 units/ml in all cases (including those with undetectable kinase I activity) except for W, where it was 420. The clones were all derived from the culture of Fig. 7, series 2. At the bottom of the table are listed the traits of the original wild-type (W) and I⁻ cells.
[b] Incidence of LRB-Glu permeable interfaces (mean ± SE). The confidence level of the difference between the incidences of I⁺ and I⁻ clones was, in all cases, $P < 0.0001$. (From Wiener and Loewenstein, 1983.)

where cAMP, by way of the type-I regulatory subunit of the protein kinase, promotes the phosphorylation of a target protein (T) determining the number of (open) cell-to-cell channels. Integrated into cell physiology, the scheme becomes

$$h + r \longrightarrow cAMP + R^I C \longrightarrow C \longrightarrow T\text{-}P \longrightarrow \text{channels} \uparrow \quad (4)$$

where h is a hormone and r its receptor.

The scheme now has broad heuristic value. It predicts that, with the right receptor and protein-kinase endowment, a cell community can regulate its communication by hormones. This regulatory capacity is finely graded and so affords a wide range of cellular controls. We have already seen that certain hormones play such a regulatory role, and, no doubt, more hormonal examples will soon become known.

Furthermore, by merely adding a (closed) loop between the channels and hormones, the scheme acquires predictive value regarding the hor-

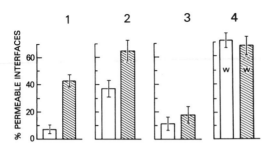

Figure 7. Cyclic-AMP-dependent protein kinase corrects communication defect in mutant I⁻ cells. The cells were loaded with the catalytic enzyme subunit, and the incidence of LRB-Glu-permeable cell interfaces was evaluated 4–4.5 hr after the loadings (crosshatched bars). Open bars give the corresponding values of the unloaded controls. In series 1 and 2, I⁻ cells were loaded with active enzyme; in 3, with inactivated enzyme; in series 4, wild-type cells (W) were loaded with active enzyme. Confidence levels of the differences between the values of enzyme-loaded cells and unloaded controls: series 1, $P < 0.002$; series 2, $0.01 > P > 0.005$; series 3, $0.10 > P > 0.05$. (The values of series 2 and 4 can be compared with each other; the cultures had equal densities.) (From Wiener and Loewenstein, 1983.)

monal form of communication: it predicts a self-reinforcement of the hormone response in certain conditions. This follows from the fact that cAMP fits through the channel. Thus, in tissues where this nucleotide is the second messenger of the hormone action, the action may be amplified: the cellular response to the hormone may spread from cell to cell where the phosphodiesterase action is not excessive. The possibility of self-reinforcement comes about because the instruments of such spread, the cell-to-cell channels, are themselves responsive to the nucleotide; they proliferate. Hence, under persistent hormone action, the hormonal communication between cells may be facilitated. Thus, we get a self-regulating cellular amplifier as a product of the interaction between two basic forms of cellular communication. Such variable amplifiers have interesting physiological potential not only for the dissemination of the hormonal response of organs and tissues but also as low-pass filters of receptor noise and as devices for hierarchical cellular interactions (Loewenstein, 1981).

The primary target of phosphorylation (T in the scheme) remains to be identified. It is possible that the channel itself is that target. Indeed, the protein of lens gap junction can be phosphorylated *in vitro* by a cAMP-dependent protein kinase (Johnson and Johnson, 1982), and it should soon become known whether this *in vitro* finding is physiologically meaningful. At present it seems equally possible that the critical target of phosphorylation is a step or more removed from the channel; the permeability regulation is relatively slow (hours), and one component, at least, depends on protein synthesis.

3. THE CELL-TO-CELL CHANNELS ARE DOWN-REGULATED BY TYROSINE PHOSPHORYLATION

I turn now to experiments in which we tried to interfere with the cAMP-dependent channel regulation. Will the normal channel regulation

Cell-to-Cell Channels and Phosphorylation

be perturbed when phosphorylations of other sorts are activated in the cells? We asked this for the case in which tyrosine residues, instead of serine or threonine (the targets of cAMP-dependent phosphorylation), are phosphorylated. Tyrosine phosphorylation is relatively rare in normal cells, but it can be induced massively by certain transforming viruses, such as the Rous sarcoma virus. The *src* gene of this RNA virus codes for a tyrosine-phosphorylating protein kinase in the host cells, the pp60^{src} protein (Collet *et al.*, 1980; Hunter, 1980; Bishop, 1982).

3.1. The *src* Protein Causes Reversible Down-Regulation

We used cells containing temperature-sensitive mutant *src* gene. Their pp60^{src} is thermolabile, so that its protein kinase activity can be switched from inactive to active form by merely shifting the culture temperature from 41°C (the nonpermissive temperature for the virus) to 33–35°C, and vice versa. Tyrosine phosphorylation sets in within a half hour of a temperature downshift (Radke and Martin, 1979). We found that the junctional permeability is down-regulated within that time (Azarnia and Loewenstein, 1984a).

Figure 8 shows the down-regulation for quail embryo cells *ts*68 as well as the reversibility of the effect: LRB-Glu junctional transfer rose on temperature upshift, fell on downshift, and rose again on upshift (Fig. 8I). These effects were not caused by the temperature changes themselves: the junctional transfer was not significantly changed over that temperature range in normal cells or in cells infected with wild-type virus. Changes in permeability of nonjunctional membrane were ruled out as sources of the effects: the loss rates of LRB-Glu were typically unchanged (Fig. 8IIA) or, in leakier cells, were in the wrong direction for that (Fig. 8IIB). Thus, as in the case of cAMP-dependent phosphorylation, the change in junctional transfer represents a change in junctional permeability.

The junctional response is fast. The reduction of junctional permeability ensued within 25 min of lowering the temperature to the permissive level—one of the fastest effects of the *src* gene known. The reversal of the effect was just about as fast and was not affected by inhibitors of protein synthesis. Thus, if we may assume that the permeability loss and reversal reflect symmetrical processes, it is likely that the action of pp60^{src} is by way of a modification of the channel open state rather than by way of a modification of channel protein synthesis. This is not to say that the process of channel assemblage—an assemblage from precursor channel halves or subunits—could not be the target for modification but that it is unlikely that the synthesis (and degradation) of the precursors be the target.

A plausible mechanism is a modification of the channel protein or of

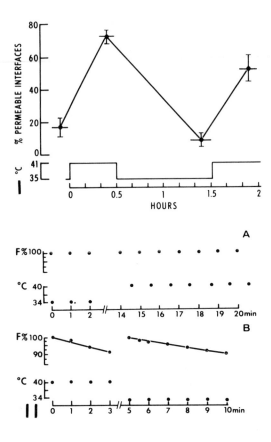

Figure 8. The *src* protein downregulates junctional permeability reversibly. Quail embryo cells infected with temperature-sensitive Rous sarcoma virus. The temperature (lower ordinates) was shifted between the permissive and the nonpermissive levels of the virus, that is, the levels at which the tyrosyl kinase of the *src* protein is activated and inactivated, respectively. I gives the incidence of LRB-Glu-permeable cell interfaces ($\pm$ S.E.) determined just before each temperature shift (upper ordinates). IIA shows the LRB-Glu loss rates at the two temperature levels of a typical tight cell; B of a leakier cell, a relatively rare case. The rate constants of the curve are 1.8 and 2.7 min^{-1} at the permissive and the nonpermissive temperature levels, respectively. F is the cellular fluorescence. (From Azarnia and Loewenstein, 1984a.)

an immediate regulator protein by (tyrosine) phosphorylation. A direct mechanism of this sort would be consistent with the swiftness of the junctional response to the pp60^{src} kinase activity. Whether the primary protein target of this viral phosphorylation is the same as that in the normal cAMP-dependent phosphorylation remains to be seen.

In this regard, I point out that the *src* protein kinase mechanism seems to override the cAMP-dependent protein kinase mechanism. When the virus-infected cells were treated with cAMP, phosphodiesterase inhibitors or the adenylate cyclase activator forskolin (treatments that powerfully raise junctional permeability in normal cells), the junctional downregulation by the *src* protein took place just the same. These treatments then failed to raise the junctional permeability even at the nonpermissive temperature (Azarnia and Loewenstein, 1984a). This dominance is explainable if the thermosensitive mutant were slightly "leaky," giving a low level of phosphorylation at the nonpermissive temperature, enough to prevent the cAMP response but not to produce the other phenotypic changes.

The effect of the *src* protein seems quite general. We obtained effects on junctional permeability similar to those of the quail embryo cell *ts*68 with chick embryo cell *ts*LA90 and mouse 3T3 cell *ts*LA90 (and vole cell *wt*) (Azarnia and Loewenstein, 1984a,b). Not in all cases was the junctional response as fast as in the quail embryo cell *ts*68. In the 3T3 cell *ts*LA90, the response took more than 1 hr. This may reflect a lower tyrosyl kinase activity; the kinase activity of this ("tighter") mutant is about one-tenth that of *ts*68 (J. Brugge, unpublished results).

3.2. The Channel Response Is Independent of the Cytoskeleton

Among the various conspicuous cellular alterations caused by the *src* protein that go under the portmanteau of transformation are alterations in the cytoskeleton. These might well have a bearing on the formation and stability of cell junctions. Thus, the question naturally arises as to whether the change in junctional communication is secondary to the cytoskeletal change. Vinculin, the protein that anchors the microfilaments to the cell membrane, is tyrosine phosphorylated (Chen and Singer, 1982), and this phosphorylation is thought to lead to microfilament disorganization and hence to the cells' rounding up (Sefton *et al.*, 1981). These cytoskeletal effects lag hours behind the effect on junctional permeability, but this does not rid one of the doubt that earlier, subtler (undetected) cytoskeletal changes might critically interfere with the cell membrane contact needed for channel formation.

This question was set to rest by experiments with *src*-containing vole cells, namely, with a clone of genetic revertants isolated by Anthony Faras. This clone had reverted to the normal cytoskeletal phenotype yet had retained the high levels of pp60^{src} kinase (and the tumorigenicity) of the fully transformed parent clone (Lau *et al.*, 1979; Nawrocki *et al.*, 1984). Its junctional permeability turned out to be lower than that of normal cells, as low as that of the fully transformed cells (Azarnia and Loewenstein, 1984b). The alteration of cell-to-cell channels determined by the *src* gene, thus, is independent of the alteration of the cytoskeleton.

3.3. The Channel Response Is Highly Sensitive to Tyrosyl Kinase

The results obtained with another revertant clone of vole cells—a revertant that had lost most of the tyrosyl kinase activity—gave an idea of the relative sensitivity of the junctional response. This clone had a remnant of only 2–3% of the pp60^{src} kinase activity of the fully transformed parent and was cytoskeletally normal (and lacked tumorigenicity) (Nawrocki *et al.,* 1984). It turned out to have a higher junctional LRB-Glu permeability than the fully transformed parent, but its permeability was still lower than that of the normal cells (Azarnia and Loewenstein, 1984b). The small kinase remnant apparently sufficed to depress junctional permeability but not to produce the other alterations of transformation. The relatively high sensitivity implied by this result echoes the discussion above of the result of the *src* dominance over the cAMP-dependent channel regulation. It is interesting in this connection that pp60^{src} is found close to the membrane of cell junction (Willingham *et al.,* 1979; Schriver and Rohrschneider 1981; Nigg *et al.,* 1982).

3.4. Down-Regulation by the Large and Small *T* Proteins Encoded by SV-40

A down-regulation of the channel also was obtained by action of the DNA virus SV-40. Here two genes are involved in the cellular transformation process: the *A* gene that codes for the large *T* protein, the *T* antigen, and the *F* gene that codes for the little *t* protein, the *t* antigen (Lai and Nathans, 1975; Tooze, 1981).

We used viruses with a mutation on the *A* gene that renders *T* thermolabile. The overall effect of shifting the temperature from the nonpermissive to the permissive level was not unlike that obtained with the thermolabile pp60^{src}: the junctional permeability fell (reversibly). But the effect was slower. It had lag periods of 4 hr or more. Such junctional down-regulation in the T^+ state was found in experiments with epithelial cells—primary rat pancreas islet cells—and fibroblasts—rat embryo cells and NREF cells infected with different viruses—*ts*A58, *ts*A239, and *ts*209—all with *A* mutations located between nucleotides 3476 and 3733 on the SV-40 genome (Azarnia and Loewenstein, 1984c).

These experiments showed that the *T* antigen causes cell-to-cell channel alteration. Is it a sufficient cause for this? To answer this question, one needs a mutant of the *F* gene. Since there are no temperature-sensitive mutants available for that gene, we made do with a deletion. We used a viral DNA constructed by Sompayrac and Danna (1983) by ligating a DNA segment containing a temperature-sensitive mutation on the *A* gene (*ts*A58 or *ts*2009) with a segment containing an *F* gene deletion (*dl*54-59), and a DNA prepared by William Topp by creating an *F* gene deletion at the Tag I site in *ts*A58 DNA.

Incorporated into rat pancreas islet cells and mouse 10T½ cells, these DNAs provided us with systems that could be switched $T^-t^- \rightleftarrows T^+t^-$ by temperature shifts. The design was straightforward: a junctional response, namely, a reduction of junctional permeability, is expected here if and only if *T* is a sufficient cause. The answer was negative. A response ensued only in the T^+t^+ condition, the condition of the pure temperature-sensitive A mutations above (Azarnia and Loewenstein, 1984c).

In conclusion, both T proteins are required for the cell-to-cell channel alteration. This requirement is distinct from the requirement for the cytoskeletal alteration, for which t is sufficient (Frisque et al., 1980). So once again, the alteration of the channel is genetically independent of the alteration of the cytoskeleton.

There is little to guide us as yet to the general mechanism of action of these two SV-40 channel-modifying proteins. T binds to a cellular phosphoprotein (Tevethia et al., 1980) and thus conceivably might act through phosphorylation, but neither the target nor the mechanism of phosphorylation is known.

REFERENCES

Azarnia, R., and Loewenstein, W. R., 1984a, Intercellular communication and the control of growth. X. Alteration of junctional permeability by the src gene. A study with temperature-sensitive mutant Rous sarcoma virus, J. Membr. Biol. **82**:191–205.

Azarnia, R., and Loewenstein, W. R., 1984b, Intercellular communication and the control of growth. XI. Alteration of junctional permeability by the src gene in a revertant cell with normal cytoskeleton, J. Membr. Biol. **82**:207–212.

Azarnia, R., and Loewenstein, W. R., 1984c, Intercellular communication and the control of growth. XII. Alteration of junctional permeability by simian virus 40, J. Membr. Biol. **82**:213–222.

Azarnia, R., Dahl, G., and Loewenstein, W. R., 1981, Cell junction and cyclic AMP. III. Promotion of junctional membrane permeability and junctional membrane particles in a junction-deficient cell type, J. Membr. Biol. **63**:133–146.

Bishop, J. M., 1982, Retroviruses and cancer genes, Adv. Cancer Res. **37**:1–37.

Browne, C. L., and Wiley, H. S., 1979, Oocyte–follicle cell gap junctions in *Xenopus laevis* and the effects of gonadotropin on their permeability, Science **203**:182–183.

Chen, W.-T., and Singer, S. J., 1982, Immunoelectron microscopic studies of the sites of cell–substratum and cell–cell contacts in cultured fibroblasts, J. Cell Biol. **95**:205–222.

Collett, M. S., Purchio, A. F., and Erikson, R. L., 1980, Avian sarcoma virus-transforming protein, $pp60^{src}$ shows protein kinase activity specific for tyrosine, Nature **285**:167–169.

Flagg-Newton, J. L., 1979, Cyclic AMP increases junctional permeability in cultured mammalian cells, Biophys. J. **25**(2):297a.

Flagg-Newton, J. L., and Loewenstein, W. R., 1981, Cell junction and cyclic AMP. II. Modulation of junctional membrane permeability, dependent on serum and cell density, J. Membr. Biol. **63**:123–131.

Flagg-Newton, J. L., Dahl, G., and Loewenstein, W. R., 1981, Cell junction and cyclic AMP. I. Upregulation of junctional membrane permeability and junctional membrane particles by administration of cyclic nucleotide or phosphodiesterase inhibitor, J. Membr. Biol. **63**:105–121.

Frisque, R. J., Rifkin, D. B., and Topp, W. C., 1980, Requirement for the large T and small t proteins of SV40 in the maintenance of the transformed state, Cold Spring Harbor Symp. Quant. Biol. **44**:325–331.

Gottesman, M. M., Singh, T., LeCam, A., Roth, C., Nicolas, J.-C., Cabial, F., and Pastan,

I., 1981, Cyclic-AMP-dependent phosphorylation in cultured fibroblasts: A genetic approach, *Cold Spring Harbor Conf. Cell Prolif.* **8A**:195–209.

Hunter, T., 1980, Proteins phosphorylated by the RSV transforming function, *Cell* **22**:647–648.

Ito, S., Sato, E., and Loewenstein, W. R., 1974, Studies on the formation of a permeable membrane junction. II. Evolving junctional conductance and junctional insulation, *J. Membr. Biol.* **19**:339–355.

Johnson, K. R., and Johnson, R., 1982, Bovine lens MP26 is phosphorylated *in vitro* by an endogenous cAMP-dependent protein kinase, *Fed. Proc.* **41**:755.

Johnson, R., Hammer, M., Sheridan, J., and Revel, J. P., 1974, Gap junction formation between reaggregated Novikoff hepatoma cells, *Proc. Natl. Acad. Sci. U.S.A.* **71**:4536–4543.

Kerrick, W. G. L., and Krasner, B., 1975, Disruption of the sarcolemma of mammalian skeletal muscle fibers by homogenization, *J. Appl. Physiol.* **39**:1052–1055.

Krebs, E. G., 1972, Protein kinases, *Curr. Top. Cell. Regul.* **5**:99–133.

Kuo, J. F., and Greengard, P., 1969, Cyclic nucleotide-dependent protein kinases, IV. Widespread occurrence of adenosine 3′,5′-monophosphate-dependent protein kinase in various tissues and phyla of the animal kingdom, *Proc. Natl. Acad. Sci. U.S.A.* **64**:1349–1355.

Lai, C.-J., and Nathans, D., 1975, A map of temperature-sensitive mutants of simian virus 40, *Virology* **66**:70–81.

Lau, A. F., Krzyzek, R. A., Brugge, J. S., Erikson, R. L., Schollmeyer, J., and Faras, A. J., 1979, Morphological revertants of an avian sarcoma virus-transformed mammalian cell line exhibit tumorigenicity and contain pp60^{src}, *Proc. Natl. Acad. Sci. U.S.A.* **76**:3904–3908.

Loewenstein, W. R., 1966, Permeability of membrane junctions, *Ann. N.Y. Acad. Sci.* **137**:441–472.

Loewenstein, W. R., 1974, Cellular communication by permeable junctions, in *Cell Membranes: Biochemistry, Cell Biology and Pathology* (G. Weissmann and R. Claiborne, eds.), H.P. Publishing, New York, pp. 105–114.

Loewenstein, W. R., 1981, Junctional intercellular communication. The cell-to-cell membrane channel, *Physiol. Rev.* **61**:829–913.

Loewenstein, W. R., Kanno, Y., and Socolar, S. J., 1978, Quantum jumps of conductance during formation of membrane channels at cell–cell junction, *Nature* **274**:133–136.

Nawrocki, J. F., Lau, A. F., and Faras, A. J., 1984, Correlation between phosphorylation of a 34,000-molecular weight protein, pp60^{src}-associated kinase activity, and tumorigenicity in transformed and revertant vole cells, *Mol. Cell Biol.* **4**:212–215.

Nigg, E. A., Sefton, B. M., Hunter, T., Walter, G., and Singer, S. J., 1982, Immunofluorescent localization of the transforming protein of Rous sarcoma virus with antibodies against a synthetic *src* peptide, *Proc. Natl. Acad. Sci. U.S.A.* **79**:5322–5326.

Oliveira-Castro, G. M., and Loewenstein, W. R., 1971, Junctional membrane permeability: Effects of divalent cations, *J. Membr. Biol.* **5**:51–77.

Radke, K., and Martin, G. S., 1979, Transformation by Rous sarcoma virus: Effects of *src*-gene expression on the synthesis and phosphorylation of cellular polypeptides, *Cold Spring Harbor Symp. Quant. Biol.* **44**:975–982.

Radu, A., Dahl, G., and Loewenstein, W. R., 1982, Hormonal regulation of cell junction permeability: Upregulation by catecholamine and prostaglandin E_1, *J. Membr. Biol.* **70**:239–251.

Rose, B., and Loewenstein, W. R., 1976, Permeability of a cell junction and the local cytoplasmic free ionized calcium concentration. A study with aequorin, *J. Membr. Biol.* **28**:87–119.

Rosen, O. M., and Krebs, E. G. (eds.), 1981, *Protein Phosphorylation. Cold Spring Harbor Conference on Cell Proliferation,* Vol. 8A,B, Cold Spring Harbor Laboratory, New York.

Schriver, K., and Rohrschneider, L., 1981, Organization of pp60^{src} and selected cytoskeletal proteins within adhesion plaques and junctions of Rous sarcoma virus-transformed cells, *J. Cell Biol.* **89**:525–535.

Sefton, B. M., Hunter, T., Ball, E. H., and Singer, S. J., 1981, Vinculin: A cytoskeletal target of the transforming protein of Rous sarcoma virus, *Cell* **24**:165–174.

Sompayrac, L., and Danna, K. J., 1983, A simian virus 40 dl884/tsA58 double mutant is temperature sensitive for abortative transformation, *J. Virol.* **46**:620–625.

Tevethia, S. S., Greenfield, R. S., Flyer, D. C., and Tevethia, M. J., 1980, SV-40 transplantation antigen: Relationship to SV-40 specific proteins. *Cold Spring Harbor Symp. Quant. Biol.* **44**:235–242.

Tooze, J. (ed.), 1981, *DNA Tumor Viruses: Molecular Biology of Tumor Viruses,* Part 2, 2nd edition, Cold Spring Harbor Laboratory, New York.

Unwin, P. N. T., and Ennis, P. D., 1983, Calcium-mediated changes in gap junction structure: Evidence from the low angle X-ray pattern, *J. Cell Biol.* **97**:1459–1466.

Unwin, P. N. T., and Zampighi, G., 1980, Structure of the junction between communicating cells, *Nature* **283**:545–549.

Wiener, E. C., and Loewenstein, W. R., 1983, Correction of cell–cell communication defect by introduction of a protein kinase into mutant cells, *Nature* **305**:433–435.

Willingham, M. C., Jay, G., and Pastan, I., 1979, Localization of the ASV *src* gene product to the plasma membrane of transformed cells by electron microscopic immunocytochemistry, *Cell* **18**:125–134.

Chapter 22

The β-Cell Bursting Pattern and Intracellular Calcium

Illani Atwater and John Rinzel

1. INTRODUCTION

Mammalian islets of Langerhans are made up principally of insulin-secreting β cells but also contain glucagon-secreting α cells, somatostatin-secreting δ cells, and polypeptide-secreting PP cells as well as sympathetic and parasympathetic nerve terminals and capillary-associated endothelial cells. The islets of Langerhans from mouse are convenient for electrophysiological studies because they are large (up to 1 mm in diameter), easily distinguished from the surrounding pancreatic tissue, and contain a large proportion of β cells (about 85%), which are mainly located in the center of the islet.

The glucose-induced electrical activity in β cells has been the subject of extensive experimental studies over the last 15 years (see Atwater *et al.*, 1983). The main observations have been the following:

1. Glucose induces β-cell depolarization, which is associated with an increase in input resistance (Atwater *et al.*, 1978).

ILLANI ATWATER • Laboratory of Cell Biology and Genetics, NIADDK, National Institutes of Health, Bethesda, Maryland 20205.   JOHN RINZEL • Mathematical Research Branch, NIADDK, National Institutes of Health, Bethesda, Maryland 20205.

2. Glucose metabolism, rather than glucose *per se*, is required for glucose to be translated into this stimulatory signal (Atwater, 1980).
3. Each β cell responds to glucose in a graded fashion, and there is also variability among islets with respect to the concentration of glucose required to reach the threshold of excitation as well as to reach maximal response (Atwater *et al.*, 1980; Beigelman *et al.*, 1977).
4. At concentrations of glucose between threshold (around 5 mM) and maximal (around 20 mM), the β-cell electrical activity occurs in a spontaneous, regular pattern of bursts (Atwater *et al.*, 1980; Meissner, 1976).
5. The β-cell membrane permeability during the bursting pattern is controlled by at least three well-defined ionic channels. A Ca^{2+}-gated K^+ channel (Atwater *et al.*, 1979b), voltage-gated K^+ channel (Atwater *et al.*, 1979a), and a voltage-gated Ca^{2+} channel (Atwater *et al.*, 1981).

2. ROLE OF $[Ca^{2+}]_i$: DEPENDENCE ON GLUCOSE

Although the Ca^{2+} influx during the action potential is undoubtedly a fundamental trigger for the secretion process, a number of observations make the underlying burst pattern an interesting phenomenon to study. In particular, the burst pattern occurs in response to glucose metabolism (and certain other metabolized substrates), whereas alternative stimulants such as the sulfonylureas, which act directly on the β-cell membrane, induce depolarization and continuous spike activity (Atwater *et al.*, 1983). Also, the burst pattern is responsive to variable glucose concentrations, whereas the membrane potentials during bursting and the action potentials are not. Thus, the burst pattern reflects the link between glucose metabolism, the principal physiological secretagogue, and insulin release. The relative duration of the active phase of the burst pattern is correlated with the glucose concentration required to reach threshold for activation of that islet. The burst pattern is variable from islet to islet while remaining essentially identical for most cells within an islet (Meda *et al.*, 1984). Thus, the burst pattern recorded from a single β cell reflects the characteristics of that individual islet.

The interplay between β-cell membrane permeability and the intracellular concentration of free calcium, $[Ca^{2+}]_i$, has been proposed to regulate the characteristic glucose-induced burst pattern (Fig. 1A). During a burst of electrical activity, the membrane depolarizes, and Ca^{2+} channels are activated (active phase), leading to the generation of action potentials and an increased Ca^{2+} influx into the cell. During this phase, $[Ca^{2+}]_i$

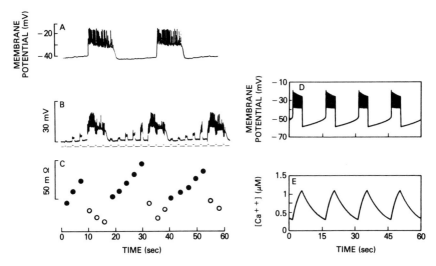

Figure 1. Spontaneous oscillations in membrane potential measured in mouse β cell in the presence of 11.1 mM glucose (A). Voltage deflections in response to current injection during the course of the membrane potential oscillations (B); the current pulses (0.1 nA) are shown just below the trace in B. Relative resistance changes during the burst pattern as measured from the voltage deflections (C). Time course profiles of membrane potential (D) and intracellular calcium concentration (E) predicted from theoretical Chay–Keizer (1983) model. Parameters of this five-variable model are as in Table 1 of Chay and Keizer (1983); here, $k_{Ca} = 0.032$ msec^{-1}.

builds up, K^+ channels are activated, and input resistance decreases (Fig. 1B,C). When the increased K^+ permeability tips the balance, then the membrane switches to the hyperpolarized value (silent phase). As the increased $[Ca^{2+}]_i$ is buffered or pumped out of the cell, the membrane slowly depolarizes, and input resistance increases. Note that the measured increase in input resistance parallels the predicted decrease in $[Ca^{2+}]_i$ (compare Figs. 1C and 1E). With sufficient depolarization, the Ca^{2+} channels are again activated, and the next burst is initiated. Thus, it has been proposed that glucose metabolism alters membrane potential by controlling the buffering of intracellular $[Ca^{2+}]_i$.

Changes in the time course of the burst pattern occur between about 5 and 20 mM glucose (Fig. 2). As glucose is increased, the relative duration of the active phase increases. The ability of glucose to modulate the burst pattern without changing membrane potential is peculiar to the β cell and different from stimulus-induced modulations in other bursting cells (see Levitan, Chapter 20). It was this property that led to the hypothesis that the Ca^{2+}-gated K^+ channel (a membrane permeability that is not voltage dependent at negative potentials and physiological $[Ca^{2+}]_i$) was the un-

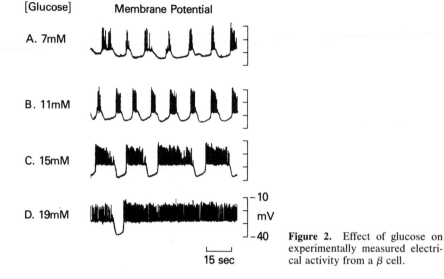

Figure 2. Effect of glucose on experimentally measured electrical activity from a β cell.

derlying link between glucose concentration and the β-cell electrical activity.

3. A BIOPHYSICAL/MATHEMATICAL MODEL

Recently, Chay and Keizer (1983) developed a mathematical model to describe the β-cell electrical activity. The model incorporates an adapted and expanded version of the classical Hodgkin and Huxley (1952) description of nerve membrane excitability. It includes four membrane permeabilities: voltage-gated K^+ and Ca^{2+} channels, Ca^{2+}-gated K^+ channels, and a leak permeability. It also describes the dynamic changes in intracellular free calcium that result from flux through the membrane and uptake (disappearance) of $[Ca^{2+}]_i$. The model reproduces well the observed electrical burst activity (Fig. 1D). Moreover, it predicts explicitly the time course of the changes in $[Ca^{2+}]_i$ (Fig. 1E), which has not yet been measured experimentally. The theoretical dependence of the burst pattern on glucose concentration is accounted for by varying only one of the parameters, k_{Ca}, the rate of Ca^{2+} uptake. As found experimentally (Beigelman et al., 1977), increasing glucose concentration (i.e., increasing k_{Ca} in the model) leads to longer active phases and shorter silent phases (Chay and Keizer, 1983); sufficiently large k_{Ca} leads to continuous spike activity—a state of depolarization uninterrupted by silent phases.

The qualitative features of this model have been exposed (Rinzel,

1985) by utilizing phase-space analysis (see FitzHugh, 1961, for an introduction to such concepts applied to models of membrane excitability). In the five-variable Chay–Keizer model, $[Ca^{2+}]_i$ changes slowly. This is because only a small fraction, f, of the total intracellular calcium is free and therefore able to affect such changes at any given time; most of the calcium is attached to high-affinity, rapid binding sites of cytoplasmic components. To explain the model we exploit the slow rate of change in $[Ca^{2+}]_i$ by first considering $[Ca^{2+}]_i$ as a parameter and studying its influence on the faster spike-generating membrane variables.

Over a certain range of $[Ca^{2+}]_i$, the theoretical membrane has an N-shaped steady-state current–voltage relationship with up to perhaps three steady-state or rest potentials (Fig. 3A, note open circles). The dependence of these steady-state potentials on $[Ca^{2+}]_i$ is indicated by the Z-shaped curve in Fig. 3B. For large $[Ca^{2+}]_i$, the Ca^{2+}-activated K^+ conductance is dominant, and there is only a single rest state of hyperpolarization. For low $[Ca^{2+}]_i$, this conductance is negligible, and the depolarized steady-state potential (about 30 mV) is determined primarily by the balance of the inward and outward V-dependent Ca^{2+} and K^+ currents. For intermediate levels of $[Ca^{2+}]_i$ and appropriate parameter values in the model, the membrane potential could remain at either the upper or lower steady-state value; each is stable. The middle potential (shown dashed) corresponds to an unstable state that is not physically realizable.

In such a case of bistability, the model can exhibit a special type of bursting: slow-wave activity in which the active phase has no spikes. To understand this, we append to the excitation subsystem the slow dynamics of $[Ca^{2+}]_i$. Each long-dashed curve in Fig. 3B indicates (for each of two different values of k_{Ca}) the voltage at which the (steady) Ca^{2+} influx just balances the uptake of Ca^{2+}. For values of V and $[Ca^{2+}]_i$ above such a curve, $[Ca^{2+}]_i$ is increasing. Below the curve, $[Ca^{2+}]_i$ decreases. Where this curve (called the Ca^{2+}-nullcline) intersects the Z curve (filled circle), neither $[Ca^{2+}]_i$ nor V is changing; this corresponds to a steady state of the full model. For low glucose (small k_{Ca}), this intersection occurs at the more negative V as in a resting β cell. At high glucose, we find a steady state of depolarization (the analogue, in this parameter range, of continuous spiking).

For an intermediate range of k_{Ca}, the intersection occurs on the middle branch of the Z curve. Since this is not a stable steady state, V and $[Ca^{2+}]_i$ do not come to rest there. Instead, they execute a slow-wave oscillation (Fig. 3C), visiting alternatively the stable upper and lower branches. During the "active" phase of depolarization, $[Ca^{2+}]_i$ increases, and therefore so does the Ca^{2+}-activated K^+ conductance. When the latter increases sufficiently, the membrane hyperpolarizes, and the silent phase ensues, during which Ca^{2+} uptake dominates. As the Ca^{2+}-gated K^+ conductance

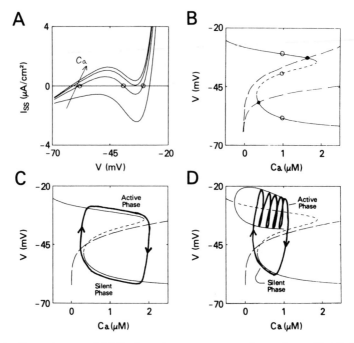

Figure 3. Dependence on $[Ca^{2+}]_i$ of steady-state current–voltage relationship (A) for four-variable HH excitable subsystem of Chay–Keizer model. From lowermost to uppermost curve, $[Ca^{2+}]_i$ = 0, 1, 2, 3 μM. Steady-state or rest potentials, obtained from zero-current crossings in A (i.e., open circles) versus $[Ca^{2+}]_i$, produce Z curve (B). For these parameters, upper and lower states are stable; middle state is unstable, a saddle. Calcium nullcline (long dashes), along which $dCa/dt = 0$, is shown for two values of k_{Ca} (0.005 msec^{-1} for lower curve and 0.06 msec^{-1} for upper curve). Intersections (filled circles) correspond to stable steady-state response of theoretical β cell. If $k_{Ca} = 0.04$ msec^{-1} (as in C), intersection corresponds to unstable steady state. Observed theoretical response is slow wave oscillation, i.e., bursting with spikes, shown schematically by heavy closed curve (C). For different parameter values, upper steady-state V on Z curve is unstable (dashed in D) and surrounded by oscillation (maximum and minimum V values form a "fork"). Observed response is bursting with spikes; trajectory is shown schematically by heavy closed curve; time course of V and $[Ca^{2+}]_i$ is similar to that in Fig. 1D,E.

decreases, the membrane slowly depolarizes until, eventually, the active phase is reentered, and the cycle repeats. Note that when the trajectory leaves one stable branch of the membrane subsystem, it rapidly moves vertically to another. This is because changes in $[Ca^{2+}]_i$ are relatively much slower.

In a parameter range more appropriate to the β cell, the upper branch of the Z curve does not correspond to a stable steady state of the fast membrane dynamics (now shown dashed in Fig. 3D). Rather, this upper steady state is surrounded by a small (15–20 mV) oscillation (the "fork" in Fig. 3D). In this case, the slow dynamics of intracellular Ca^{2+} handling lead to the typical burst pattern with spikes, which reflects the membrane's basic bistability (a low-V steady state and a high-V spikelike oscillation).

To understand the dependence of the active and silent phase durations on glucose concentration, we return briefly to Fig. 3C. For k_{Ca} just above the threshold for "bursting," the steady state is on the middle branch of the Z curve but close to the left knee. Therefore, the burst trajectory moves very slowly as it rounds this knee and passes close to the steady state. This means that the silent phase duration is very long. Similarly, we see why the active phase duration is very long for larger k_{Ca}, just below the value for which the steady state passes onto the stable upper branch.

4. BURST FREQUENCY DEPENDS ON THE RATIO [FREE $Ca^{2+}]_i$/[TOTAL Ca]$_i$

Another experimental observation is that from islet to islet, the burst period is variable, and this is independent of the relative duration of the active phase. For example, as illustrated in Fig. 4, cells from two different islets that are both responding to 11 mM glucose with an active phase of about 50% duration may have variable burst frequencies ranging between about two per minute (upper part of Fig. 4A) and about six per minute (lower part of Fig. 4A).

In the Chay–Keizer model, the rate of Ca^{2+} uptake is modulated by the ratio of free to total intracellular calcium, the parameter f. We have been able to reproduce the experimental findings of variable burst frequency for a constant relative burst duration by varying f. As f is decreased (by increasing the concentration of bound Ca), burst period lengthens (also noted by Chay and Keizer, 1983). Under these conditions, the absolute duration of each active phase is increased, although the relative duration of the active phase remains constant (illustrated in Fig. 4B). These findings indicate that the islets showing the slower oscillations may contain more Ca^{2+}-binding capacity (for example, they may have more endoplasmic reticulum).

The model also predicts the time course and magnitude of the oscillations in intracellular Ca^{2+} concentration (shown in Fig. 4C). It is notable that the absolute levels of $[Ca^{2+}]_i$ are not altered; the range of the oscillation remains constant. Only the time course is altered. The burst trajectory in Fig. 3D is similar for different f values but traversed at a different rate. Preliminary data indicate that islets that show a higher frequency in their burst pattern release higher levels of insulin (with respect to their basal release) than those with the slower periodicity, even though the

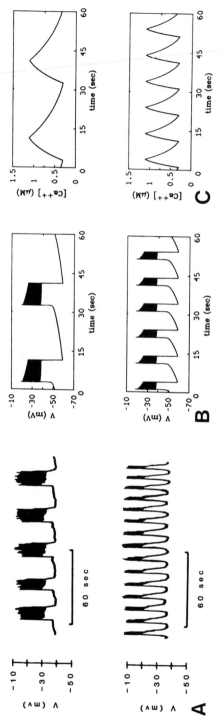

Figure 4. Burst patterns recorded from different islets but with same glucose concentration (11.1 mM) exhibit low (A, upper) or high (A, lower) burst frequency but same relative duration of active phase. Similar theoretical V (B) and $[Ca^{2+}]_i$ (C) time courses correspond to different values of f, the ratio of free to total intracellular calcium concentrations; $f = 0.016$ (upper) and $f = 0.048$ (lower). The total number of spikes per minute in B (upper) is the same as in B (lower).

islets show the same relative active phase duration and the same average spike frequency. These results may indicate that the rate of rise of Ca^{2+} is more important for secretion than the absolute levels obtained, or this may reflect an increased number of times during which a threshold level of $[Ca^{2+}]_i$ is attained, even though the average $[Ca^{2+}]_i$ levels are the same for both types of bursting behavior (fast or slow). Also of interest, islets taken from diabetic mice show considerably longer burst periods than do normal islets (Rosario *et al.*, 1985), which may implicate an excessive total calcium store as a contributing factor in diabetes.

5. SUMMARY

We have outlined the major experimental observations that lead to a model for the glucose-induced electrical activity of the β cell. Intracellular calcium, by activating a Ca^{2+}-gated K^+ channel, plays a key role in determining the burst pattern. Glucose sensitivity is treated theoretically by varying the intracellular uptake rate of calcium. The theoretical model has yielded insights on the experimental/physiological conditions that affect the electrical activity. It predicts the glucose-induced changes in cytosolic free Ca^{2+} concentration, which are not yet possible to measure in intact islets. The model has also been used to demonstrate that even though it is deterministic, there are parameter regimes for which it exhibits aperiodic ("chaotic") responses (Chay and Rinzel, 1985). These results were used to understand, in terms of ionic channel-gating properties, the chaotic bursting pattern observed in islets from mice fed a higher-lipid diet (Lebrun and Atwater, 1985). By interpreting experimental findings in the context of a quantitative model, one can explore various alternative hypotheses. As experiment and theory continue to interact, one can anticipate that the model will evolve and lead to new experiments, insights, and interpretations.

REFERENCES

Atwater, I, 1980, Control mechanisms for glucose induced changes in the membrane potential of mouse pancreatic β-cells, *Ciencia Biol.* **5:**299–314.

Atwater, I., Ribalet, B., and Rojas, E., 1978, Cyclic changes in potential and resistance of the β-cell membrane induced by glucose in islets of Langerhans from mouse, *J. Physiol. (Lond.)* **278:**117–139.

Atwater, I., Ribalet, B., and Rojas, E., 1979a, Mouse pancreatic β-cells: Tetraethylammonium blockage of the potassium permeability increase induced by depolarization, *J. Physiol. (Lond.)* **288:**561–574.

Atwater, I., Dawson, C. M., Ribalet, B., and Rojas, E., 1979b, Potassium permeability activated by intracellular calcium ion concentration in the pancreatic β-cell, *J. Physiol. (Lond.)* **288**:575–588.

Atwater, I., Dawson, C. M., Scott, A., Eddlestone, G., and Rojas, E., 1980, The nature of the oscillatory behavior in electrical activity for pancreatic β-cell, *J. Horm. Metab. Res. [Suppl.]* **10**:100–107.

Atwater, I., Dawson, C. M., Eddlestone, G. T., and Rojas, E., 1981, Voltage noise measurements across the pancreatic β-cell membrane: Calcium channel characteristics, *J. Physiol. (Lond.)* **314**:195–212.

Atwater, I., Rosario, L., and Rojas, E., 1983, Properties of the Ca-activated K^+ channel in pancreatic β-cells, *Cell Calcium* **4**:451–461.

Beigelman, P. M., Ribalet, B., and Atwater, I., 1977, Electrical activity of mouse pancreatic beta-cells II. Effects of glucose and arginine, *J. Physiol. (Paris)* **73**:201–217.

Chay, T. R., and Keizer, J., 1983, Minimal model for membrane oscillations in the pancreatic β-cell, *Biophys. J.* **42**:181–190.

Chay, T. R., and Rinzel, J., 1985, Bursting, beating, and chaos in an excitable membrane model, *Biophys. J.* **47**:357–366.

FitzHugh, R., 1961, Impulses and physiological states in models of nerve membrane, *Biophys. J.* **1**:445–466.

Hodgkin, A. L., and Huxley, A. F., 1952, A quantitative description of membrane current and its application to conduction and excitation in nerve, *J. Physiol. (Lond.)* **117**:500–544.

Lebrun, P., and Atwater, I., 1985, Chaotic and irregular bursting electrical activity in mouse pancreatic β-cells, *Biophys. J.* **48**:529–531.

Meda, P., Atwater, I., Goncalves, A., Bangham, A., Orci, L., and Rojas, E., 1984, The topography of electrical synchrony among β-cells in the mouse islet of Langerhans, *Q. J. Exp. Physiol.* **69**:719–735.

Meissner, H. P., 1976, Electrical characteristics of the beta-cells in pancreatic islets, *J. Physiol. (Paris)* **72**:757–767.

Rinzel, J., 1985, Bursting oscillations in an excitable membrane model, in *Ordinary and Partial Differentiated Equations* (B. D. Sleeman and R. J. Jarvis, eds.) Springer–Verlag, New York, pp. 304–316.

Rosario, L. M., Atwater, I., and Rojas, E., 1985, Membrane potential measurements in islets of Langerhans from *ob/ob* obese mice suggest an alteration in $[Ca^{++}]_i$-activated K-permeability, *Q. J. Exp. Physiol.* **70**:137–150.

Chapter 23

Neurotrophic Effects of *in Vitro* Innervation of Cultured Muscle Cells. Modulation of Ca^{2+}-Activated K^+ Conductances

Benjamin A. Suarez-Isla and Stanley I. Rapoport

1. INTRODUCTION

The trophic influence of nerve on muscle is expressed as several complex changes of the muscle plasma membrane (McArdle, 1983). Muscle denervation studies (Luco and Eyzaguirre, 1955; Thesleff, 1974) showed that cutting of the nerve decreased the membrane resting potential, increased the specific membrane resistance, and induced the appearance of hyperpolarizing afterpotentials. Denervation also decreases the sensitivity to blockage by tetrodotoxin of the action potential mechanism (Thesleff, 1974) and promotes the insertion of acetylcholine receptors into extrajunctional regions (Berg and Hall, 1975).

A dramatic example of neural influence on properties of muscle cells is observed after denervation of frog slow muscle fibers (Schmidt and Stefani, 1977), which acquire the action potential mechanism 2 weeks after denervation or crushing of the nerve. In addition, these fibers recover their "slow" characteristics after long-term reinnervation.

These changes are probably the expression of modifications of prop-

---

BENJAMIN A. SUAREZ-ISLA AND STANLEY I. RAPOPORT • Laboratory of Neurosciences, National Institute on Aging, National Institutes of Health, Bethesda, Maryland 20892. *Present address for B.A.S.:* Centro de Estudios Científicos de Santiago, Santiago, Chile.

erties of channel and/or receptor macromolecules that underlie conductance changes during development or alteration of normal innervation patterns. They have been shown to be preceded by specific changes in protein synthesis (Thesleff, 1974; McArdle, 1983).

Preliminary studies (Suarez-Isla *et al.*, 1982, 1983) showed that coculture of rat myotubes with spinal cord neurons for only 24 hr elicited several significant changes in the electrical membrane properties of the muscle cells, among them decrements in specific membrane resistance, resting membrane potential, and incidence of the typical slow hyperpolarizing potential (slow HAP; Barrett *et al.*, 1981). *In this chapter we show that this coculture system can be used to study in detail the changes elicited after in vitro innervation.* These changes closely resemble those that take place during maturation of the muscle cells *in vivo*. In particular, the presence of the neurons affects the two types of Ca^{2+}-activated K^+ conductances found in rat myotubes (Romey and Lazdunski, 1984). However, only one of them (apamin sensitive, TEA insensitive) seems to underlie the slow HAP that is blocked after innervation. Part of this work has been presented before as preliminary communications (Suarez-Isla *et al.*, 1982, 1983).

2. METHODOLOGICAL CONSIDERATIONS

2.1. Preparation of Muscle Cultures

Muscle cell cultures were prepared by dissociating striated muscle cells from hindlimbs of newborn Wistar rats as described by Ruffolo *et al.* (1979), using trypsin and gentle trituration. Cells were plated at low density (about 100,000 cells/ml to avoid extensive confluency and formation of branched myotubes) on collagen-coated plastic dishes. Cultures were kept at 37°C under 10% CO_2–air in Dulbecco's modified Eagle medium with L-glutamine, 4.5 g/liter glucose, and without sodium pyruvate. The medium was changed every second day. Cytosine arabinoside, 1 μM, was given for 24 hr at day 3. Under these conditions, clearly delineated spindle-shaped myotubes appropriate for intracellular recording are obtained (Fig. 1).

2.2. Dissociation of Spinal Cord Neurons

Spinal cords were dissected in Ca^{2+}/Mg^{2+} free Tyrode's solution from 6-day-old white Leghorn chick embryos. After removal of dorsal root ganglia, the spinal cords were incubated for 15 min at 37°C in 0.05% w/v trypsin dissolved in the previous medium and later dissociated by 12

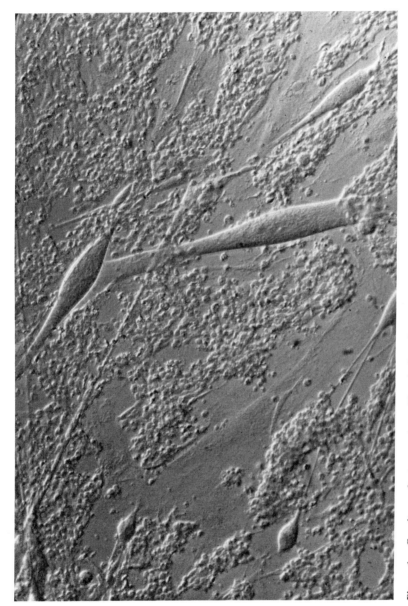

Figure 1. Coculture of rat myotubes with spinal cord neurons from 6-day chick embryos as seen under Nomarski optics. Large central myotube is 30 μm wide at its center.

passages through a pipette with a 1.5-mm opening. Five million cells were plated on myotubes previously cultured for 5 to 10 days. Intracellular recording indicated that 80 to 100% of the muscle cells showed spontaneous miniature end-plate potentials (m.e.p.p.s) after 24 hr in coculture (Suarez-Isla *et al.*, 1984; cf. Fig. 2c). These m.e.p.p.s were taken as criteria of *in vitro* innervation (Ruffolo *et al.*, 1979, Suarez-Isla *et al.*, 1984).

2.3. Electrophysiological Techniques

Spindle-shaped myotubes (Fig. 1) (average dimensions 300 µm length and 40 µm maximum width) were selected for patch and intracellular recording. Intracellular recording was done with microelectrodes of 30 MΩ resistance filled with 3 M KCl using a high-input-impedance amplifier (Model M-707, WP Instruments, New Haven, CT) equipped with a bridge circuit that allowed current injection and stimulation through the same microelectrode. Electrical contact was achieved via Ag/AgCl pellets. Cells could be easily impaled by gentle tapping of the table surface. All intracellular recordings were made at 37°C in Eagle's MEM, and a stream of 5% CO_2–air was gently flushed over the medium surface to help maintain a pH of 7.4 ± 0.1. Action potentials were elicited by anode-break excitation at the break of a 200-msec hyperpolarizing current pulse (<5 nA amplitude). Input resistance was measured from the steady-state voltage deflections at the end of 200-msec hyperpolarizing pulses of 0.1, 0.2, 0.5, 1.0, and 2.0 nA. Specific membrane resistance was calculated by dividing input resistance by the area of the cell. The latter was estimated by using the longest and shortest axes of the cell as measured with a grid mounted in the ocular piece at 200× total magnification (minimum measurable distance 2.5 µm). The cell shape could be approximated by a rhomboid. Thus, area = $l \times w \times 2$. Average myotube dimensions were 200 × 30 µm. This quick approximation was in good agreement with measurements obtained from micrographs.

Patch-clamp recordings were done as described in detail by Hamill *et al.* (1981) with an extracellular patch-clamp amplifier (EPC-5, List Electronic, Darmstadt, Federal Republic of Germany) set at a filter setting of 10 kHz low pass. The signal coming from the clamp was recorded on FM tape. For data analysis, the signals from intracellular or patch-clamp recordings were digitized at 100 µsec/point with a digital oscilloscope (Nicolet 4094, Nicolet Instrument Corp., Madison, WI). Action potentials were analyzed for amplitude, duration, and dV/dt using subroutines controlled by the oscilloscope. Single-channel recordings were filtered at 1 or 2 kHz low pass with an eight-pole Bessel filter and analyzed for amplitude and open and closed times with a version of a program kindly provided by Dr. Osvaldo Alvarez, implemented for a VAX 780 computer (Digital Equipment Corp.).

3. INNERVATION AND MUSCLE CELL ELECTRICAL ACTIVITY

The results described in this section were obtained after 24-hr coculture of muscle cells with spinal cord neurons, and intracellular recordings were done in paired control and experimental culture dishes. Duplicates were used in each case, and at least 10 muscle cells were tested in each dish.

3.1. Coculture with Spinal Cord Neurons Reduces the Incidence of Slow Hyperpolarizing Afterpotentials in Rat Myotubes

Action potentials elicited from myotubes kept for 5 to 10 days in control dishes showed a distinct slow HAP of up to 20 mV in amplitude that lasted up to 250 msec at 37°C (Fig. 2a). These action potentials resemble those that can be elicited from denervated (Thesleff, 1974) or dysgenic muscle (Albuquerque, 1970). In some cases subthreshold stim-

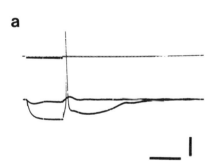
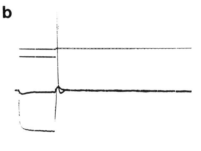
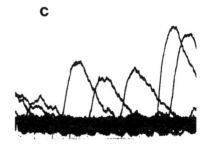

Figure 2. Effect of *in vitro* innervation on the action potential of a rat myotube evoked by anode-break excitation. a: The control action potential displays a prominent slow hyperpolarizing afterpotential (slow HAP) of 16 mV amplitude and duration at 50% amplitude of 130 msec (at 37°C). b: The action potential elicited from a cocultured muscle cell lacks the slow HAP. In this case it was necessary to apply a stronger pulse to elicit the action potential. Vertical calibrations: upper trace, 5 nA; lower trace, 20 mV. Horizontal calibration: 50 msec. The trace depicting the current pulse has been located at 0 mV membrane potential. c: Spontaneous miniature end-plate potentials recorded from an innervated myotube. Vertical calibration, 0.5 mV; horizontal calibration, 10 msec.

ulation produced damped oscillations. Stronger stimulations could also generate trains of repetitive action potentials triggered from the depolarizing phase of the slow HAP. The percentage of cells with slow HAPs in control dishes was 86.9 ± 3.1 (mean ± S.E.M.) in 17 pairs of dishes obtained from six primary dissociations.

In paired dishes containing myotubes that had been cocultured with spinal cord neurons from 6-day chick embryos for 24 hr, the percentage of cells with slow HAPs decreased significantly ($P < 0.01$) to 52.2 ± 5.0 (mean ± S.E.M.) (Fig. 2b). The effect was present beyond 24 hr in culture (Fig. 3) but was more pronounced at 24 hr.

The slow hyperpolarizing afterpotential has been associated with the activation of a calcium-dependent potassium conductance (Barrett *et al.*, 1981) that could result in spontaneous oscillatory electrical activity before innervation takes place *in vivo*, thereby maintaining contractile activity. Possible mechanisms underlying the blockage of the slow HAP after *in vitro* innervation are the subject of experiments discussed in Section 4.3.

3.2. The Distribution of Specific Membrane Resistance Does Not Change when All Cells in Control and Cocultured Dishes Are Compared

Frequency distribution histograms of specific membrane resistance (Rm) (bin size = 200 Ω × cm²) were not significantly different (Kol-

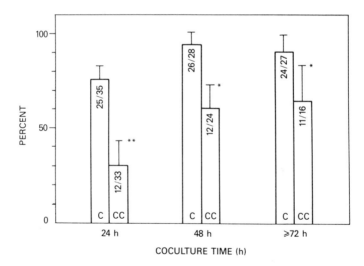

Figure 3. Effect of coculture time of the incidence of slow HAPs. Vertical axis is the percentage of cells with HAPs ± S.E.M. Data have been obtained from 17 matched pairs of culture dishes prepared from six primary dissociations. The number of cells with HAPs and the total number of cells tested are given inside the respective columns. **$P < 0.01$; *$P < 0.1$. C, controls; CC, cocultured myotubes.

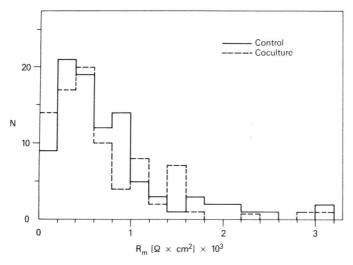

Figure 4. The distribution of R_m values is not significantly different among control and cocultured dishes when the presence of slow HAPs is not taken into consideration (Kolmogorov–Smirnov test; bin size 200 $\Omega \times cm^2$).

mogorov–Smirnov test) (Fig. 4) between cells in paired control and cocultured dishes. This also was the case when only the subset of innervated myotubes in cocultured dishes was compared with controls.

3.3. The Specific Membrane Resistance of Cells without HAPs Is Decreased

The distribution histogram of specific membrane resistances of cells without slow HAPs was significantly shifted to lower values ($P < 0.01$; Kolmogorov–Smirnov test) (Fig. 5). Mean Rm for cells without HAPs was 608 $\Omega \times cm^2$ ($n = 86$) as compared to 918 $\Omega \times cm^2$ in cells with HAPs ($n = 98$). Similar shifts were observed when cells with and without HAPs were compared separately within control and coculture sets. In other words, the neurotrophic-induced difference between control and cocultured cells was seen primarily as a decrease in the incidence of slow HAPs and not as a decrease in specific membrane resistance.

3.4. Cocultured Myotubes Are Significantly More Sensitive to Tetrodotoxin Than Control Cells

Overshooting action potentials could be elicited by anode-break excitation from up to 90% of control myotubes in the presence of 2 μM TTX, a toxin that specifically blocks sodium currents in adult nerve and muscle cells in the nanomolar concentration range (Figs. 6a,c, 7a). After

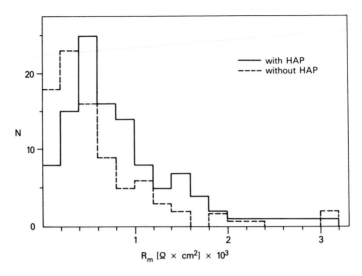

Figure 5. The distribution of R_m values is significantly lower in cells devoid of slow HAPs. Mean R_m decreased from 918 $\Omega \times cm^2$ to 608 $\Omega \times cm^2$ (Kolmogorov–Smirnov test; bin size 200 $\Omega \times cm^2$).

24-hr coculture, 2 µM TTX was sufficient to block the action potential mechanism in up to 90% of the cells (Figs. 6b,d, 7a). However, the maximal rate of rise was similarly decreased in control and cocultured cells (Fig. 7b). It is known that the development of TTX resistance and decrease in dV_{max}/dt can occur independently (Cullen *et al.*, 1975). In fact, chicken muscle does not acquire TTX resistance after denervation, but dV_{max}/dt is reduced.

These results can be summarized as follows: *in vitro* innervation by spinal cord neurons elicits several changes in the electrical membrane properties of the target muscle cells. These changes qualitatively resemble those taking place during muscle maturation *in situ* or after reimplantation of nerves in previously denervated adult muscles. In particular, *in vitro* innervation significantly decreases the percentage of muscle cells with slow HAPs, a prominent feature in the action potential of immature or denervated muscle (Thesleff, 1974).

The next section presents experiments conducted to investigate which conductance underlies the slow afterpotential and how it is modulated by *in vitro* innervation.

3.5. The Slow HAP Is Not Blocked by Tetraethylammonium

As shown in Fig. 8, addition of 20 mM tetraethylammonium (TEA) chloride significantly increased the duration of the action potential but

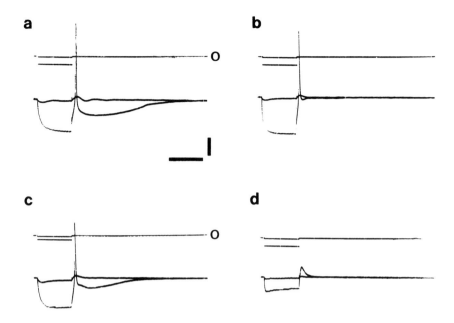

Figure 6. *In vitro* innervation significantly increases the TTX sensitivity of the action potential mechanism. a: Control myotube. b: Myotube cocultured for 24 hr with spinal cord neurons. c: Control cell in the presence of 2 μM TTX; note the generation of an overshooting action potential. d: After 24 hr coculture, 2 μM TTX blocks the action potential mechanism. Vertical calibrations, 5 nA for the upper trace (injected current) and 20 mV for the lower trace (membrane potential); horizontal calibration, 100 msec.

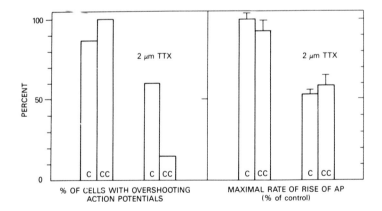

Figure 7. a: Percentage of cells with overshooting action potentials (as referred to controls, C, in the absence of TTX) in the absence and presence of 2 μM TTX. C, control cells; CC, cocultured cells. Data obtained from three primary dissociations with 10 to 20 cells tested per condition. b: Maximal rate of rise of the action potential expressed as percentage of control (C) value in the absence of 2 μM TTX. This was 200 V/sec. Values belong to the same set of data as in panel a and are expressed as means ± S.E.M.

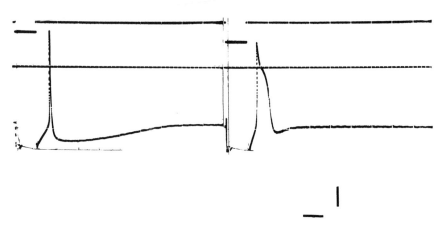

Figure 8. The slow HAP is not blocked by tetraethylammonium. Application of 20 mM TEA increases the duration of the fast repolarization phase, but the slow HAP is still present. Vertical calibration, 2 nA for the upper trace (current injected) and 20 mV for the lower trace (membrane potential); horizontal calibration, 100 msec.

was not sufficient to mask the remaining afterhyperpolarization. The action potential duration increased from 1.4 ± 0.4 msec (mean ± S.E.M.; nine determinations) to 22.4 ± 0.6 msec (six determinations). At lower TEA concentrations (not shown), the remaining slow HAP had a larger amplitude, and the action potential duration increment was correspondingly smaller. However, addition of nanomolar concentrations of the bee venom polypeptide apamin, a known blocker of $I_{K,Ca}$ conductances in other preparations (Romey and Lazdunski, 1984; Burgess et al., 1981), elicited a rapid block of the slow HAP without a concomitant increase in the action potential duration measured at 0 mV.

3.6. The Slow HAP Is Blocked by Nanomolar Concentrations of Apamin

As shown in Fig. 9, bath application of 12 nM apamin blocked completely the slow HAP within 15 to 40 sec after addition. Action potential duration was 1.6 ± 0.6 msec (mean ± S.E.M.) for eight determinations and was not significantly different from control values (see above). These results suggest that the neurotrophic effect blocks selectively the apamin-sensitive $I_{K,Ca}$ conductance without significantly affecting the ionic channels underlying the fast repolarization phase. These possibilities were investigated with the patch-clamp technique.

Barrett et al. (1981) have shown that in rat myotubes the slow HAP

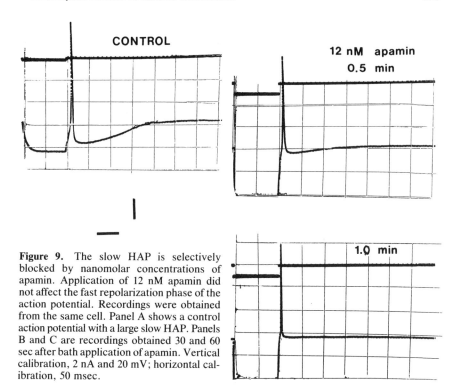

Figure 9. The slow HAP is selectively blocked by nanomolar concentrations of apamin. Application of 12 nM apamin did not affect the fast repolarization phase of the action potential. Recordings were obtained from the same cell. Panel A shows a control action potential with a large slow HAP. Panels B and C are recordings obtained 30 and 60 sec after bath application of apamin. Vertical calibration, 2 nA and 20 mV; horizontal calibration, 50 msec.

is associated with activation of a Ca^{2+}-dependent K^+ conductance and have proposed that this conductance participates in the maintenance of spontaneous electrical and contractile activity before innervation takes place *in vivo*, being suppressed at later stages of development and maturation.

To investigate whether the decrease in the incidence of slow HAPs resulted from modulation of Ca^{2+}-dependent K^+ conductances, we compared the gating and kinetic properties of single $I_{K,Ca}$ channels in intact membrane patches of paired control and cocultured myotubes using the extracellular patch clamp technique (Hamill *et al.*, 1981).

Single-channel recordings were made with the cell-attached patch configuration only in well-developed myotubes that had high resting potentials (> -55 mV) and that responded with an overshooting action potential on intracellular stimulation. We compared single-channel recordings from control myotubes that displayed prominent slow HAPs (amplitude > 10 mV) with those from cocultured myotubes with completely blocked slow HAPs. In addition, the analysis of single-channel data was restricted to patches in which only one channel was apparently active at any given time and to those that showed very infrequent overlapping of two channels (less than 2% of the total recording time) at depolarizations higher than

+50 mV. Data presented here were obtained from a total of six patches formed in control myotubes and from four patches formed in cocultured myotubes that belonged to paired culture dishes. Paired recordings were made the same day, and results are exemplified here with data obtained from one paired experiment.

3.7. The High-Conductance Ca^{2+}-Activated K^+ Channel Is Active in Both Control and Innervated Myotubes

Figure 10 summarizes several aspects of the main single-channel activity pattern observed in both control and cocultured myotubes. The absolute patch potential was measured with an independent intracellular

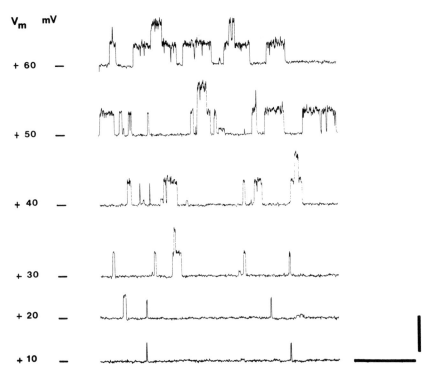

Figure 10. Effect of membrane potential on the Ca^{2+}-activated K^+ channel in an intact membrane patch from an innervated myotube. Voltages on the left-hand side represent absolute membrane potential sensed by the patch. Lines indicate the closed state, and upward current deflections represent outward flow of K^+ ions. Vertical calibration, 10 pA; horizontal calibration, 100 msec. Low-pass filter set at 2 kHz. Seal resistance was 15 Ω. Pipette contains Earle's minimum essential medium with HEPES buffer, 250 nM acetylcholine, and 2 μM TTX, pH 7.2, 21°C. Smaller outward deflections correspond to the opening of ACh channels.

microelectrode kept in the cell during the patch recording or by sampling the resting membrane potential before and after the patch recording. In this recording from an innervated myotube, single-channel events appeared on depolarization of the patch pipet above -20 mV. Increasing depolarizations increased in the frequency of events and lengthened the time spent in the channel-open state. This was accompanied by short closing events and apparent enhancement of the noise level during openings. Channel activity was not random, and long periods of rather regular channel activity at a steady polarization of the patch pipet could be followed by quiescent periods of up to several minutes. The fractional open time spent by the channel was always different after the quiescent periods. These qualitative aspects of channel activity were not different between control and cocultured cells.

Separate control experiments carried out with excised inside-out patches in the presence of K^+ gradients and different bath Ca^{2+} concentrations (not shown here) confirm that these channels correspond to the large Ca^{2+}-dependent K^+ channel described by Pallotta et al. (1981) and Methfessel and Boheim (1982) in rat myotubes.

The single-channel conductance was estimated from the slope of the current–voltage relation as shown in Fig. 11. For depolarizations below $+20$ mV, the I/V curves were linear, giving a single-channel conductance of 110 pS in both control and cocultured myotubes. However, a strong sublinearity appeared with increasing depolarizations. Furthermore, the potential at which this sublinearity became apparent was not identical for control myotubes and cocultured cells: the estimated "rollover" potential was always higher in control cells ($+70$ mV) than in cocultured cells ($+40$ mV). The phenomenon of sublinearity has been observed in adrenal chromaffin cells (Yellen, 1984) and has been attributed to a rapid voltage-dependent blockage of the Ca^{2+}-activated K^+ channel by an intracellular cation.

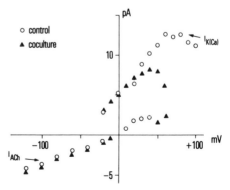

Figure 11. Comparison of current–voltage relationships between control (open circles) and innervated myotubes (closed triangles). Points in the upper two quadrants represent Ca^{2+}-activated channels. Lower points represent ACh-activated channels recorded from the same patch. Note that in innervated myotubes, the sublinearity of the I/V curve appears at less depolarized potentials. In contrast, no difference is observed in ACh-activated channels. Vertical axis, single-channel current amplitude in picoamperes; horizontal axis, absolute membrane potential in millivolts.

These results suggest that the neurotrophic effect is expressed as a change in the permeation characteristics of the channel. This was analyzed in terms of a voltage-dependent ion block of the channel (Woodhull, 1973; Coronado and Miller, 1982). In this steady-state approach, the voltage dependence of the apparent time-averaged current is given by equation 1:

$$I / I_0 = (1 + [B]/K_d)^{-1} \qquad (1)$$

where $K_d = K_0 \exp\{-z\delta FV/RT\}$, [B] is the blocker concentration, z its valence, K_0 its dissociation constant at zero applied potential, I_0 is the unblocked channel current, and δ is the fraction of the voltage gradient "sensed" by the binding site for the blocker. As shown in Fig. 12a, the

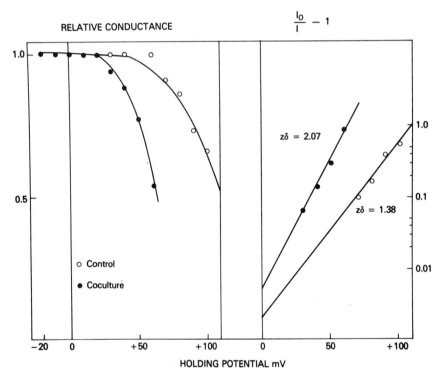

Figure 12. The alteration of the I/V curve shape can be explained in terms of a voltage-dependent ion-block model. Left panel depicts the relative conductance for control (open circles) and cocultured (closed circles) myotube patches. Note the large shift in the hyperpolarizing direction of the cocultured patch. Right panel represents the linearized conductances obtained from equation 1 (see text). The slope of the line fitted to the conductance values of the cocultured cell (closed circles; $z\delta = 2.07$) is significantly increased with respect to control values (open circles; $z\delta = 1.38$).

plot of relative conductances versus membrane potential is markedly shifted in the hyperpolarizing direction in a patch formed in cocultured myotubes. Linearization of currents in terms of Woodhull's treatment (1973) (Fig. 12b) is given by equation 2:

$$\ln [I_0/(I_0 - 1)] = \ln \{[B]/K_0\} - z\delta FV/RT \qquad (2)$$

This is a convenient way to determine K_0 as the y-intercept at zero applied voltage and the "effective valence" of the blocking reaction ($z\delta$) from the slope of the linearized conductance curve. This indicated that the voltage sensitivity of the hypothetical ion block increased significantly after innervation. In particular, the fraction of the field sensed by the ion-binding site (as given by $z\delta$) increased from 1.38 to 2.07 in this particular paired experiment. However, the ratio between the dissociation constants at zero voltage for cocultured (CC) and control (C) patches, K_{cc}/K_c, was 0.89. This suggests that no affinity change took place after innervation. Similar results were obtained in the other three paired comparisons.

Furthermore, the neurotrophic effect was not expressed as a significant change in channel kinetics. As exemplified in Fig. 13, the voltage dependence of the open and closed time constants did not change after innervation *in vitro*. Similarly, the fraction of time that the channel spent in the open state, $f(V)$, increased e-fold with 20 mV of depolarization for low levels of channel activity in both control and cocultured cells (Fig. 14).

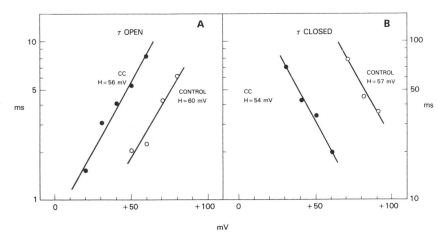

Figure 13. The voltage dependence of the open and closed time constants does not change after innervation *in vitro*. Vertical calibration, time constants in milliseconds on log scale; horizontal calibration, absolute membrane potential in millivolts. Same paired cell patches as in Figs. 11 and 12.

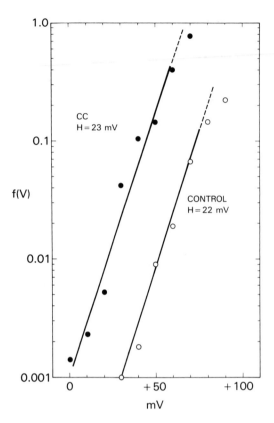

Figure 14. The voltage dependence of the fractional open time is not modified after *in vitro* innervation. Data obtained from the same paired patches as in Figs. 11–13.

The reason for the large shifts observed in the voltage axis is unknown. They occurred among control cells and even in a single patch during recording time. This heterogeneity has been observed by others in rat myotubes (Pallotta *et al.*, 1981) and in transverse tubular membranes from rabbit muscle (Vergara *et al.*, 1983).

These results illustrate several important points. (1) The cell membrane of cocultured myotubes that do not show slow HAPs still possesses functional 110-pS Ca^{2+}-activated K^+ channels with gating and kinetic properties indistinguishable from channels located in control rat myotubes. (2) The only significant and reproducible change brought about after *in vitro* innervation is a change in the permeation properties of the channel. (3) This is expressed as a sublinearity in the current–voltage relationship that becomes significant at lower membrane potentials in cocultured cells devoid of slow HAPs as compared to controls with slow HAPs. (4) The voltage range at which the *I/V* curve is affected is above

+20 mV and well outside the voltage range in which the slow HAPs develop (i.e., near the resting potential). This is consistent with the assumption that the 110-pS channel is not associated with the activation of the currents underlying the slow HAP. In addition, this channel is not active at the resting potential. However, a transient increase in intracellular calcium could theoretically activate this channel at potentials nearer to the resting potential (cf. Barrett *et al.*, 1982).

To test this possibility, we studied the effect of a propagated action potential on the activation of the 110-pS channel. A myotube was impaled by an intracellular microelectrode, and, after a stable recording was achieved, a cell-attached patch was formed within 100 μm of the intracellular electrode. Then a patch pipette polarization was selected that elicited infrequent and short channel openings. After 1 min of base-line recording, action potentials were generated through the intracellular electrode, and the patch current signal was continuously monitored. As shown in Fig. 15, upper panel, the experiment indicated that: (1) the propagated action potential was able to increase significantly the probability of opening of the channel above the control level; (2) the activation was transient but outlasted the duration of the slow HAP as recorded by the intracellular electrode; (3) the transient stimulation could be enhanced if the patch pipette was previously depolarized to a higher positive potential (Fig. 15, lower trace), but no transient activation was observed when the patch was held at the resting potential. These observations show that the 110-pS Ca^{2+}-dependent K^+ channel does not participate in the generation of the slow afterhyperpolarization.

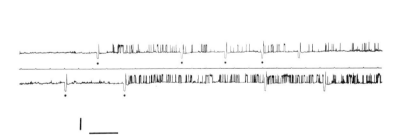

Figure 15. The probability of opening of the $I_{K,Ca}$ channel increases transiently after a propagated action potential. A conventional microelectrode was used to inject current pulses to evoke an action potential while simultaneously recording from a patch electrode. Upper trace is the patch current at 0 mV. In the lower trace, the patch was depolarized to +60 mV. Channels appeared in bursts, probably activated by increased intracellular calcium. No activation was seen when the patch was held at the resting potential (−55 mV). Seven-day-old control myotube.

4. CONCLUSIONS

These studies indicate that short-term coculture of muscle cells with spinal cord neurons elicits changes within 24 hr in the electrical membrane properties of the muscle cells. These changes include (1) blockage of the slow afterhyperpolarization, (2) modification of the open-channel current–voltage relationship of the 110-pS Ca^{2+}-activated K^+ channel, (3) sensitivity increase of the action potential mechanisms to blockage by tetrodotoxin, and (4) decrease of the specific membrane resistance. Similar treatment of the muscle cells also produces a significant increase in the clustering of acetylcholine receptors in the cell membrane (Thompson, 1984). Taken together, these changes mimic qualitatively those that take place during muscle maturation *in vivo* or after reinnervation of previously denervated adult muscles. In addition, they mirror the consequences of muscle denervation (Thesleff, 1974).

4.1. Large-Conductance Ca^{2+}-Activated K^+ Channels Are Not Involved in the HAP

Of particular interest is the finding of the neurotrophic influence on the 110-pS Ca^{2+}-activated K^+ conductance. This is the first demonstration of a neurotrophic modulation of a membrane conductance at the single-channel level, expressed as an increase in the voltage dependence of internal cation block. This change in single-channel properties does not elicit a significant alteration of the action potential trajectory, and it could not have been disclosed solely by analysis of the action potential characteristics. These findings also confirm and extend a previous report by Romey and Lazdunski (1984) that presented evidence of two types of Ca^{2+}-activated K^+ conductances in rat myotubes. They confirm that the 110-pS channel does not underlie the activation of the slow afterhyperpolarization.

4.2. Apamin-Sensitive Channels Remain to Be Resolved

Pennefather *et al.* (1985) have observed two types of Ca^{2+}-dependent K^+ currents in voltage-clamped bullfrog sympathetic ganglion cells. One of them, I_{ahp}, activates with a time course similar to the slow HAP in those neurons. In our experiments on the effect of a propagated action potential on single-channel conductances (Fig. 15), we have been unable so far to find other channels that are activated by a propagated action potential while the patch membrane is held at the resting potential. Two possibilities can tentatively explain these negative results. (1) The single-channel conductance of the putative I_{ahp} channel is too small to be resolved. (2) The channels are present in the myotube membrane at a very low density, and our number of samples is too small to detect one of them. Measurement of macroscopic currents with intracellular voltage clamp in the presence of blockers of the large conductance will have to be applied to elucidate whether there is a current that could account for the activation of the slow HAP.

4.3. Innervation and Large-Conductance Ca^{2+}-Activated K^+ Channels

The neurotrophic effect on the 110-pS channel can be interpreted as a change in the permeation characteristics of the channel, and it can be described in terms of a voltage-dependent ion block of the channel (Woodhull, 1973). Within this framework, it can be concluded that the neurotrophic influence acts by increasing the voltage dependence of the ion block by an internal cation. These results (Fig. 12) suggest that innervation does not increase significantly the affinity of the blocking ion for its binding site. They do indicate, however, that the ion-binding site is sensing a larger fraction of the applied field after the neurotrophic effect sets in.

Sublinearities of the current–voltage relationship of Ca^{2+}-activated K^+ channels have been observed in smooth muscle cells (Singer, 1982) and in adrenal chromaffin cells (Yellen, 1984). It has recently been proposed (Läuger, 1985) that rapid voltage-dependent transitions between subconductance states of a channel could give rise to time-averaged determinations of current amplitudes that are an average of the current amplitude values of the open state and other subconductive states. This question should be addressed using excised inside-out patches of control and treated myotubes to elucidate whether the increased sublinearity after innervation reflects an intrinsic change of the channel macromolecule leading to the appearance of new unresolved subconductance states or to the concentration increase of a putative intracellular blocking ion.

REFERENCES

Albuquerque, E. X., and McIsaac, R. J., 1970, Fast and slow mammalian muscle after denervation, *Exp. Neurol.* **26**:183–202.

Barrett, J. N., Barrett, E. F., and Dribin, L. B., 1981, Calcium-dependent slow potassium conductance in rat skeletal myotubes, *Dev. Biol.* **82**:258–266.

Barrett, J. N., Magleby, K. L., and Pallotta, B. S., 1982, Properties of single calcium-activated potassium channels in cultured rat muscle, *J. Physiol. (Lond.)* **331**:211–230.

Berg, D. K., and Hall, Z. W., 1975, Increased extrajunctional acetylcholine sensitivity produced by chronic post-synaptic neuromuscular blockade, *J. Physiol.* **244**:659–676.

Burgess, G. M., Claret, M., and Jenkinson, D. H., 1981, Effects of quinine and apamin on the Ca-dependent K permeability of mammalian hepatocytes and red cells, *J. Physiol. (Lond.)* **317**:67–90.

Coronado, R., and Miller, C., 1982, Conduction and block by organic cations in a K-selective channel from sarcoplasmic reticulum incorporated into planar phospholipid bilayers, *J. Gen. Physiol.* **79**:529–547.

Cullen, M. J., Harris, J. B., Marshall, M. W., and Ward, M. W., 1975, An electrophysiological study of normal and denervated chicken latissimus dorsi muscles, *J. Physiol. (Lond.)* **245**:371–385.

Hamill, O. P., Marty, A., Neher, E., Sakman, B., and Sigworth, F., 1981, Improved patch-clamp techniques for high resolution current recording from cells and cell-free membrane patches, *Pfluegers Arch.* **391**:85–100.

Läuger, P., 1985, Ionic channels with conformational substates, *Biophys. J.* **47**:581–591.

Luco, J. V., and Eyzaguirre, C., 1955, Fibrillation and hypersentivity of ACh in denervated muscle: Effect of length of degenerating nerves, *J. Neurophysiol.* **18**:65–73.

McArdle, J. J., 1983, Molecular aspects of the trophic influence of nerve on muscle, *Prog. Neurobiol.* **21**:135–198.

Methfessel, C., and Boheim, G., 1982, The gating of single calcium-dependent potassium channels is described by an activation/blockade mechanism, *Biophys. Struct. Mech.* **9**:355–60.

Pallotta, B. S., Magleby, K. L., and Barrett, J. N., 1981, Single channel recordings of Ca^{2+}-activated K^+ currents in rat muscle cells in culture, *Nature* **293**:471–474.

Pennefather, P., Lancaster, B., Adams, P. R., and Nicoll, R. A., 1985, Two distinct Ca-dependent K currents in bullfrog sympathetic ganglion cells, *Proc. Natl. Acad. Sci.* **82**:3040–3044.

Romey, G., and Lazdunski, M., 1984, The coexistence in rat muscle cells of two distinct classes of Ca^{+2}-dependent K^+ channels with different pharmacological properties and different physiological functions, *Biochim. Biophys. Res. Comm.* **118**:669–674.

Ruffolo, R. R., Jr., Eisenbarth, G. E., Thompson, J. M., and Nirenberg, M., 1979, Synapse turnover: A mechanism for acquiring synapse specificty, *Proc. Nat. Acad. Sci. U.S.A.* **75**:2281–2285.

Singer, J. J., and Walsh, J. W., 1984, Large conductance Ca^{+2}-activated K^+ channel in smooth muscle cell membrane, *Biophys. J.* **45**:68–70.

Schmidt, H., and Stefani, E., 1977, Action potentials in slow muscle fibres of the frog during regeneration of motor nerves, *J. Physiol.* **270**:507–517.

Suarez-Isla, B. A., and Rapoport, S. I., 1983, *In vitro* innervation modulates a calcium-activated potassium conductance in rat myotubes, *Soc. Neurosci. Abstr.* **9**:22.

Suarez-Isla, B. A., Thompson, J. M., and Rapoport, S. I., 1982, Effect of coculture and *in vitro* (re)innervation on the electrical membrane properties in rat myotubes, *Soc. Neurosci. Abstr.* **8**:125.

Suarez-Isla, B. A., Pelto, D. J., Thompson, J. M., and Rapoport, S. I., 1984, Blockers of calcium permeability inhibit neurite extension and formation of neuromuscular synapses in cell culture, *Dev. Brain Res.* **14**:263–270.

Thesleff, S., 1974, Physiological effects of denervation of muscle, *Ann. N.Y. Acad. Sci.* **228**:89–104.

Thompson, J. M., and Rapoport, S. I., 1982, Absence of stable retina–muscle synapses is related to absence of acetylcholine receptor aggregation factor in retina neurons, *Soc. Neurosci. Abstr.* **8**:185.

Vergara, C., and Latorre, R., 1983, Kinetics of Ca-activated K channels from rabbit muscle incorporated in lipid bilayers: Evidence for Ca and Ba blockade, *J. Gen. Physiol.* **82**:543–568.

Woodhull, A. M., 1973, Ionic blockage of sodium channels in nerve, *J. Gen. Physiol.* **61**:687–708.

Yellen, G., 1984, Ionic permation and blockage in Ca^{2+}-activated K^+ channels in bovine chromaffin cells, *J. Gen. Physiol.* **84**:157–186.

Part V

Ionic Channel Structure, Functions, and Models

Chapter 24

Correlation of the Molecular Structure with Functional Properties of the Acetylcholine Receptor Protein

Francisco J. Barrantes

Dedicated to Dr. Luis F. Leloir
on the occasion of his 80th
birthday, September 6, 1986

1. INTRODUCTION

The nicotinic acetylcholine receptor (AChR) is an integral membrane protein that transduces the binding of acetylcholine into a local depolarization of the postsynaptic membrane through the transient opening of a cation-selective channel. The AChR is a glycoprotein of about a quarter of a million molecular weight and is by far the best known of all neurotransmitter receptors from both a functional and a structural point of view. This is because the AChR macromolecule is very abundant in tissues such as the electric organ of fish, thus providing an excellent biological source for biochemical studies, and because of the availability of appropriate chemical tools such as α-toxins, which are quasiirreversible competitive antagonists of AChR function. These toxins have extremely high affinity for the receptor and have not only enabled the characterization of some pharmacological properties of the latter but have helped, among other

---

FRANCISCO J. BARRANTES • Instituto de Investigaciones Bioquimicas, Universidad Nacional del Sur, 8000 Bahia Blanca, Argentina.

things, to localize the protein in cells and tissues, to study state transitions elicited by agonists, and to purify the AChR by affinity chromatography (see reviews by Barrantes, 1983; Changeux et al., 1983).

The AChR 9 S monomer is composed of two quasiidentical subunits (α) and three additional subunits (β, γ, δ) in a mole ratio of $\alpha_2\beta\gamma\delta$ (Reynolds and Karlin, 1978; Raftery et al., 1980; Lindstrom et al., 1982). Apparent molecular weights of the α, β, γ, and δ chains determined by SDS gel electrophoresis are 39,000, 48,000, 58,000, and 64,000, respectively, whereas the exact molecular weights derived from the nucleotide sequences of the cDNA coding for each subunit are 50,116, 53,681, 56,279, and 57,565, respectively. The nature of the discrepancies (of varying signs) is not yet known. The α subunits carry the recognition site for agonists and competitive antagonists. All subunits are glycopolypeptides (Lindstrom et al, 1982) and transverse the membrane at least once.

2. THE AChR MACROMOLECULE

The detergent-solubilized AChR monomer is an asymmetrical, roughly cylindrical body with a radius of gyration of about 4.6 nm (see Barrantes, 1983). In agreement with its subunit stoichiometry (Lindstrom et al., 1982), the AChR monomer appears to lack symmetry in the plane of the membrane (Kistler et al., 1982; Zingsheim et al., 1982a,b). Application of low-dose (<10 e$^-$/Å^2) scanning transmission electron microscopy (STEM) and image-averaging techniques specially developed to analyze noncrystalline specimens enabled us to produce the first low-resolution maps of the average AChR particle in the native receptor-rich membranes at a resolution of 15–20 Å (Zingsheim et al., 1980). Briefly, three distinct regions were identified around the central pit of the AChR in end-on view micrographs, and to our surprise the distribution of mass lacked any symmetry. Such observations challenged models postulating a symmetrical, hexameric organization of subunits; more importantly, they called for the search for the symmetry properties implicit in the postulated allosteric behavior of the AChR in features other than the subunit arrangement in the monomer.

In the axis normal to the membrane surface, the AChR protein is also unevenly distributed with respect to the lipid bilayer. The mass of the AChR molecule extends about 5.5 nm towards the extracellular space and about 1.5 nm into the cytoplasmic compartment (Klymkowsky and Stroud, 1979; Zingsheim et al., 1982a). On removal of nonreceptor proteins from the cytoplasmic face of the membrane, a different profile of the AChR particle becomes visible (Barrantes, 1982). This is presumably the cytoplasmic-facing end of the molecule. Thus, the AChR monomer displays asymmetry along axes both parallel and perpendicular to the membrane plane.

Symmetrical bodies normally pack in two or three dimensions into regular arrays. The AChR exists in the membrane as closely but not maximally packed particles (Barrantes *et al.*, 1980; Zingsheim *et al.*, 1980). Ordered arrays (Brisson, 1980; Klymkowsky and Stroud, 1979) occur exceptionally in the native membranes (another reflection of the lack of symmetry of the individual molecule), and the type of symmetry displayed in such ordered lattices is variable (see review in Barrantes, 1979). The usual case is that of disordered particles, although rows of particles have been observed (Heuser and Salpeter, 1979; Barrantes, 1982b). We have used a combination of (1) depletion of nonreceptor proteins by alkaline treatment and (2) a mild physical shearing of the resulting fragile membranes to observe oligomeric forms of the AChR in the membrane by negative-contrast electron microscopy (Barrantes, 1982b). The 9 S AChR monomer predominant in reduced membranes, the 13 S species in normal or alkylated membranes, and higher oligomeric forms in both normal or oxidized membranes can be correlated with the AChR species revealed by gradient centrifugation and with the presence or absence of some nonreceptor proteins (Barrantes, 1982a,b).

3. ARRANGEMENT OF SUBUNITS IN THE AChR MACROMOLECULE

The localization of individual AChR subunits is the subject of many current biochemical, immunologic, and structural studies. Lactoperoxidase iodination (St. John *et al.*, 1982), controlled proteolysis in sealed, intact vesicles (Strader and Raftery, 1980), cross linking with neurotoxins (see references in Conti-Tronconi and Raftery, 1982), and chemical analysis of AChR chains indicate exposure of all subunits to the extracellular millieu. This is corroborated by the use of antisubunit antibodies (see Lindstrom *et al.*, 1982), which also make clear the exposure of all subunits to the intracellular space (Froehner, 1981). The AChR subunits can also be reached from the intramembranous phase with lipophilic reagents such as iodonaphtylazide (Tarrab-Hazdai *et al.*, 1980) or pyrene sulfonylazide (Sator *et al.*, 1979). Hidden antigenic determinants of the AChR can be exposed after alkaline extraction, which removes essentially the conspicuous nonreceptor ν-proteins (Froehner, 1981). The latter have also been localized at the cytoplasmic face of the membrane (Barrantes, 1982b; St. John *et al.*, 1982). This type of evidence indicates that all subunits transverse the membrane at least once. In fact (see below), current models suggest four to five transmembrane helices for each chain, thus bringing the total number of helices within the bilayer to 20–25. The AChR is an integral transmembrane glycoprotein.

Direct structural studies aimed at localizing individual subunits have also been carried out in recent times. The comparison of native membranes with those subjected to cleavage of the disulfide bonds by reducing agents has yielded the putative average position of the δ subunits (Zingsheim *et al.*, 1982a) in the native membrane-bound AChR (Fig. 1). Similar results were obtained by using an indirect approach based on the measurement of angles subtended by biotinylated toxin–avidin complexes in AChR trimers (Wise *et al.*, 1981). Fairclough *et al.* (1983) measured the angles subtended by Fab fragments of monoclonal antibodies directed against the α subunits on AChR particles. They concluded that the δ–δ bonded dimers are a mixed population of translationally related particles (R–R) and particles related by a C2 type of symmetry (R–Я). The angular distributions of Fab fragments are broadened by the contribution arising from the flexibility about the δ–δ bond, which was calculated to be ±22°. The reason for the different occurrence of symmetrical dimers related by a twofold axis in native membranes (Zingsheim *et al.*, 1982a) and asymmetric dimers in lipid-enriched or reconstituted vesicles (Bon *et al.*, 1982; Fairclough *et al.*, 1983) is not known. The extensive protein–protein interactions operative in the former case might bear on the discrepancy.

In spite of the ambiguities mentioned in the preceding section, knowledge of the location of the δ–δ bond linking two monomers in a dimer provides a useful landmark for establishing the relative topography of other AChR subunits in the population of dimers displaying C2 symmetry

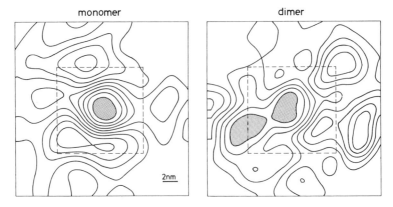

Figure 1. Average end-on views of AChR monomer and dimer obtained from low-dose electron microscopy of membrane-bound, dithiothreitol-reduced (left) and unreduced (right) preparations. Resolution is 6.4 nm. (From Zingsheim *et al.*, 1982a.)

in the native AChR membranes (Zingsheim et al., 1982a). Since the α subunits carry the recognition site for agonists and antagonists, the latter can be used in an attempt to localize the α chains by structural methods. Studies using the external tag avidin bound to biotinylated α-toxin enabled the calculation of the angles separating the α subunits (Holtzman et al., 1982). The conclusion was reached that the two α chains in a monomer cannot be contiguous. We have made use of native α-bungarotoxin for the direct localization of its recognition sites on the AChR molecule by low-dose electron microscopy and single-particle image-averaging techniques (Zingsheim et al., 1982b). Attachment of a single native toxin to its recognition site on each α chain produces a significant increase of the mass contributing to the average image of the AChR, roughly one-quarter of the mass per α subunit. A difference map between the toxin-tagged and untreated membranes yields two statistically significant peaks of stain-excluding density. The two α chains exhibit contacts with a different set of subunits; i.e., they possess different local environments. This provides a structural framework within which one can rationalize the functional nonequivalence of the recognition sites. More recently, Lindstrom et al. (1983) have produced evidence indicating that only one of the two α chains is glycosylated, a further manifestation of asymmetry in the AChR, in this case at the level of chemical features in individual subunits within the receptor monomer.

4. THE AChR PRIMARY STRUCTURE, cDNA RECOMBINANT TECHNIQUES, AND MODELING RECEPTOR STRUCTURE

The sequence of the first 54 amino acid residues from the NH terminus of the four subunits has been determined by conventional chemical sequencing techniques (Raftery et al., 1980). Sequence homologies between the amino termini of the subunits were found, and the evolutionary origin of all chains from a common ancestral gene, followed by divergence through gene duplication, has been postulated (Raftery et al., 1980). Calculations of the secondary structure of the AChR subunits emerging from the above studies have also been attempted (Guy, 1981). The information obtained from such limited sequence data indicates considerable α-helical and β-pleated structure in all subunits.

With the advent of DNA recombinant techniques, the DNA sequences of all subunits became available (Noda et al., 1982, 1983; Numa et al., 1983; Sumikawa et al., 1982; Claudio et al., 1983; Devillers-Thiery et al., 1983), and hence the complete primary structure of the AChR chains has been elucidated. A comparison of the amino acid sequences, as aligned by Noda and co-workers (1983), shows that the homology between chains extends through most of the primary structure of the subunits with the exception of small stretches, particularly between residues 342 and 425 (Fig. 2). This is indicative of the existence of similar structural features in each subunit and hence of similar contributions of each subunit to the overall AChR structure. It also reinforces the view (Raftery *et al.*, 1980) that the subunits could have evolved from a common ancestral gene.

The distribution of amino acid residues along different subunit polypeptide chains can be compared and also analyzed in terms of the occurrence of hydrophobic and hydrophilic stretches, from which the location of segments relative to the membrane lipid bilayer can be postulated. This is most conveniently done by application of algorithms such as those developed by Kyte and Doolittle (1982) or Hopp and Woods (1981). This type of analysis made apparent the occurrence of four segments of hydrophobic amino acid residues in all receptor subunits. The remarkable alignment of similar stretches along the chains, which brings into register

```
               -24                    -1
ALPHA       MILCSYWHVGLVLLLFSCCGLVLG
BETA        MENVRRMALGLVVMMALALSGVGA
GAMMA              MVLTLLLIICLALEVRS
DELTA       MGNIHFVYLLISCLYYSGCS   G

1
SEHETRLVANLL  EN YNKVIRPVEHHTHFVDITVGLQLIQLISVDEVNQIVETNVRLRQQWIDVRLRWNPADYGGIKKIRLPSDDV
SVMEDTLLSVLF  ET YNPKVRPAQTVGDKVTVRVGLTLTNLLILNEKIEEMTTNVFLNLAWTDYRLQWDPAAYEGIKDLRIPSSDV
ENEEGRLIEKLL  GD YDKRIIPAKTLDHIIDVTLKLTLTNLISLNEKEEALTTNVWIEIQWNDYRLSWNTSEYEGIDLVRIPSELL
VNEEEERLINDLLIVNKYNKHVRPVKHNNEVVNIALSLTLSNLISLKETDETLTSNVWMDHAWYDHRLTWNASEYSDISILRLPPELV
                      100
WLPDLVLYNNADGDFAIVHMTKLLLDYTGKIMWTPPAIFKSYCEIIVTHFPFDQQNCTMKLGIWTYDGTKVSIS        PESDRP
WQPDIVLMNNNDGSFEITLHVNVLVQHTGAVSWQPSAIYRSSCTIKVMYFPFDWQNCTMVFKSYTYDTSEVTLQ      HALDAKGERE
WLPDVVLENNVDGQFEVAYYANVLVYNDGSMYWLPPAIYRSTCPIAVTYFPFDWQNCSLVFRSQTYNAHEVNLQLS       AEEGEA
WIPDIVLQNNNDGQYHVAYFCNVLVRPNGYVTWLPPAIFRSSCPINVLYFPFDWQNCSLKFTALNYDANEITMDLMTDTIDGK DYP
                         200
          DLSTFMESGEWVMKDYRGWKH WVYYTCCPD TPYLDITYHFIMQRIPLYFVVNVIIPCLLFSFLTGLVFYLPTDSG EKM
VKEIVINKDAFTENGQWSIEH KPSRKNW RSD DP  S YEDVTFYLIIQRKPLFYIVYTIIPCILISILAILVFYLPPDAG EKM
VEWIHIDPEDFTENGEWTIRH RPAKKNYNWQLTKDD TDFQEIIFFLIIQRKPLFYIINIIAPCVLISSLVVLVyFLPAQAGGQKC
IEWIIIDPEAFTENGEWEIIH KPAKKN IYPDKFPNGTNYQDVTFYLIIRRKPLFYVINFITPCVLISFLASLAFYLPAESG EKM
                         300
TLSISVLLSLTVFLLVIVELIPSTSSAVPLIGKYMLFTMIFVISSIIITVVVINTHHRSPSTHTMPQWVRKIFIDTIPNV
SLSISALLAVTVFLLLLADKVPETSLSVPIIIRYLMFIMILVAFSVILSVVVLNLHHRSPNTHTMPNWIRQIFIETLPPFLWIQRPV
TLSISVLLAQTIFLFLIAQKVPETSLNVSLIGKYLIFVMFVSMLIVMNCVIVLNVSLRTPNTHSLSEKIKHLFLGFLPKYLGMQLEP
STAISVLLAQAVFLLLTSQRLPETALAVPLIGKYLMFIMSLVTGVIVNCGIVLNFHFRTPSTHVLSTRVKQIFLEKLPRILHMSRAD
                         400
                   MFFSTMKRASKEKQENKIFADDIDISDISGKQVTGEVIFQTPLIKNP DV
TTPSPD        SKPTIISRANDEYFIRKPA     GDFVCPVDNARVAVQPERLFSEMKWHLNG        LTQPVTLPQDL
SEETPEKPQ  PR RRSSFGIMI  KA  EEYILKKPRSELMFEEQKDRHGLKRVNK  MTSDIDI        GTTVDLYKDLANFAP  EI
ESEQPDWQNDLKLRRSSSVGYISKAQ  EYFNIKSR   SELMFEKQSERHGLVP  RVTPRIGFGNNNEN         IAASDQLHDEI
                             500
KSAIEGVKYIAEHMKSDEESSNAAEEWKYVAMVIDHILLCVFMLICIIGTVSVFAGRLIELSQEG
KEAVEAIKYIAEQLESASEFDDLKKDWQYVAMVADRLFLYVFFVICSIGTFSIFLDASHNVPPDNPFA
KSCVEACNFIAKSTKEQNDSGSENENWVLIGKVIDKACFWIALLLFSIGTLAIFLTGHFNQVPEFPFFPGDPRKYVP
KSGIDSTNYIVKQIKEKNAYDEEVGNWNLVGQTIDRLSMFIITPVMVLGTIFIFVMGNFNHPPAKPFEGDPFDYSSDHPRCA
```

Figure 2. Aligned amino acid sequence of the four acetylcholine receptor subunits (Noda *et al.*, 1982, 1983; Sumikawa *et al.*, 1982; Claudio *et al.*, 1983; Devilliers-Thiery *et al.*, 1983).

the hydrophobic regions of the different subunits, led Noda *et al.* (1982), Claudio *et al.* (1983), and Devilliers-Thiery *et al.* (1983) to postulate models of the subunit arrangement with respect to the lipid bilayer in which each of the four hydrophobic domains transverses the membrane once and displays α-helical configuration. These postulations are based on (1) the observed length of the hydrophobic stretches; (2) the analogy with other transmembrane proteins such as bacteriorhodopsin and glycophorin A; and (3) favorable length of the observed stretches in α-helix as opposed to, say, β-sheets to cover the bilayer thickness. These models predict that the amino acid and carboxyl termini of the subunits face the extracellular millieu. This is in contrast to the data obtained on other integral membrane proteins studied to date, in which the N terminus is expected to remain on the extracellular space after cleavage of the leader sequence, and the C terminus is exposed to the intracellular face of the membrane (Sabatini *et al.*, 1982). Furthermore, this type of model, the so-called four-helix model, implies the presence of a relatively large portion of the AChR protein (*ca.* 30%) on the cytoplasmic face of the membrane.

An alternative type of model (the five-helix model) places a much smaller (*ca.* 20%) portion of the protein on the intracellular face of the membrane and predicts that in addition to the four transmembrane helices there is a fifth, amphipathic transmembrane helix (Guy, 1984; Fairclough *et al.*, 1983). These models place the C terminus on the cytoplasmic face of the membrane. The aforementioned authors also produce evidence suggesting that the amphipathic helix possesses a continuous hydrophobic face on one side and a hydrophilic face on the other, thus providing a structural substratum for the lining of the internal walls of the ionic channel by the latter. Furthermore, the continuous hydrophilic face has the appropriate length to span the lipid bilayer (about 38 Å) and a distribution of charges (21 positive, 19 negative, 10 uncharged) in its residues (Fairclough *et al.*, 1983) that makes it ideally suited for intersubunit ion pairing.

The regions between residues 70 and 160 in the four subunit sequences are also highly homologous (27 residues are totally conserved). Both the four- and five-helix models postulate that this region is on the synaptic cleft. Furthermore, this stretch would carry two important features: a site for N-glycosylation (nucleotides 143–145) and the disulfide bond that reacts after reduction with affinity reagents such as 4-(N-maleimido)benzyltrimethylammonium (MBTA) (C130, C144) on the α subunit (Noda *et al.*, 1982). This stretch is also postulated to be rich in amphipathic β-sheets common to all subunits in the form of a six- or eight-strand β-barrel domain. The "solvent" side of the β-barrel could line the central well of the receptor. In the α subunit, a β-bend in this barrel near disulfides 130–144 could form part of the ACh-binding site, as suggested by Noda *et al.* (1982). According to Fairclough *et al.* (1983), the predominant negative charges in the "vestibule" of the receptor might contribute to the cation specificity of the receptor.

5. IMMUNOCHEMISTRY OF AChR AND THE TESTING OF MODELS

Poly- and monoclonal antibodies of different degrees of specificity (antitotal AChR, antisubunit, etc.) have been successfully used to map regions of the receptor molecule inaccessible to other techniques such as affinity labeling or chemical modification. A main immunogenic region (MIR, Lindstrom et al., 1982) consisting of at least two adjacent determinants located outside the ACh recognition site on the α subunit appears to be a phylogenetically preserved immunologic feature of the AChR. Antibodies aganst the MIR cross react between species and passively transfer the acute form of experimental autoimmune myasthenia gravis (EAMG). Within the "library" of antibodies available to date, some types noncompetitively inhibit the $^{22}Na^+$ flux, and others are able to distinguish between the AChR monomer and dimer or cross link two δ subunits (Lindstrom et al., 1982). By using monoclonal antibodies against synthetic peptides prepared on the basis of the known sequence data, Lindstrom et al. (1984) have demonstrated that the C termini of the AChR chains are exposed to the cytoplasmic face of the membrane, thus favoring the five-helix type of model.

The recent studies of Juillerat et al. (1984) and Lindstrom et al. (1984) also served to test other predictions about the putative location of the MIR (Noda et al., 1982) based on the sequence data obtained from the cDNA nucleotide structure. The latter authors had suggested that the most hydrophilic domain (residues 161–166) of the α subunits formed the main immunogenic region. These predictions were based on the knowledge that the most hydrophilic regions of the amino acid sequence in a protein are prominent antigenic determinants (Hopp and Woods, 1981). Lindstrom et al. (1984) have recently deomonstrated that the 161–166 stretch of the α-chains is not the MIR and is not even contiguous to this region. Numa et al. (1983), however, have alternatively located the MIR in the α-196–207 stretch.

6. VOLTAGE-GATED AND AGONIST-GATED CHANNELS: A COMPARISON

The fact that the sodium channel for the eel electrocyte has recently been cloned and its complete nucleotide and amino acid sequences determined (Noda et al., 1984) prompts comparisons with the recently acquired knowledge of the structure of the AChR. In contrast to the AChR oligomeric, multisubunit complex, the sodium channel consists of a single polypeptide chain of more than 260,000 M_r with some additional smaller polypeptides of M_r 37,000–45,000. The cDNA sequence of the main polypeptide reveals the occurrence of four domains of 250–300 amino acid residues that are roughly 50% homologous with each other. Each domain has a similar overall structure; six subdomains characteristically occur in them all. Two subdomains presumably correspond to hydrophobic, membrane-transversing segments, presumably in α-helical conformation; two are negatively charged, and the two remaining are positively charged and neutral, respectively.

Numa et al. (1984) speculate on the structure–function relationship of the predicted domains. The transmembrane segments are postulated to be involved directly or indirectly in the formation of the voltage-gated ionic channel. At least one helical segment from each repetitive domain is supposed to participate in the lining of the channel, whereas the negatively charged domain is assumed to be related to the ionic selectivity of the channel. The positively charged domains are suggested to be involved in the activation–inactivation gates, which are located near the cytoplasmic membrane surface (Armstrong, 1975). They further hypothesize that each of the four positively charged domains of the sodium channel, together with the corresponding four negatively charged domains, acts as a voltage sensor in the activation process.

The coupling between membrane field and channel gating is believed to result from the change in the dipole moment of the sodium channel macromolecule associated with the transition(s) between closed and open channel states. The dipole moment change is very large for the sodium channel (hundreds of Debyes) as compared with the gating of the receptor-controlled gating operating in the case of the agonist-activated AChR (in the order of tens of Debyes). No unusual accumulation of charges appears to occur in the sodium channel as depicted in the model presented by Noda et al. (1984) except for the distribution of positively and negatively charged domains described above. The sodium channel also appears to differ from the AChR molecule in that both amino and carboxyl termini are exposed to the cytoplasmic face of the membrane.

7. DYNAMICS OF AChR AND LIPIDS IN THE MEMBRANE

The highly specialized function of the AChR calls for an organization of the postsynaptic membrane—the natural site of the AChR—optimally suited to the requirements of signal transduction within the operational time scale of the nicotinic synapse. The AChR is closely packed at densities of more than 10,000 particles/μm^2 in the lipid bilayer. A lipid matrix interrupted by integral membrane proteins such as that contemplated by the "fluid mosaic model" is certainly inadequate to describe the postsynaptic membrane. Instead, and on the basis of available evidence, a protein matrix with lipid filling the interstices has been suggested as a closer approximation (Barrantes, 1979).

There is broad agreement that lipid motion in monolayers and bilayers follows in general terms the predictions of the so-called free-area theories, neglecting interactions of the diffusant phospholipid head group with the aqueous medium and of the fatty acyl chains with the opposing monolayer. Essentially, a lipid molecule undergoes a diffusive step when a free volume (a sort of a "hole") exists adjacent to it, and solvent fluctuations "close" the free volume around the diffusant by jumping into that volume. The free-area theories are used to explain the lateral diffusion coefficient of fluorescent lipid analoges in lipid bilayers similar to those used to measure the diffusion of reconstituted AChR. NBD-Phosphatidylethanolamine exhibited values of D, the lateral diffusion coefficient, of 8×10^{-8} cm^2 sec^{-1} (Criado et al., 1982).

When large intergral membrane proteins like the AChR—whose dimensions generously exceed those of the surrounding lipids—are reconstituted into artificial bilayers at relatively low protein:lipid ratios, the lateral diffusion of the protein can be approximated by the Saffman and Delbrück (1975) formalism. This is based on the Stokes–Einstein treatment of motion in a solvent medium, in which lateral motion can be envisaged as originating in random collisions of the macromolecule with the solvent molecules and is opposed by frictional forces with the same. The diffusion coefficient, D, is given by $D = kT/f$, where k is the Boltzmann constant and f is the frictional coefficient. In the case of a spherical molecule of radius r diffusing in a medium of viscosity η, $D = kT/6\pi\eta r$ (the Stokes–Einstein equation). Saffman and Delbrück applied this formalism to the case of a cylinder of radius r oriented perpendicular to and diffusing in a thin viscous sheet of viscosity η and thickness h (equal to the height of the embedded cylinder), a treatment that is more relevant to that of a protein transversing a membrane. The diffusion coefficient becomes:

$$D = (kT/4\pi\eta_m h) \ln(\eta_m h/\eta_w r - 0.5772)$$

where η_m is the viscosity of the viscous sheet ("the membrane") and η_w ($\simeq 10^{-2}$ poise) is the viscosity of the aqueous medium in which the sheet is immersed. The sheet is treated as a continuum of solvent molecules. The formalism predicts a weak dependence of D on protein radius, a prediction that appears to be followed roughly by the few integral membrane proteins studied to date, including the AChR (Criado et al., 1982).

It should be noted that the theory finds validity only in the case of very diluted proteins (as with proteins reconstituted in lipid bilayers); the rate of diffusion of the AChR monomers and dimers lies in the range of $1–3 \times 10^{-8}$ cm² sec^{-1} in lipid bilayers in the liquid–crystalline phase and within the 14–37°C range (Criado *et al.*, 1982). The presence of cholesterol does not appear to modify D.

In the plasmalemma of developing rat myotubes, the situation is somewhat different. Two populations of AChR molecules are observed: a diffusely distributed population having $D = 5 \times 10^{-11}$ cm² sec^{-1} and a localized, patched population with $D \approx 10^{-12}$ cm² sec^{-1} (Axelrod *et al.*, 1976). In blebs of myoblast sarcolemma, Tank *et al.* (1982) found D to be 3×10^{-9} cm² sec^{-1}, whereas Poo (1982) estimated D to be 2.6×10^{-9} cm² sec^{-1} in *Xenopus* embryonic myoblasts. D in reconstituted membranes is at least 50-fold higher than the highest values found in native membranes. That is, the AChR in synthetic lipid membranes undergoes essentially free lateral diffusion. It does not exhibit a large decrease in mobility as compared to the lipids. Reconstituted membranes differ from natural membranes in their lipid composition, concentration of protein, and absence of peripheral proteins. The former factor was not found to influence the lateral diffusion of the AChR at very high lipid:protein ratios. It is likely that the extensive protein–protein interactions occurring both between AChR molecules on the one hand and between AChR molecules and other nonreceptor proteins on the other are mainly responsible for the relative immobilization of AChR in the native membrane and particularly in the synaptic region.

Considerations similar to those outlined above for the translational mobility of the AChR are also applicable to the rotational motion of the protein. This has been studied by means of electron spin resonance and phosphorescence anisotropy techniques. A spectrum of relaxation times is observed with both approaches; the rotational correlation time of the monomeric AChR is about 10–25 μsec (Bartholdi *et al.*, 1981), in full agreement with the expected value. Slower relaxation times are probably related to the existence of higher oligomeric AChR species. Since the translational motion of the AChR in reconstituted lipid bilayers can be practically considered a free, unrestricted lateral diffusion, D can be related to the rotational relaxation time by $D \approx 4\, r^2/\phi$.

8. ACETYLCHOLINE-RECEPTOR-CONTROLLED CHANNEL PROPERTIES

The detailed description of the molecular mechanisms leading from agonist binding to gating of the AChR-controlled channel is still one of the most challenging open questions in this field. It is now possible to directly observe the elementary currents associated with ion permeation through a single AChR-controlled channel in the living cell. This is accomplished by the application of novel electrophysiological recording techniques: the patch-clamp method (Neher and Sakmann, 1976). More recently, the same type of single-current measurements has been obtained in reconstituted systems, in which (1) the chemical composition of lipid and protein, (2) the thermodynamic state of the host membranes, (3) the degree of purity of the AChR, (4) its oligomeric state, and (5) the sidedness of the membrane can be controlled more easily than in the native membrane (Boheim et al., 1981). Thus, one can now characterize discrete open and closed states of the individual AChR channels (Neher and Sakmann, 1976) under normal, blocked, and desensitized states, as schematically represented in Fig. 3. The traces obtained by patch-clamp recordings from living cells or membrane fragments excised therefrom are idealized in this

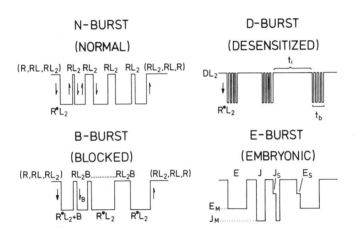

Figure 3. Diagramatic representation of single-channel behavior. A burst of openings (downward deflection) is defined as a sequence of periods in which the channel, having reached its open state, is interrupted by short closed periods. Three types of bursts can be observed (see review by Barrantes, 1983). A normal ("N") burst is believed to result from the transition from the biliganded closed (RL_2) AChR conformation to the biliganded open conformation (R^*L_2). The AChR channel spends short periods in the closed state ("gaps") before finally closing for a much longer period, i.e., before losing one (RL) and two agonist molecules to reach the resting nonliganded state R. D bursts occur in the presence of desensitizing agonist concentrations. Bursts of this type have long average durations and are separated by much longer silent intervals having durations of hundreds of milliseconds to seconds. B bursts ("blocked") are observed in the presence of channel blockers (e.g., local anesthetics). Some blockers (B) are believed to bind the open channel conformation (R^*L_2) and drive the AChR into an open but nonconducting, blocked state (RL_2B). The ternary complex has to dissociate before the channel can finally close. Embryonic muscle cells have AChR channels displaying heterogeneity of open times and amplitudes. The "embryonic" E burst is probably a mixture of two types of channels (J_M = junctional, adult type, and E_M = embryonic type) operating simultaneously. These channels display substates (J_S, E_S) of smaller conductance.

diagram. It followed from these studies that the M_r-250,000 monomer, with its pentameric structure, is the minimal functional unit possessing ligand recognition ability and controlling the ionic permeability (see Barrantes, 1983, and references therein). It is also possible to tentatively relate the AChR channel states to corresponding states of ligation of the AChR recognition site (see Fig. 3.).

In spite of this progress, many fundamental problems remain unanswered, such as the lack of a coherent explanation for the apparent cooperativity of the response. Similarly, there is no comprehensive, wholistic model covering simultaneously (1) ligand binding, (2) AChR conformational transitions, and (3) channel response. Information is also lacking on the relationship between AChR oligomeric states and channel-gating kinetics. The correlation of function and structure of the channel itself is also an area that will be rapidly developed as our knowledge of the AChR molecule reaches the necessary depth. The impact of cDNA recombinant techniques will be particularly appreciated in this field. Selective mutagenesis, analysis of protein synthesis conducted with altered mRNAs or lacking the message for one or more subunits, etc., are possible paths for the near future. Experiments of this sort will also enable the testing of models derived from analysis of sequence data aimed at deciphering the putative regions of the AChR involved in channel structure. Determination of the topography of individual subunits and the supramolecular arrangement of the AChR in the membrane should also be accompanied by the description of the functional counterparts associated with such structural forms. The actual findings on the structure and mechanisms operating in the nicotonic AChR may not be necessarily (or fully) applicable in a direct manner to many other neurotransmitter and hormonal receptors. Some of its basic properties (ligand recognition, adequate three-dimensional organization for insertion in the membrane with the correct vectorial orientation, and the capacity to effect a meaningful cellular response), however, are common to all endogenous receptors. The rapid advances in our knowledge of the nicotinic ACh receptor will at least help us in the design of more correct experiments and strategies for studying other, more complicated receptor systems.

REFERENCES

Armstrong, C. M., 1975, Ionic pores, gates, and gating currents, *Q. Rev. Biophys.* **7**:179–210.

Axelrod, D., Koppel, D. E., Schlessinger, J., Elson, E., and Webb, W. W., 1976, Mobility measurements by analysis of fluorescence photobleaching recovery kinetics, *Biophys. J.* **16**:1055–1069.

Barrantes, F. J., 1979, Endogeneous chemical receptors: Some physical aspects, *Annu. Rev. Biophys. Bioeng.* **8**:287–321.

Barrantes, F. J., 1982a, Interactions of the membrane-bound acetylcholine receptor with the non-receptor, peripheral ν-peptide, in *Neuroreceptors* (F. Hucho, ed.), W. de Gruyter, Berlin, New York, pp. 315–328.

Barrantes, F. J., 1982b, Oligomeric forms of the membrane-bound acetylcholine receptor disclosed upon extraction of the M_r 43,000 nonreceptor peptide, *J. Cell Biol.* **92**:60–68.

Barrantes, F. J., 1983, Recent developments in the structure and function of the acetylcholine receptor, *Int. Rev. Neurobiol.* **24**:259–341.

Barrantes, F. J., Neugebauer, D.-C., and Zingsheim, H. P., 1980, Peptide extraction by alkaline treatment is accompanied by rearrangement of the membrane-bound acetylcholine receptor from *Torpedo marmorata*, *FEBS Lett.* **112**:73–78.

Bartholdi, M., Barrantes, F. J., and Jovin, T. M., 1981, Rotational molecular dynamics of the membrane-bound acetylcholine receptor revealed by phosphorescence spectroscopy, *Eur. J. Biochem.* **120**:389–397.

Boheim, G., Hanke, W., Barrantes, F. J., Eibl, H., Sakmann, B., Fels, G., and Maelicke, A., 1981, Agonist-activated ionic channels in acetylcholine receptor reconstituted into planar lipid bilayers, *Proc. Natl. Acad. Sci. U.S.A.* **78**:3586–3590.

Bon, F., Lebrun, E., Gomel, J., Van Rapenbusch, R., Cartaud, J., Popot, J.-L., and Changeux, J.-P., 1982, Orientation relative des deux oligomeres constituant la forme lourde du recepteur de l'acetylcholine chez la *Torpille marbree*, *C.R. Acad. Sci. (Paris) [D]* **295**:199–204.

Brisson, A., 1980, *These Doctorat d'Etat*, Grenoble, France.

Changeux, J.-P., Bon, F., Cartaud, J., Devilliers-Thiéry, A., Giraudat, J., Heidmann, T., Holton, B., Nghiem, H.-O, Popot, J.-L., Van Rapenbusch, R., and Tzartos, S., 1983, Allosteric properties of the acetylcholine receptor protein from *Torpedo marmorata*, *Cold Spring Harbor Symp. Quant. Biol.* **48**:35–52.

Claudio, T., Ballivet, M., Patrick, J., and Heinemann, S., 1983, Nucleotide and deduced amino acid sequences of *Torpedo californica* acetylcholine receptor γ subunit, *Proc. Natl. Acad. Sci. U.S.A.* **80**:1111–1115.

Conti-Tronconi, B. M., and Raftery, M. A., 1982, The nicotinic cholinergic receptor: Correlation of molecular structure with functional properties, *Annu. Rev. Biochem.* **51**:491–530.

Criado, M., Vaz, W. L. C., Barrantes, F. J., and Jovin, T. M., 1982, Translational diffusion of acetylcholine receptor (monomeric and dimeric forms) of *Torpedo marmorata* studied by fluorescence recovery after photobleaching, *Biochemistry* **21**:5750–5755.

Devilliers-Thiery, A., Giraudat, J., Bentavoulet, M., and Changeux, J. -P., 1983, Complete mRNA coding sequence of the acetylcholine binding α-subunit of *Torpedo marmorata* acetylcholine receptor: A model for the transmembrane organization of the polypeptide chain, *Proc. Natl. Acad. Sci. U.S.A.* **80**:2067–2071.

Fairclough, R. H., Finer-Moore, J., Love, R. A., Kristofferson, D., Desmeules, P. J., and Stroud, R. M., 1983, Subunit organization and structure of an acetylcholine receptor, *Cold Spring Harbor Symp. Quant. Biol.* **48**:9–20.

Froehner, S. C., 1981, Identification of exposed and buried determinants of the membrane-bound acetylcholine receptor from *Torpedo californica*, *Biochemistry* **20**:4905–4915.

Guy, H. R., 1981, Structural models of the nicotinic acetylcholine receptor and its toxin-binding site, *Cell. Mol. Neurobiol.* **1**:231–258.

Guy, H. R., 1984, A structural model of the acetylcholine receptor channel based on partition energy and helix packing calculations, *Biophys. J.* **45**:249–261.

Heuser, J. E., and Salpeter, S. R., 1979, Organization of acetylcholine receptors in quick-frozen, deep-etched, and rotary-replicated *Torpedo* postsynaptic membrane, *J. Cell Biol.* **82**:150–173.

Holtzman, E., Wide, D., Wall, J., and Karlin, A., 1982, Electron microscopy of complexes of isolated acetylcholine receptor, biotinyltoxin, and avidin, *Proc. Natl. Acad. Sci. U.S.A.* **79**:310–314.

Hopp, T. P., and Woods, K. R., 1981, Prediction of protein antigenic determinants from amino acid sequences, *Proc. Natl. Acad. Sci. U.S.A.* **78**:3824-3828.

Juillerat, M. A., Barkas, T., and Tzartos, S. J., 1984, Antigenic sites of the nicotinic acetylcholine receptor cannot be predicted from the hydrophilicity profile, *FEBS Lett.* **168**:143-148.

Kistler, J., Stroud, R. M., Klymkowsky, M. W., Lalancette, R. A., and Fairclough, R. H., 1982, Structure and function of an acetylcholine receptor, *Biophys. J.* **37**:371-383.

Klymkowsky, M. W., and Stroud, R. M., 1979, Immunospecific localization and three-dimensional structure of a membrane-bound acetylcholine receptor from *Torpedo californica*, *J. Mol. Biol.* **128**:319-334.

Kyte, J., and Doolittle, R. F., 1982, A simple method for displaying the hydropathic character of a protein, *J. Mol. Biol.* **157**:105-132.

Lindstrom, J., Tzartos, S. J., and Gullick, W., 1982, Structure and function of acetylcholine receptors studied using monoclonal antibodies, *Ann. N.Y. Acad. Sci.* **377**:1-19.

Lindstrom, J., Tzartos, S., Gullick, W., Hochschwender, S., Swanson, L., Sargent, P., Jacob, M., and Montal, M., 1983, Use of monoclonal antibodies to study acetylcholine receptors from electric organs, muscle, and brain and the autoimmune response to receptor in myasthenia gravis, *Cold Spring Harbor Symp. Quant. Biol.* **48**:89-99.

Lindstrom, J., Criado, M., Hochschwender, S., Fox, J. L., and Sarin, V., 1984, Immunochemical tests of acetylcholine receptor subunit models, *Nature* **311**:573-575.

Neher, E., and Sakmann, B., 1976, Single channel currents recorded from membrane of denervated frog muscle cells, *Nature* **260**:799-802.

Noda, M., Takahashi, H., Tanabe, T., Toyosato, M., Furutani, Y., Hirose, T., Asai, M., Inayama, S., Miyata, T., and Numa, S., 1982, Primary structure of α-subunit precursor of *Torpedo californica* acetylcholine receptor deduced from cDNA sequence, *Nature* **299**:793-797.

Noda, M., Takahashi, H., Tanabe, T., Toyosato, M., Kikyotani, S., Hirose, T., Asai, M., Takashima, H., Inayama, S., Miyata, T., and Numa, S., 1983, Primary structures of BB- and DD- subunit precursors of *Torpedo-californica* acetylcholine receptor deduced from cDNA sequences, *Nature* **301**:251-255.

Noda, M., Shimizu, S., Tanabe, T., Takai, T., Kayano, T., Ikeda, T., Takahashi, H., Nakayama, H., Kanaoka, Y., Minamino, N., Kangawa, K., Matsuo, H., Raftery, M. A., Hirose, T., Inayama, S., Hayashida, H., Miyata, T., and Numa, S., 1984, Primary structure of *Electrophorus electricus* sodium channel deduced from cDNA sequence, *Nature* **312**:121-127.

Numa, S., Noda, M., Takahashi, H., Tanabe, T., Toyosato, M., Furutani, Y., and Kikyotani, S., 1983, Molecular structure of the nicotinic acetylcholine receptor, *Cold Spring Harbor Symp. Quant. Biol.* **48**:57-69.

Poo, M., 1982, Rapid lateral diffusion of functional ACh receptor in embryonic muscle cell membranes, *Nature* **295**:332-334.

Raftery, M. A., Hunkapiller, M. W., Strader, C. D., and Hood, L. E., 1980, Acetylcholine receptor: Complex of homologous subunits, *Science* **208**:1454-1457.

Reynolds, J. A., and Karlin, A., 1978, Molecular weight in detergent solution of acetylcholine receptor from *Torpedo californica*, *Biochemistry* **17**:2035-2038.

Sabatini, D. D., Kreibich, G., Morimoto, T., and Adesnik, M., 1982, Mechanisms for the incorporation of proteins in membranes and organelles, *J. Cell. Biol.* **92**:1-22.

Saffman, P. G., and Delbrück, M., 1975, Brownian movement in biological membranes, *Proc. Natl. Acad. Sci. U.S.A.* **72**:3111-3113.

St. John, P. A., Froehner, S. C., Goodenough, D. A., and Cohen, J. B., 1982, Nicotinic postsynaptic membranes from *Torpedo*: Sidedness, permeability to macromolecules, and topography of major polypeptides, *J. Cell. Biol.* **92**:333-342.

Sator, V., González-Ros, J. M., Calvo Fernández, P., and Martínez-Carrion, M., 1979,

Pyrenesulfonyl azide: A marker of acetylcholine receptor subunits in contact with membrane hydrophobic environment, *Biochemistry* **18**:1200–1206.

Strader, C. D., and Raftery, M. A., 1980, Topographic studies of *Torpedo* acetylcholine receptor subunits as a transmembrane complex, *Proc. Natl. Acad. Sci. U.S.A.* **77**:5807–5811.

Sumikawa, K., Houghton, M., Smith, J. C., Bell, L., Richards, B. M., and Barnard, E., A., 1982, The molecular cloning and characterization of cDNA coding for the α subunit of the acetylcholine receptor, *Nucleic Acids Res.* **10**:5809–5822.

Tank, D. W., Wu, E.-S., and Webb, W. W., 1982, Enhanced molecular diffusibility in muscle membrane blebs: Release of lateral constraints, *J. Cell Biol.* **92**:207–212.

Tarrab-Hazdai, R., Bercovici, T., Goldfarb, V., and Gitler, C., 1980, Identification of the acetylcholine receptor subunit in the lipid bilayer, *J. Biol. Chem.* **155**:1204–1209.

Wise, D. S., Wall, J., and Karlin, A., 1981, Relative locations of the β and δ chains of the acetylcholine receptor determined by electron microscopy of isolated receptor trimer, *J. Biol. Chem.* **256**:12624–12627.

Zingsheim, H. P., Neugebauer, D.-C., Barrantes, F. J., and Frank, J., 1980, Structural details of membrane-bound acetylcholine receptor from *Torpedo marmorata*, *Proc. Natl. Acad. Sci. U.S.A.* **77**:952–956.

Zingsheim, H. P., Neugebauer, D.-C., Frank, J., Hanicke, W., and Barrantes, F. J., 1982a, Dimeric arrangement and structure of the membrane-bound acetylcholine receptor studied by electron microscopy, *EMBO J.* **1**:541–547.

Zingsheim, H. P., Barrantes, F. J., Frank, J., Hanicke, W., and Neugebauer, D.-C., 1982b, Direct structural localization of two toxin-recognition sites on an ACh receptor protein, *Nature* **299**:81–84.

Chapter 25

Amiloride-Sensitive Epithelial Sodium Channels

Dale J. Benos

1. INTRODUCTION

Electrically high-resistance epithelia actively transport sodium from the luminal side to the blood (Macknight *et al.*, 1980). The first step in this transepithelial movement of Na^+ is the facilitated diffusion of this ion across the luminal or apical membrane down its electrochemical potential energy gradient. This particular transport pathway is rate limiting for the overall transport, is regulated hormonally, and is inhibited by the diuretic drug amiloride. Single-site turnover numbers deduced from current-noise experiments (10^6 ions/sec) are consistent with a channel or pore-type mechanism (Lindemann and Van Driessche, 1977). Fuchs *et al.* (1977) and Van Driessche and Lindemann (1979) found that under their experimental conditions, Na^+ permeation through these channels could be adequately described by an electrodiffusion model in which the passive movement of Na^+ obeys the independence principle.

In this chapter, I first summarize briefly the history of the use of amiloride as a Na^+ transport inhibitor and then compare the known properties of the two general categories of amiloride-sensitive Na^+ transport

---

DALE J. BENOS • Department of Physiology and Biophysics, Laboratory of Human Reproduction and Reproductive Biology, Harvard Medical School, Boston, Massachusetts 02115; *present address:* Department of Physiology and Biophysics, University of Alabama at Birmingham, Birmingham, Alabama 35294.

systems. Then, I discuss the Na^+ entry mechanism in epithelia in some detail. Specifically, I present results of experiments designed to elucidate some of the characteristics of amiloride/Na^+ channel receptors, namely, amiloride analogue structure–activity relationships, a flux ratio analysis of Na^+ movement through this channel, the voltage dependence of the amiloride block, and the effect of interfacial potential changes on the rate of Na^+ entry. I conclude this chapter by summarizing some of the recent attempts to isolate this channel functionally by incorporation into planar bilayers and by patch-clamp studies.

In the early 1960s, chemists at the pharmaceutical firm of Merck, Sharp and Dohme synthesized a new K^+-sparing diuretic called amiloride (Bickling *et al.*, 1965; Glitzer and Steelman, 1966; Baer *et al.*, 1967; Fig. 1). This drug potently and specifically inhibits Na^+ transport in a wide variety of cellular and epithelial preparations ranging from erythrocytes to kidney tubules to human and chicken rectum, although notably not in nerve (see Benos, 1982, for a review). The observation that a single synthetic molecule can inhibit Na^+ transport in such an incredibly large number of diversified tissues obtained from nearly every phylum of the animal kingdom (with the exception of protozoans, sponges, and coelenterates) underscores the existence of a fundamental process or processes in the plasma membranes of most animal cells.

Before 1978, most investigators thought that amiloride only blocked Na^+ entry into high-resistance Na^+-transporting epithelia. There were scattered reports demonstrating that amiloride could inhibit Na^+ transport in some nonepithelial systems, but the significance of these observations was not appreciated or realized (Aceves and Cereijido, 1973; Johnson *et al.*, 1976). Since 1979, well over 200 papers have been published describing amiloride-sensitive Na^+ transport systems in over 25 nonepithelial cells as well as in electrically low-resistance epithelia such as renal proximal tubule and gallbladder. These amiloride-sensitive Na^+ transport systems have been implicated in important physiological functions such as salt balance, pH regulation, fertilization, growth and differentiation, and cell volume control.

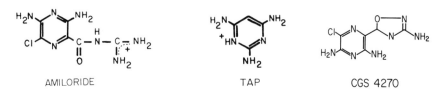

Figure 1. Chemical structures of amiloride, TAP, and CGS-4270.

Amiloride-Sensitive Sodium Channels

In general, amiloride acts rapidly and reversibly. A survey of the literature reveals that amiloride-sensitive transport processes can be arbitrarily divided into two main groups based on their sensitivity to the drug: those in which the apparent equilibrium dissociation constant (K_i) is less than 1 μM and those in which the K_i is greater than 1 μM (as determined from dose–response curves done at high external [Na^+]). The Na^+ transport mechanism in this first group (high amiloride affinity) consists of a conductive pathway, presumably an ion channel, whereas that in the second category (low amiloride affinity) consists of an electroneutral exchange like a Na^+/proton or Na^+/Ca^{2+} antiport process (Fig. 2). In general, the Na^+ channel is found in high-resistance epithelia, whereas Na^+–proton exchangers are found in the low-resistance epithelia and cellular systems, although exceptions to this statement are known (Moran and Moran, 1984).

2. AMILORIDE-SENSITIVE NA$^+$ TRANSPORT PROCESSES

Table I compares and contrasts the general characteristics of the amiloride-sensitive Na^+ channel and Na^+/proton exchanger. Several points should be made. First, there appears to be a high degree of selectivity with regard to the cations transported by both systems: only Na^+, Li^+, or H^+ can pass through either system. The concentration of external Na^+ required to produce half-maximal activation of transport (K_s) is likewise similar for both systems. However, Villereal (1981) in human fibroblasts

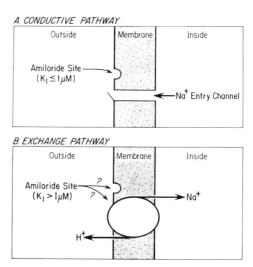

Figure 2. Amiloride-sensitive ion transport systems.

Table I. Characteristics of the Two Major Modes of Amiloride-Sensitive Na^+ Entry

	Na^+ channel	Na^+/H^+ exchanger
Selectivity	Na^+, Li^+, H^+	Na^+, Li^+, H^+
K_s for Na^+	5–50 mM	3–14 mM
Mode of transport	Conductive	Electroneutral
Location	Apical membrane	Apical, basolateral, or plasma membrane
Physiologically inducible	Yes	Yes
Amiloride K_I	<1 μM (at Na^+ > 100 mM)	>1 μM (at Na^+ > 100 mM)
Amiloride inhibition	Competitive/noncompetitive	Competitive/mixed
Other inhibitors	Benzamil	Harmaline
	6-Bromoamiloride	
	CGS 4270	N-Amidino-3-amino-5-dimethylamino-6-chloropyrazine
	TAP	

and Benos and Sapirstein (1983) in cultured rat glioma found that the amiloride-sensitive Na^+/proton exchanger did not saturate with external Na^+ (up to 140 mM), in contrast to the results of Erickson and Spring (1982) in *Necturus* gallbladder, Moolenaar et al. (1981) in mouse neuroblastoma, Kinsella and Aronson (1981) in rabbit renal microbillus membrane vesicles, and Pouyssegur et al. (1982) in hamster lung fibroblasts.

Both systems are physiologically inducible; i.e., each transport system can be functionally induced by appropriately chosen conditions. For example, Will et al. (1980) demonstrated in isolated rat colon that raising mineralocorticoid levels would induce an amiloride-sensitive Na^+ conductive pathway. This effect could be prevented by prior spironolactone treatment. *In vitro* colon preparations obtained from control rats display no amiloride sensitivity whatsoever to their transepithelial rates of Na^+ transport. The stimulation of an amiloride-sensitive Na^+/proton exchange system has been demonstrated in many different types of growth-arrested cells in culture after the readdition of serum or specific growth factors (Villereal, 1981; Moolanaar et al., 1982; Owens and Villereal, 1982; Benos and Sapirstein, 1983) or in *Amphiuma* (Cala, 1983) or dog (Parker, 1983) erythrocytes and mammalian lymphocytes (Grinstein et al., 1984) in response to a volume perturbation.

The inhibition of the Na^+ entry channel by amiloride in high-resistance epithelia has been shown to be either competitive, noncompetitive, or mixed with respect to Na^+ (see Benos, 1982, for a discussion of this matter). For the Na^+/H^+ exchanger, amiloride has been shown to be either competitive (Kinsella and Aronson, 1981; Chan and Giebisch, 1980; Paris and Pouyssegur, 1983) or mixed (Ives et al., 1983). These observations make it crucial that the exact Na^+ concentration be specified when assigning amiloride inhibitory constants. For example, in growth-arrested fibroblasts, the serum-stimulated component of Na^+ influx has an amiloride K_i of 3 µM at 20 mM external Na^+ (Pouyssegur et al., 1982) and a K_i of 200 µM at 140 mM Na^+ (Villereal, 1981).

The structure–activity relationships of amiloride and some of its analogues have been studied for the Na^+ conductive pathway (Benos et al., 1976; Cuthbert and Fanelli, 1978) as well as for the exchanger (L'Allemain et al., 1984). Chemical modifications of the parent molecule on the 5-pyrazine ring amino group render the molecule ineffective in inhibiting the Na^+ channel. In fact, if alkyl groups (e.g., methyl, ethyl,) are substituted for one or both of the 5-position amino hydrogens in amiloride, stimulation of the Na^+ conductive pathway occurs (Li and deSousa, 1978). On the other hand, modifications in the acylguanidinium portion of the molecule are more permissive; i.e., inhibitory activity can be retained or even enhanced (e.g., benzamil). The converse situation holds for the Na^+/proton exhanger. L'Allemain et al. (1984) found that although the 5-amino group was not essential for inhibition, replacement of the protons of this group by alkyl derivatives would yield compounds with up to 100-fold more potency than amiloride. Modifications in the acylguanidine moiety usually result in a diminution of inhibitory activity (but see O'Donnell et al., 1983). The hallucinogen harmaline reversibly and competitively inhibits Na^+/H^+ exchange with a K_i of 0.1–0.2 mM at 1 mM Na^+ in renal proximal tubule vesicles (Aronson and Bounds, 1980), whereas in toad urinary bladder epithelium, external addition of 0.1–1 mM harmaline stimulates the Na^+ channel (de Sousa and Grosso, 1978). These results clearly indicate that the drug sensitivities of these two major modes of amiloride-sensitive Na^+ entry are different. Amiloride has been reported to inhibit four other transport systems, namely, renal collecting duct HCO_3^- absorption (McKinney and Burg, 1978), transepithelial urea movement across toad bladder (Eggena, 1981), serum-stimulated, Na^+-dependent amino-isobutyric acid uptake in cultured chicken myotubes (Vandenburgh and Kaufman, 1982), and Na^+/Ca^{2+} exchange (Schellenberg and Swanson, 1982; Smith et al., 1982; Sordahl et al., 1984).

3. CHARACTERIZATION OF AMILORIDE-SENSITIVE Na^+ CHANNELS IN INTACT EPITHELIA

3.1. Structure–Activity Studies

The interaction between amiloride and the apical Na^+ entry channel in frog skin or toad bladder epithelia is very specific. Slight alterations in the chemical structure of amiloride result in a tremendous change in activity. For example, in frog skin, amiloride itself must be protonated, i.e., positively charged, to inhibit Na^+ transport (Benos et al., 1976). Balaban et al. (1979) compared the inhibitory characteristics of 2,4,6-triaminopyrimidine (TAP, Fig. 1) to amiloride in frog skin epithelium. Amiloride and TAP share many similarities in terms of their inhibitory actions: both inhibit rapidly and reversibly only from the external solution; both are noncompetitive with Na^+; and both are only active in their monoprotonated form. The major difference between these drugs lies in their inhibitory constants, TAP having a K_i some four orders of magnitude higher than amiloride (1 mM versus 0.1 µM). By constructing a model based on Michaelis–Menten kinetics in which the interaction between two inhibitors of the same transport system can be studied, information concerning the interaction between these two inhibitors can be obtained. The two simplest possibilities are that TAP and amiloride bind to the same molecular locus and thus act in a mutually exclusive fashion or that TAP and amiloride bind to spatially distinct sites that do not interact. To investigate these possibilities, a kinetic model based on noncompetitive inhibition of the Na^+ flux was derived (Segel, 1975; Balaban et al., 1979). If these drugs interact at the same site, a Dixon plot ($1/Na^+$ influx versus [TAP] at different fixed concentrations of amiloride) would result in a family of parallel lines; if they interact at different sites, the lines would not be parallel and would intersect. Expectedly, TAP and amiloride are competitive with each other, as evidenced by parallel Dixon plots (Fig.3).

This same analysis was applied to another inhibitor of the epithelial Na^+ channel, CGS-4270 (Fig. 1; Benos and Watthey, 1981). CGS-4270 is also a potassium-sparing diuretic similar in structure to amiloride except that the conformationally mobile acylguanidine group of amiloride was replaced by a rigid oxidiazol ring system. Like TAP, this molecule displays

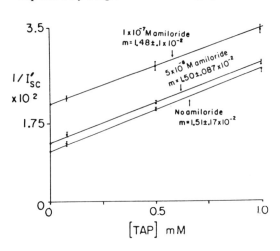

Figure 3. Dixon plot of $1/I_{SC}$ versus TAP concentration at three different amiloride concentrations. (From Balaban et al., 1979.) In both Figs. 3 and 4, the luminal [Na^+] was 110 mM, and the I_{SC} represents the short-circuit current (a measure of Na^+ entry) normalized to the current measured in the absence of inhibitors.

an inhibitory profile similar to amiloride: it is noncompetitive with Na^+, reversible, and works only from the external solution. Its K_i is 50 μM, some two orders of magnitude higher than amiloride and 50 times lower than TAP. But unlike TAP and amiloride, CGS-4270 is uncharged. Most interestingly, Dixon plot analysis reveals intersecting lines (Fig. 4), indicating that CGS-4270 and amiloride act at separate loci. Certain other molecules such as benzimidizole stimulate Na^+ entry into frog skin, and Dixon plot analysis indicates that benzimidizole and amiloride are competitive with each other (Benos and Watthey, 1981). Apparently, there are multiple regions in the vicinity of this Na^+ entry channel where these aminoguanidine compounds can interact, ultimately inhibiting or stimulating the Na^+ channel.

3.2. Flux Ratio Analysis of the Amiloride-Sensitive Na^+ Channel

One necessary criterion of independent ion movement through a particular transport pathway is that the relationship between unidirectional Na^+ influx, J_i, and efflux, J_o, obey the Behn–Teorell–Ussing flux ratio equation. Ussing and Zerahn (1951) first proposed that the ratio of the unidirectional transepithelial Na^+ fluxes through the active Na^+ transport pathway provided a measure of the effective driving force of the sodium pump in frog skin. But the transepithelial flux ratio cannot be used to conclude much about the apical Na^+ entry channel because a significant fraction of these fluxes may traverse the skin via paracellular shunts. In fact, because of the relatively high degree of basolateral membrane im-

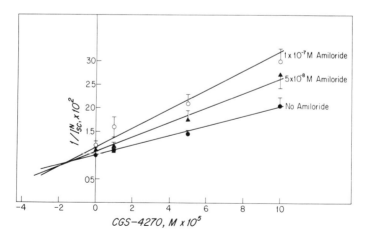

Figure 4. Dixon plot of $1/I_{SC}$ versus CGS-4270 concentration at three different concentrations of amiloride. (From Benos and Watthey, 1981.)

permeability to Na^+, the very small intracellular Na^+ transport compartment, and the electrochemical gradient across the apical membrane favoring Na^+ entry, a tracer flux ratio analysis through the amiloride-sensitive Na^+ channel in intact epithelia has only recently been accomplished (Benos et al., 1983).

Our experiments were performed on abdominal skin epithelia from the bullfrog *Rana catesbeiana* in sulfate Ringer's solution. Because most of the knowledge concerning the kinetic and electrophysiological characteristics of the Na^+ entry channel has been derived from current fluctuation measurements made under conditions of elevated serosal potassium (see Lindemann, 1983, for a review), we wanted to perform our experiments under comparable conditions. Under conditions of elevated serosal $[K^+]$, the relatively large surface area of the basolateral membrane and its high permeability to K^+ ions insures both depolarization and a high conductance of this membrane relative to the luminal membrane. The transepithelial voltage thus becomes identical to the voltage across the luminal membrane even when current is flowing across the entire epithelium. As mentioned above, a crucial problem in terms of data interpretation involves the assumptions concerning the Na^+ electrochemical potential gradient across the apical membrane. Consequently, microelectrode measurements of apical membrane potential under all experimental conditions were made. We found that this potential was always equal to the transepithelial clamp potential.

Because of very low permeability of the basolateral membrane of frog skin to Na^+, our approach was to render this membrane permeable to Na^+ by incorporating the cation ionophore nystatin. Figure 5 shows the results of experiments in which $^{22}Na^+$ efflux was measured in the absence and presence of 0.1 mM apical amiloride at different concentrations of serosally added nystatin. It is apparent that in the absence of added ionophore, there is no amiloride-sensitive component to $^{22}Na^+$ efflux. However, when nystatin is present, a significant ($P < 0.005$) inhibition of Na+ efflux by amiloride can be measured.

We also assume in our analysis that the intracellular (Na+) is equal to 10 mM under all experimental conditions (the serosal bathing solution always contained 10 mM Na^+). This value of intracellular Na^+ is close to that measured by ion-specific microelectrodes (Nagel et al., 1981) and electron probe (Rick et al., 1978). Our last assumption is that 0.1 mM amiloride inhibits only the Na^+-specific apical membrane channel and has no effect on any parallel shunt pathways. Experimental support for the validity of this last assumption is that amiloride has no effect on unidirectional $^{22}Na^+$ efflux in the absence of serosally added nystatin even if external [NaCl] or [urea] is raised to 300 mM, conditions that increase shunt pathway permeability.

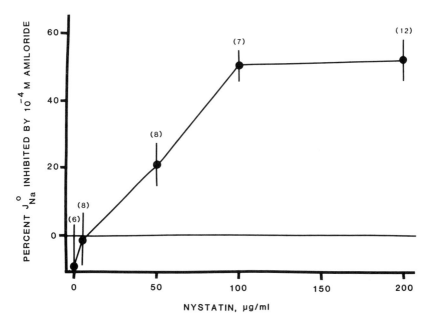

Figure 5. The effect of serosal nystatin addition on amiloride-sensitive $^{22}Na^+$ efflux from high-serosal-potassium-treated frog skin epithelia. The points represent the mean of (N) experiments, and the vertical lines represent 1 S.E.M. (From Benos *et al.*, 1983).

Figure 6 plots the natural logarithm of the measured flux ratio versus that predicted by the flux ratio equation. The excellent agreement between the measured and predicted flux ratio over a very wide range of electrochemical potential gradients indicates that Na^+ movement through this amiloride-sensitive channel obeys independence under these experimental conditions. Palmer (1982a) reached essentially the same conclusions for sodium movements through the apical membrane channels in toad urinary bladder. In both sets of experiments, the flux ratio analysis was performed in potassium-depolarized epithelial preparations (see below).

3.3. Voltage Dependence of the Amiloride Block

Previous studies have suggested the differences exist among species of frog skin as well as other epithelial preparations with regard to the interaction between amiloride and Na^+. Amiloride has been found to be a competitive inhibitor of Na^+ entry in *R. temporaria* skin (Cuthbert and Shum, 1974; Li and Lindemann, 1984), rabbit descending colon (Gottlieb *et al.*, 1978), toad urinary bladder (Sudou and Hoshi, 1977), the skin of the Mexican axolotl (Cuthbert, 1981), and the chicken chorioallantoic

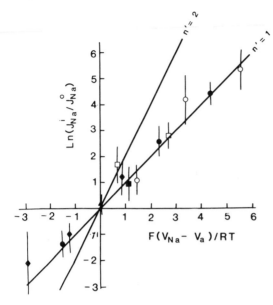

Figure 6. The natural logarithm of the observed amiloride-sensitive Na$^+$ flux ratio across frog skin epithelia versus the predicted flux ratio at different electrochemical potentials. (From Benos et al., 1983.)

membrane (Cuthbert, 1981) and noncompetitive in the skins of *R. pipiens* and *R. catesbeiana* (Benos et al., 1979), the hen coprodeum (Cristensen and Bindslev, 1982), and the toad urinary bladder (Bentley, 1968). Mixed inhibition has been reported in toad skin (Benos et al., 1979), frog skin (Salako and Smith, 1970), and toad urinary bladder (Takada and Hayashi, 1981).

A major criticism of all the studies cited above is that changes in apical membrane potential when external Na$^+$ and amiloride concentrations were varied were not considered. The electric field across the apical membrane could easily influence the binding of amiloride, especially because amiloride must carry a net positive charge in order to inhibit Na$^+$ entry in frog skin. With F. Abrahamchek, J. L. Tang, and S. I. Helman, we found that the magnitude of the apparent inhibitory dissociation constant for amiloride in frog skin depends on voltage. We also estimate that the fraction of the electric potential felt by amiloride is 0.43. By incorporating these parameters into a purely noncompetitive inhibitory model, we contend that previously observed deviations from straightforward inhibitory kinetics (Benos et al., 1979; Fig. 7) and, indeed, competitive or mixed inhibition can be explained. The competitive inhibitory kinetics between amiloride and Na$^+$ in the European frogs as compared to the noncompetitive behavior in the American frog species can occur because of differences in the magnitude of the change in apical membrane potential brought about by reducing external [Na$^+$], hence resulting in larger differences in the apparent K_i. In the European variety of frog, the initial

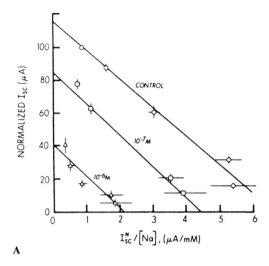

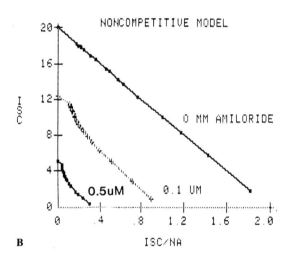

Figure 7. (A) Single-reciprocal plots of short-circuit current (I_{SC}) versus [Na$^+$] data obtained from isolated frog skin epithelia in the absence and presence of 0.1 and 1 μM amiloride (From Benos et al., 1979.) Note the slight upturn in the data in the presence of amiloride at high [Na$^+$], i.e., near the y-axis (B) Predicted single-reciprocal plots of computer-generated I_{SC} versus [Na$^+$] data at 0, 0.1, and 0.5 μM amiloride concentrations. The data were generated from a purely noncompetitive amiloride inhibitory model, allowing for a voltage-dependent blockade by amiloride. In this model, the apparent inhibitory dissociation constant for amiloride (K_i) is given by the equation $K_i(V) = K_i(0) \exp(z\delta F V_a/RT)$, where $K_i(0)$ is the inhibitory constant at zero apical membrane potential (V_a), i.e., 0.56 μM, δ is the electrical distance, namely 0.43, and z, F, R, and T have their usual meanings. The data were normalized to I_{SC} of 20 μA/cm^2 at 100 mM Na$^+$ in the absence of amiloride. The K_S for Na$^+$ was assigned a value of 10 mM.

apical membrane potential at 110 mM external Na^+ can be as low as -15 mV, hyperpolarizing to -100 mV at 3 mM external Na^+ (Harvey and Kernan, 1984), whereas in the American species, the apical membrane potential at 110 mM is usually -90 to -100 mV (Helman and Fisher, 1977) with only a 10- to 30-mV change on reducing external $[Na^+]$ to 3 mM. However, it is important to point out that this entire voltage-dependent amiloride block analysis (as well as the determination of macroscopic inhibitory constants) assumes that Na^+ channel density and single-channel currents remain constant. These assumptions may turn out to be invalid (S. I. Helman, personal communication). Recently, Palmer (1984) published results showing a voltage-dependent block by amiloride of apical Na^+ channels in the urinary bladder of the toad. However, in that tissue amiloride senses only 12% of the electric field.

3.4. Interfacial Potentials and Apical Na^+ Entry Channels

Benos, Latorre, and Reyes (1981) investigated the effects that alterations in interfacial potentials have on Na^+ entry in bullfrog skin epithelium. The results of that study demonstrate that changes in the ionic strength of the external solution do not affect the rate of Na^+ transport through the amiloride-sensitive entry channels. Also, we found that uranyl ions do not change Na^+ transport through these channels in spite of the fact that this molecule at 2.5 mM can induce a $+145$ mV increase in the interfacial potential of phosphatidylserine monolayers. Therefore, it is likely that the surface charge density in the vicinity of the Na^+-selective channel located in the apical membrane is small (i.e., 1 $e^-/600$ Å^2) and that this Na^+ entry channel is insensitive to changes in apical surface potential.

4. INCORPORATION OF AMILORIDE-SENSITIVE Na^+ CHANNELS INTO PLANAR BILAYERS

The application of intracellular microelectrode recording techniques, current fluctuation analysis, and tissue culture techniques to the study of epithelial transport problems has permitted a more detailed view of the molecular processes involved in transepithelial Na^+ transport. However, the acquisition of specific data concerning the amiloride-sensitive Na^+ entry channel has been difficult, in large measure because of the inherent geometrical complexity of the tissue. Until now, all of the kinetic properties of these channels have been inferred from macroscopic measurements of amiloride-induced current noise under voltage-clamped condiions (Lindemann, 1980). Indeed, this Na^+ entry channel was first identified as such because of the very high rate of turnover of each transport site (Lindemann and Van Driessche, 1977).

In order to observe and characterize amiloride-sensitive Na^+ channels directly, we have incorporated apical membrane vesicles containing these channels into planar lipid bilayer membranes (Sariban-Sohraby et al., 1984). These vesicles were made from a cultured toad kidney cell line, A6, grown on Millipore® filters. These cells express amiloride-sensitive Na^+ channel activity only when grown on permeable supports (Sariban-Sohraby et al., 1983). The major evidence that Na^+ channels are present in the cells and in the apical membrane vesicle fraction is that amiloride inhibited the $^{22}Na^+$ influx into both with a K_i of 0.07 μM (at a $[Na^+]$ of 110 mM). Incorporation into solvent-containing bilayers was achieved osmotically (see Miller, 1983; Latorre and Benos, 1985, for a discussion of these methods). An example of such single-channel activity is shown in Fig. 8.

The range of single-channel conductance was 4 to 80 pS at 200 mM NaCl with a mean value of 35 pS. Both the open-channel conductance and the probability of a channel being in the open or closed state were independent of voltage in the range ±60 mV. The channels were found to be perfectly cation selective. However, the ability of these channels to discriminate between Na^+ and K^+ was less than in other epithelia. The measured permeability ratio of Na^+ to K^+ for these single channels was 2 or 3 to 1. A selectivity ratio for Na^+ versus K^+ of at least 100:1 or 1000:1 has been reported for frog skin (Benos et al., 1980) and toad bladder (Palmer, 1982b), respectively. In contrast, Na^+/K^+ selectivity ratios for the Na^+ entry channel in rabbit urinary bladder have been found to be between 2:1 and 10:1 depending on dietary Na^+ intake. The Na^+/K^+ selectivity ratio of the Na^+ channel in these A6 cultured cells is unknown,

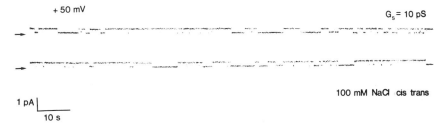

Figure 8. Cultured A6 apical membrane amiloride-sensitive Na^+ channels incorporated into planar lipid bilayers. The bilayer was symmetrically bathed with a solution of 100 mM NaCl, 10 mM MOPS–tris, pH 7.0. The arrows indicate the zero-current level (see Sariban-Sohraby et al., 1984, for details).

but the ratio of the amiloride-sensitive $^{22}Na^+$ influx to amiloride-sensitive $^{42}K^+$ influx in A6 apical membrane vesicles was 3:1 (S. Sariban-Sohraby and D. J. Benos, unpublished observations).

Amiloride interacts with this channel whether present in the *cis* or *trans* compartment of the bilayer setup. Addition of amiloride to the *trans* side produced rapid flickering between the open and closed conductance states suggestive of those observed by noise analysis in intact epithelia (Fig. 9A). *Cis* amiloride, however, did not produce flickering but rather reduced the open-state conductance (Fig. 9B), suggesting that the block was short-lived with respect to the amplifier response time. The *cis* block by amiloride was concentration dependent; the open-state conductance was reduced to 50% of its initial value by 0.1 μM amiloride. The concentration dependence of the *trans* block has not yet been studied. Nonetheless, these characteristics of the amiloride block suggest the presence of multiple binding sites for amiloride or different amiloride accessibilities to a single receptor site.

Hamilton and Eaton (1985) have made observations on a putative amiloride-sensitive Na^+ channel in intact A6 cells using the patch-clamp technique. They report that amiloride induces channel flickering when the drug is present in the pipette solution, that the channel displays a Na^+/K^+ selectivity ratio of 4:1, and that the block by amiloride is voltage dependent with amiloride sensing approximately 40% of the applied electric field.

Olans *et al.* (1984) have analyzed the variability in conductance shown by the epithelial Na^+ channel in bilayers. We studied two extremes of channel conductances. In one case, the single open-channel conductance reaches a maximum value of 4 pS, and the conductance-versus-Na^+ activity curve is well described by a rectangular hyperbola with an apparent K_s of 17 mM. The conductance-versus-[Na^+] curve for the other population of channels shows a maximum conductance of 44 pS and a K_s of 47 mM. Both the small- and large-conductance channels display similar characteristics; namely, both are amiloride sensitive (*cis* amiloride reducing open-state conductance with a K_i of 0.1 μM), both are perfectly cation selective, both show a P_{Na^+}/P_{K^+} ratio of 2, and both saturate with increasing sodium. These data do not agree with the proposed "substrate inhibition" model of macroscopic apical Na^+ entry (Van Driessche and Lindemann, 1979; Lindemann, 1980). In this model, saturation produced by self-inhibition means that the individual transport channels themselves do not saturate with [Na^+] but rather that the number of functioning channels is reduced as [Na^+] is increased.

It is important to point out that the noise experiments on which this substrate inhibition model is based were done in the presence of elevated serosal K^+. Recently, Hoshiko and Van Driessche (see Hoshiko, 1984)

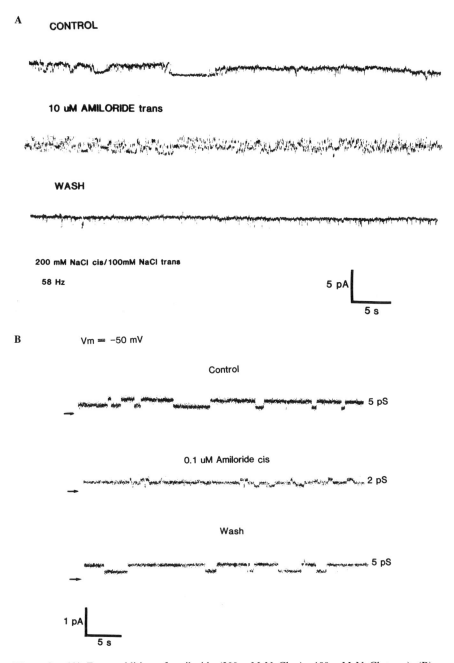

Figure 9. (A) *Trans* addition of amiloride (200 mM NaCl *cis*, 100 mM NaCl *trans*). (B) Amiloride inhibition of single-channel conductance; *cis* addition of amiloride (10 mM NaCl both sides). (From Sariban-Sohraby et al., 1984).

repeated these noise experiments in frog skins not depolarized with high serosal K^+ and reported that individual amiloride-sensitive channels do in fact saturate in the presence of serosal Na^+ rather than K^+ (see Fig. 10). These observations are in accord with measurements of the saturation behavior of epithelial Na^+ channels in bilayers by Olans *et al.* (1984). High serosal K^+, in addition to depolarizing the basolateral membrane, results in a seven- to tenfold increase in intracellular levels of cAMP in frog skin (Cuthbert and Wilson, 1981). Hence, single-channel properties measured in K^+-depolarized epithelial preparations may be altered by modifications in intracellular hormonal and/or ionic conditions.

5. CONCLUDING REMARKS

Studies concerning the molecular nature of the sodium entry pathway in the luminal membrane of electrically high-resistance epithelia have largely involved experiments in which the interaction between amiloride and this permeation process has been examined. Because of the inherent geometrical constraints imposed on the investigator by the intact epithelia, these studies have for the most part been macroscopic in nature, with the molecular details of transport being deduced. However, within the past year, single epithelial Na^+ channel activity has been observed using either the patch-clamp or planar lipid bilayer technique. These techniques have

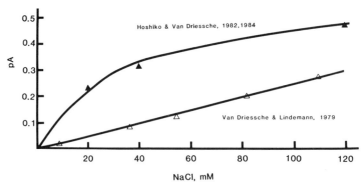

Figure 10. Current through single sodium channels in frog skin at different luminal Na^+ concentrations, as calculated from macroscopic current fluctuation analysis. Current fluctuations were, in both sets of experiments, induced by micromolar concentrations of amiloride.

allowed the direct study of single channels in an environment of controllable composition and electrochemical potential. Now, for the first time, questions concerning the molecular nature of the amiloride block, the channel's selectivity to the alkali metal cations, and the modulation of ion transport through this channel by other ions such as calcium or hormones such as vasopressin and aldosterone can be addressed in a rigorous fashion. Certainly a much clearer understanding of the nature and regulation of epithelial transport processes will emerge in the years to come.

ACKNOWLEDGMENTS. I thank Drs. Joan Bell, Lynda Rushing, and Sarah Sariban for their helpful criticisms of the manuscript. The preparation of this chapter was supported by National Institutes of Health Grant AM-25886.

REFERENCES

Aceves, J., and Cereijido, M., 1973, The effect of amiloride on sodium and potassium fluxes in red cells, *J. Physiol. (Lond.)* **229**:707–718.

Aronson, P. S., and Bounds, S. E., 1980, Harmaline inhibition of Na dependent transport in renal microvillus membrane vesicles, *Am. J. Physiol.* **238**:F210–F217.

Baer, J. E., Jones, C. B., Spitzer, S. A., and Russo, H. F., 1967, The potassium sparing and naturiuretic activity of N-amidino-3,5-diamino-6-chloropyrazine-carboxamide hydrochloride dihydrate (amiloride hydrochloride), *J. Pharmacol. Exp. Ther.* **157**:472–485.

Balaban, R. S., Mandel, L. J., and Benos, D. J., 1979, On the cross reactivity of amiloride and 2,4,6-triaminopyrimidine (TAP) for the cellular entry and tight junctional cation permeability pathways in epithelia, *J. Membr. Biol.* **49**:363–390.

Benos, D. J., 1982, Amiloride: A molecular probe of sodium transport in tissues and cells, *Am. J. Physiol.* **242**:C131–C145.

Benos, D. J., and Sapirstein, V. S., 1983, Characteristics of an amiloride-sensitive sodium entry pathway in cultured rodent glial and neuroblastoma cells, *J. Cell. Physiol.* **116**:213–220.

Benos, D. J., and Watthey, J. W. H., 1981, Inferences on the nature of the apical sodium entry site in frog skin epithelium, *J. Pharmacol. Exp. Ther.* **219**:481–488.

Benos, D. J., Simon, S. A., Mandel, L. J., and Cala, P. M., 1976, Effect of amiloride and some of its analogues on cation transport in isolated frog skin and thin lipid membranes, *J. Gen. Physiol.* **68**:43–63.

Benos, D. J., Mandel, L. J., and Balaban, R. S., 1979, On the mechanism of the amiloride–sodium entry site interaction in anuran skin epithelia, *J. Gen. Physiol.* **73**:307–326.

Benos, D. J., Mandel, L. J., Simon, S. A., 1980, Cationic selectivity and competition at the sodium entry site in frog skin, *J. Gen. Physiol.* **76**:233–247.

Benos, D., Latorre, R., and Reyes, J., 1981, Surface potentials and sodium entry in frog skin epithelium, *J. Physiol. (Lond.)* **321**:163–174.

Benos, D. J., Hyde, B. A., and Latorre, R., 1983, Sodium flux ratio through the amiloride-sensitive entry pathway in frog skin, *J. Gen. Physiol.* **81**:667–685.

Bentley, P. J., 1968, Amiloride: A potent inhibitor of sodium transport across the toad bladder, *J. Physiol. (Lond.)* **195**:317–330.

Bickling, J. E., Mason, J. W., Woltersdorf, O. W., Jr., Jones, J. H., Kwon, S. F., Robb, C. M., and Cragoe, E. J., 1965, Pyrazine diuretics. I. N-amidino-3-amino-6-halopyrazine-carboxamides, *J. Med. Chem.* **8**:638–642.

Cala, P. M., 1983, Volume regulation by red blood cells. Mechanisms of ion transport between cells and mechanisms, *Mol. Physiol.* **4**:33–52.

Chan, Y. L., and Giebisch, G., 1981, Relation between sodium and bicarbonate transport in the rat proximal convoluted tubule, *Am. J. Physiol.* **240**:F222–F230.

Christensen, O., and Bindslev, N., 1982, Fluctuation analysis of short-circuited current in a warm-blooded sodium-retaining epithelium: Site current, density, and interaction with triamterene, *J. Membr. Biol.* **65**:19–30.

Cuthbert, A. W., 1981, Sodium entry step in transporting epithelia: Results of ligand binding studies, in *Ion Transport by Epithelia* (S. G. Schultz, ed.), Raven Press, New York, pp. 181–195.

Cuthbert, A. W., and Fanelli, G. M., 1978, Effect of some pyrazine-carboxamides on sodium transport in frog skin, *Br. J. Pharmacol.* **63**:139–149.

Cuthbert, A. W., and Shum, W. K., 1974, Amiloride and the sodium channel, *Naunyn Schmiedebergs Arch. Pharmacol.* **281**:61–69.

Cuthbert, A. W., and Wilson, S. A., 1981, Mechanisms for the effect of acetylcholine on sodium transport in frog skin, *J. Membr. Biol.* **59**:65–75.

de Sousa, R. C., and Grosso, A., 1978, Vasopressin-like effects of a hallucinogenic drug—harmaline—on sodium and water transport, *J. Membr. Biol.* **40**:77–94.

Eggena, P., 1981, Inhibition of urea-linked water flux and ^{14}C-urea transport across the toad bladder by amiloride, *Proc. Soc. Exp. Biol. Med.* **167**:55–61.

Erikson, A.-C., and Spring, K. R., 1982, Volume regulation by *Necturus* gallbladder: Apical Na–H and Cl–HCO_3 exchange, *Am. J. Physiol.* **243**:C146–C150.

Fuchs, W., Hviid Larsen, E., and Lindemann, B., 1977, Current–voltage curve of sodium permeability in frog skin, *J. Physiol. (Lond.)* **267**:137–166.

Glitzer, M. S., and Steelman, S. L., 1966, N-amidino-3,5-diamino-6-chloropyrazinecarboxamide: An active diuretic in the carboxamide series, *Nature* **212**:191–193.

Gottlieb, G. P., Turnheim, K., Frizzell, R. A., and Schultz, S. G., 1978, p-Chloromercuribenzene sulfonate blocks and reverses the effect of amiloride on sodium transport across rabbit colon in vitro, *Biophys. J.* **22**:125–129.

Grinstein, S., Cohen, S., and Rothstein, A., 1984, Cytoplasmic pH regulation in thymic lymphocytes by an amiloride-sensitive Na/H antiport, *J. Gen. Physiol.* **83**:341–369.

Harvey, B. J., and Kernan, R. P., 1984, Intracellular ion activities in frog skin in relation to external sodium and effects of amiloride and/or ouabain, *J. Physiol. (Lond.)* **349**:501–507.

Hamilton, K. L., and Eaton, D. C., 1985, Single-channel recordings from amiloride-sensitive epithelial sodium channel, *Am. J. Physiol.* **249**:C200–C207.

Helman, S. I., and Fisher, R. S., 1977, Microelectrode studies of the active Na^+ transport pathway of frog skin, *J. Gen. Physiol.* **69**:571–604.

Hoshiko, T., 1984, Fluctuation analysis of apical sodium transport, *Curr. Top. Membr. Transport* **20**:3–26.

Ives, H. E., Yee, V. J., Warnock, D. G., 1983, Mixed type inhibition of the renal Na/H antiporter by Li and amiloride, *J. Biol. Chem.* **258**:9710–9716.

Johnson, J. D., Epel, D., and Paul, M. 1976, Intracellular pH activation of sea urchin eggs after fertilization, *Nature* **262**:661–664.

Kinsella, J. L., and Aronson, P. S., 1981, Interaction of NH_4 and Li with the renal microvillus membrane Na–H exchanger, *Am. J. Physiol.* **241**:C220–C226.

L'Allemain, G., Franchi, A., Cragoe, E., Jr., and Pouyssegur, J., 1984, Blockade of the Na/H antiport abolishes growth factor-induced DNA synthesis in fibroblasts, *J. Biol. Chem.* **259**:4313–4319.

Latorre, R., and Benos, D., 1985, Reconstitution of ionic channels into planar lipid bilayer

membranes, in *Transmembrane Signaling and Sensation* (F. Oosawa, ed.), Japan Scientific Societies Press, Tokyo, pp. 199–213.
Li, J. H., and de Sousa, R. C., 1979, Inhibitory and stimulatory effects of amiloride analogues on sodium transport in frog skin, *J. Membr. Biol.* **46**:155–169.
Li, J. H.-Y., and Lindemann, B., 1984, Competitive blocking of epithelial sodium channels by organic cations: The relationship between macroscopic and microscopic inhibition constants, *J. Membr. Biol.* **76**:235–251.
Lindemann, B., 1980, The beginning of fluctuation analysis of epithelial ion transport, *J. Membr. Biol.* **54**:1–11.
Lindemann, B., and Van Driessche, W., 1977, Sodium-specific membrane channels of frog skin are pores: Current fluctuations reveal high turnover, *Science* **195**:292–294.
Macknight, A. D. C., Dibona, D. R., and Leaf, A., 1980, Sodium transport across toad urinary bladder: A model "tight" epithelium, *Physiol. Rev.* **60**:615–715.
McKinney, T. D., and Burg, M. B., 1978, Bicarbonate absorption by rabbit cortical collecting tubules *in vivo*, *Am. J. Physiol.* **234**:F141–F145.
Miller, C., 1983, First steps in the reconstruction of ionic channel functions in model membranes, in *Current Methods in Cellular Neurobiology*, (J. L. Barber, ed.), John Wiley & Sons, New York, pp. 1–37.
Moolenaar, W. H., Boonstra, J., Van der Saag, P. T., and deLast, S. W., 1981, Sodium proton exchange in mouse neuroblastoma cells, *J. Biol. Chem.* **256**:12883–12887.
Moolenaar, W. H., Yarden, Y., deLaat, S. W., and Schlessinger, J., 1982, Epidermal growth factor induces electrically silent Na influx in human fibroblasts, *J. Biol. Chem.* **257**:8502–8506.
Moran, A., and Moran, N., 1984, Amiloride-sensitive sodium pathways in LLC-PK1 epithelia, *J. Gen. Physiol.* **84**:28a.
Nagel, W., Garcia-Diaz, J. F., and Armstrong, W. McD., 1981, Intracellular ionic activities in frog skin, *J. Membr. Biol.* **61**:127–134.
O'Donnell, M. E., Cragoe, E., Jr., and Villereal, M. L., 1983, Inhibition of Na influx and DNA synthesis in human fibroblasts and neuroblastoma–glioma hybrid cells by amiloride analogs, *J. Pharmacol. Exp. Ther.* **226**:368–372.
Olans, L., Sariban-Sohraby, S., and Benos, D. J., 1984, Saturation behavior of single, amiloride sensitive Na channels in planar lipid bilayers, *Biophys. J.,* **46**:831–835.
Owens, N. E., and Villereal, M. L., 1982, Evidence for a role of calmodulin in serum stimulation of Na influx in human fibroblasts, *Proc. Natl. Acad. Sci. U.S.A.* **79**:3537–3541.
Palmer, L. G., 1982a, Na transport and flux ratio through apical Na channels in toad bladder, *Nature* **297**:688–690.
Palmer, L. G., 1982b, Ion selectivity of the apical membrane Na channel in the toad urinary bladder, *J. Membr. Biol.* **67**:91–98.
Palmer, L. G., 1984, Voltage-dependent block by amiloride and other monovalent cations of apical Na channels in the toad urinary bladder, *J. Membr. Biol.* **80**:153–165.
Paris, S., and Pouyssegur, J., 1983, Biochemical characterization of the amiloride-sensitive Na/H antiport in chinese hamster lung fibroblasts, *J. Biol. Chem.* **258**:3503–3508.
Parker, J. C., 1983, Volume-responsive sodium movements in dog red blood cells, *Am. J. Physiol.* **244**:C324–C330.
Pouyssegur, J., Chambard, J. C., Franchi, A., Paris, S., and Van Obberghen-Schilling, E., 1982, Growth factor activation of an amiloride-sensitive Na/H exchange system in quiescent fibroblasts: Coupling to ribosomal protein S6 phosphorylation. *Proc. Natl. Acad. Sci. U.S.A.* **79**:3935–3539.
Rick, R., Dorge, A., Van Arnim, E., and Thurau, K., 1978, Electron microprobe analysis of frog skin epithelium: Evidence for a syncytial Na transport compartment, *J. Membr. Biol.* **39**:257–271.

Salako, L. A., and Smith, A. J., 1970, Changes in sodium pool and kinetics of sodium transport in frog skin produced by amiloride, *Br. J. Pharmacol.* **39**:99–109.

Sariban-Sohraby, S., Burg, M. B., and Turner, R. J., 1983, Apical sodium uptake in toad kidney epithelial cell line A6, *Am. J. Physiol.* **245**:C167–C171.

Sariban-Sohraby, S., Latorre, R., Burg, M., Olans, L., and Benos, D., 1984, Amiloride-sensitive epithelial Na channels reconstituted into planar lipid bilayer membranes, *Nature* **308**:80–82.

Schellenberg, G. D., and Swanson, P. D., 1982, Properties of the Na–Ca exchange transport system from rat brain: Inhibition by amiloride, *Fed. Proc.* **41**:673.

Segel, I. H., 1975, *Enzyme Kinetics*, John Wiley & Sons, New York.

Smith, R. L., Macara, I. G., Levenson, R., Housman, D., and Cantley, L., 1982, Evidence that a Na/Ca antiport system regulates murine erythroleukemia cell differentiation, *J. Biol. Chem.* **257**:773–780.

Sordahl, L. A., LaBelle, E. F., and Rex, K. A., 1984, Amiloride and diltiazem inhibition of microsomal and mitochondrial Na and Ca transport, *Am. J. Physiol.* **246**:C172–C176.

Sudou, K., and Hoshi, T., 1977, Mode of action of amiloride in toad urinary bladder, *J. Membr. Biol.* **32**:115–132.

Takada, M., and Hayashi, H., 1981, Interactions of cadmium, calcium, and amiloride in the kinetics of active sodium transport through frog skin, *Jpn. J. Physiol.* **31**:285–303.

Ussing, H. H., and Zerahn, K., 1951, Active transport of sodium as the source of electric current in the short-circuited isolated frog skin, *Acta Physiol. Scand.* **23**:110–127.

Vandenburgh, H. H., and Kaufman, S., 1982, Coupling of voltage-sensitive sodium channel activity to stretch-induced amino acid transport in skeletal muscle *in vitro*, *J. Biol. Chem.* **257**:13448–13454.

Van Driessche, W., and Lindemann, B., 1979, Concentration dependence of currents through single sodium-selective pores in frog skin, *Nature* **282**:519–520.

Villereal, M. L., 1981, Sodium fluxes in human fibroblasts: Effects of serum, Ca and amiloride, *J. Cell. Physiol.* **107**:359–369.

Will, P. C., Lebowitz, J. L., and Hopfer, U., 1980, Induction of amiloride-sensitive sodium transport in the rat colon by mineralocorticoids, *Am. J. Physiol.* **238**:F261–F268.

Chapter 26

A Channel Model for Development of the Fertilization Membrane in Sea Urchin Eggs

Gerald Ehrenstein and Richard FitzHugh

1. INTRODUCTION

The channels we consider in this chapter have not yet been observed. They are part of a model we have developed to describe the increase in cytosolic calcium that occurs within an egg following fertilization. Before describing the model, it is useful to provide some background. First, we outline the processes that occur following fertilization of a sea urchin egg, in order to put into context the role of the increase in cytosolic calcium. Then we summarize the experimental evidence that guided the development of the model. Finally, we describe the model in words and in equations and indicate some of its predictions.

2. PROCESSES FOLLOWING FERTILIZATION

Fertilization of a sea urchin egg activates a complex sequence of events quite apart from the joining of the germ plasm of sperm and egg. These events include the movement of ions across the plasma membrane,

---

GERALD EHRENSTEIN AND RICHARD FITZHUGH • Laboratory of Biophysics, NINCDS, National Institutes of Health, Bethesda, Maryland 20205.

liberation of free calcium from internal stores, exocytosis of cortical granules, formation of a fertilization membrane, activation of enzymes, a rise in oxygen consumption, an increase in intracellular pH, an increase in protein synthesis, and initiation of DNA synthesis. The exact causal relations among these many events have not yet been determined, but there is presently good experimental evidence for a general pattern involving a cascade of events, such as that shown in Fig. 1 (Epel, 1982).

Although it is subject to minor change, Fig. 1 provides a good summary of present knowledge and shows that an increase in the concentration of cytosolic calcium ions is the central event that links fertilization with activation and development. The question we address is how fertilization gives rise to the increased concentration of cytosolic calcium.

3. EXPERIMENTAL BASIS FOR MODEL

For sea urchins, it is possible to activate eggs in calcium-free seawater (Chambers, 1980), indicating that the increased cytosolic calcium comes from internal stores and not from the external solution. For our mathematical model, it is not necessary to specify the location of the internal stores of calcium. We simply require that the calcium be stored in some internal organelle. For future experimental tests of the model, however, the nature of the organelle will be important. A good candidate is the tubular reticulum recently described (Poenie et al., 1982).

The temporal variation of cytosolic calcium concentration following fertilization of sea urchin eggs has been determined by measuring the

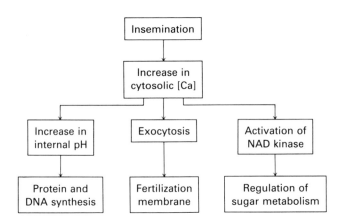

Figure 1. Block diagram of cascade of events following fertilization.

luminescence of a group of eggs injected with aequorin, a substance that fluoresces in the presence of calcium. It was found that the calcium increase starts about 30 sec after fertilization and lasts for 1–2 min (Steinhardt et al., 1977). These experiments could not be used to determine the spatial variation of the cytosolic calcium concentration because sea urchin eggs are too small relative to the spatial resolution of the method. However, in fish eggs, which are much larger than sea urchin eggs, both spatial and temporal variation of the cytosolic calcium concentration has been measured. It was found that the increased concentration of calcium first occurs at the point of sperm entry and then moves through the egg in a narrow band until it reaches the region of the egg opposite the point of sperm entry. At each point, the increase in calcium concentration is followed by a decrease back to the original calcium level (Gilkey et al., 1978).

The experimental results described above relate to the role of calcium. We now consider some very recent experimental results that relate to triggering the events that lead to the increase in cytosolic calcium. In these experiments, the assay for an increase in cytosolic calcium was not aequorin but rather formation of a fertilization membrane, which can be easily seen under a microscope. As shown in Fig. 1, the fertilization membrane is also a calcium indicator, albeit a qualitative one. In one set of experiments, a fertilization membrane was triggered by injection into sea urchin eggs of a soluble component of sperm (Dale et al., 1985), and in another set of experiments, a fertilization membrane was triggered by injection of inositol-1,4,5-triphosphate (ITP) (Whitaker and Irvine, 1985).

4. DESCRIPTION OF MODEL

A model that is consistent with the above results and that describes how a component of sperm can trigger a series of events leading to the release of free calcium is shown schematically in Fig. 2. In this model, a component of sperm enters the egg during fertilization and activates an enzyme, phospholipase C. The activated enzyme then hydrolyzes phosphatidylinositol-4,5-bisphosphate (PIP_2), forming ITP and diacylglycerol (DAG). The ITP acts as an agonist to open calcium channels in the tubular reticulum, thus allowing sequestered calcium ions to cross the reticular membrane and increase the concentration of cytosolic free calcium. This free calcium and DAG act together to stimulate a kinase, which phosphorylates, and hence activates, phospholipase C. This completes the cycle of chemical reactions, one result of which is the desired release of free calcium into the cytosol. The sperm factor, which started the whole

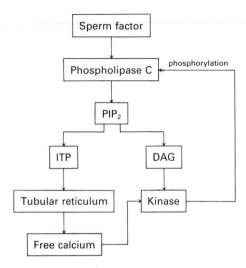

Figure 2. Schematic diagram of model to describe how fertilization leads to an increase of cytosolic free calcium.

process, is no longer needed, since phospolipase C is now activated by phosphorylation. The role of the sperm factor is simply to get the cycle started.

An important part of the model that cannot be seen in the schematic diagram is molecular diffusion within the egg. The complete mathematical model consists of a set of differential equations that describe this diffusion as well as the kinetics of the chemical reactions.

There are many experimental tests of this model that could be performed. The mathematical equations themselves allow us to perform a theoretical test by determining whether or not the model predicts a propagated wave of free calcium, in which the concentration of free calcium at each point first rises from a low level and then returns back to that low level. This theoretical test is important, since merely specifying a set of chemical reactions is by no means sufficient to insure that they lead to a propagated wave. In order to perform the theoretical test, one must show that the solution of the differential equations describing the model results in a wave with the desired properties. It is not necessary that predictions of the model accurately agree with experimental data; what is wanted initially is rather a demonstration that the desired qualitative behavior results from the model. Later, refinements of the model can be attempted from the purpose of obtaining closer quantitative agreement with experiment.

5. EQUATIONS OF MODEL

The variables representing concentrations of the different substances are as follows:

variable	concentration of
P	PIP_2
I	ITP
D	DAG
B	bound calcium
C	free calcium
E	enzyme (phospholipase C)

For brevity, the terms "bound" and "free" are used here to refer to the calcium sequestered in internal organelles and the calcium free within the cytosol. Each of the above six variables is a function of both time, t, and position within the cell. For lack of actual concentration values, they are given in arbitrary dimensionless units rather than physical units. Differential equations describing the overall kinetics of the first three of these substances are:

$$\partial P/\partial t = -bEP \quad (1)$$

$$\partial I/\partial t = k_I \nabla^2 I + bEP - hI \quad (2)$$

$$\partial D/\partial t = k_D \nabla^2 D + bEP \quad (3)$$

where b, h, k_I, and k_D are constants.

Equations 1–3 describe the reversible splitting of PIP_2 into ITP and DAG by the enzyme (terms bEP). It is assumed that resynthesis of PIP_2 is slow enough that the back reaction need not be included in these equations. Experience with equations of this form suggests that in order to get a spreading wave with recovery, something must irreversibly decay and disappear. By analogy with the hydrolysis of the agonist acetylcholine by acetylcholinesterase, we postulate the decay of ITP. This process is represented by the term $-hI$ in equation 2. In equations 2 and 3, the terms with the Laplacian operator ∇^2 denote the spread of ITP and DAG by diffusion, and k_I and k_D are their diffusion constants. Since ITP is hydrophilic and DAG is hydrophobic, they will diffuse in aqueous and lipid media, respectively, and appropriate diffusion constants should be used (here $k_D = k_I/10$). For the cortical region of an egg, aqueous diffusion occurs in a narrow band near the plasma membrane, and lipid diffusion occurs within the plasma membrane. Since we will approximate this region by a two-dimensional spherical surface, both types of diffusion should be considered.

Differential equations for the last two variables can be written as follows:

$$\partial C/\partial t = gIB - fC \qquad (4)$$

$$\partial E/\partial t = kCD - nE \qquad (5)$$

where g, f, k and n are constants.

Equation 4 represents the passive release of calcium from the tubular reticulum (term gIB) and its reuptake (term $-fC$). A diffusion term is not included in this equation because we expect that diffusion will be relatively small compared to reuptake. Equation 5 represents the activation of the enzyme by the joint action of calcium and DAG (via kinase, term kCD) and its spontaneous inactivation (term $-nE$).

Although the volume of the internal organelle containing B and the cytosolic volume for C are different, their volume ratio can be incorporated into the constant g in equation 4. Thus, without loss of generality, a conservation equation can be written in terms of normalized units for B and C:

$$B = 1 - C \qquad (6)$$

The reaction scheme described above can be simplified by eliminating the differential equations for C and E, if it is assumed (in the absence of evidence to the contrary) that these reactions are so fast that they are nearly at equilibrium. By setting the time derivatives equal to zero in equations 4 and 5 and using equation 6, one obtains:

$$C = I/(I + q) \qquad (7)$$

$$E = sCD \qquad (8)$$

where $q = f/g$ and $s = k/n$.

Thus, the model can be described by equations 1, 2, 3, 6, 7, and 8.

6. SOLUTIONS OF MODEL EQUATIONS

We are interested in solving the equations of the model for a sea urchin egg, which can be reasonably approximated by a sphere. A simpler alternative would be to solve them for a fish egg (such as medaka), which has a very large central yolk and can, therefore, be reasonably approximated by a two-dimensional spherical surface. At first, to avoid geometric complications, the equations will be solved for the even simpler one-dimensional case with one spatial variable x. It can be seen below that the solution for the spherical surface is qualitatively similar to that for one dimension.

If x represents distance in one dimension (in arbitrary dimensionless units), the Laplacian operator ∇^2 becomes $\partial^2/\partial x^2$ in equations 2 and 3. Initially, I, C, D, and E are set to zero, and B to 1. P_0, the initial value of P, is adjustable. With these initial conditions, all time derivatives would be zero, and the system would remain stationary. To start the process at $t = 0$, we increase E at one point ($x = 0$) to an arbitrary value E_0. This corresponds to the assumption that some chemical in the sperm, as the latter enters the egg, either activates the enzyme locally or disinhibits an enzyme inhibitor.

This initial condition momentarily violates equation 8 at $t = 0$, but subsequently this equation is satisfied. The momentary increase in E provides a stimulus that, according to equations 2 and 3, increases I and D suddenly to new positive values I_0 and D_0 at $x = 0$. These new values are then used as initial values for the subsequent solution of the partial differential equations.

Values for the parameters P_0, b, h, q, s, and k_1 were chosen by trial and error and are not necessarily the only ones to give satisfactory results. Further experimental data are needed to provide better choices for them.

The results of numerically solving the equations are shown in Fig. 3, where I, D, C, and E are plotted against x for different times. The curves are symmetric about $x = 0$. Variables I, C, and E show an impulse that propagates to the right, with recovery. Variable D also propagates, but without recovery. At the right end ($x = 50$), the curves become inaccurate because of the artificial boundary conditions imposed there ($B = 1$, $P = P_0$, other variables $= 0$) to keep the x interval finite. Otherwise, these pulses would continue to propagate towards $x = +\infty$.

The most important result shown in Fig. 3 is that C propagates with recovery. This indicates that the model considered in this chapter does agree with the experimental result that after fertilization, free cytosolic calcium advances along the egg as a wave, increasing from a very low value and then decreasing back to that low value.

Another interesting result is that the model with the indicated parameters shows a sharp threshold phenomenon. The stimulus value for Fig. 3 is 1 (in arbitrary units); the initial increase of E is proportional to the stimulus. It was found that decreasing the stimulus to 0.056 did not change the curves much but that decreasing it further to 0.055 produced a sudden

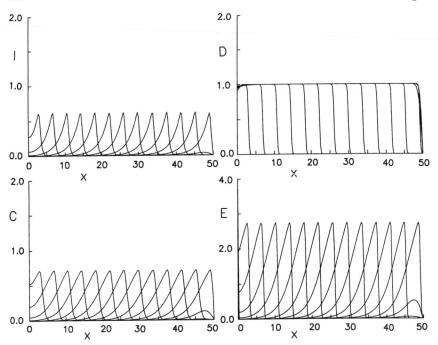

Figure 3. Predictions of model for variation of ITP (inositol-1,4,5-triphosphate), DAG (diacylglycerol), cytosolic free calcium, and phospholipase C, labeled *I*, *D*, *C*, and *E*, respectively, as functions of distance for different times increasing in steps of 2 time units. Units of time and distance are arbitrary. Parameter values: stimulus = 1, P_0 = 1, b = 5, h = 1, q = 0.25, s = 4, k_I = 1, k_D = 0.1.

change: *I*, *D*, and *C*, after small initial increases, settle back to zero, and no propagated wave occurs. We have not yet determined whether this threshold phenomenon pertains to a wide range of values of the parameters, but the existence of such a threshold in the model provides an interesting hypothesis for experimental test.

Solving the equations for the geometric condition of the spherical surface of an egg, instead of on the *x*-axis, requires changing the form of the Laplacian operator ∇^2 in equations 2 and 3. The three spherical coordinates are radial distance r, colatitude θ, and longitude ϕ. Of these, r and ϕ drop out of the expression for the Laplacian, since r is constant on the surface of a sphere, and the spreading wave is independent of ϕ by symmetry. The remaining variable θ, in radians, varies from 0 at the "north pole", where the sperm enters, to π at the "south pole". The Laplacian is given by:

$$\nabla^2 = \frac{1}{\sin \theta} \frac{\partial}{\partial \theta} \left(\sin \theta \frac{\partial}{\partial \theta} \right) \qquad (8)$$

Fertilization Membrane in Sea Urchin Eggs

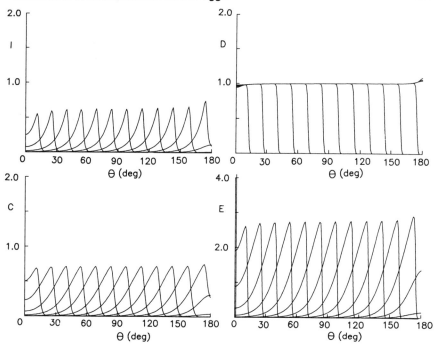

Figure 4. Same variables as Fig. 3 plotted for spherical surface. Angle θ plotted on horizontal axis in degrees. Parameter values same as in Fig. 3 except stimulus = 0.1, k_1 = 0.004, k_D = 0.0004 (appropriate for θ in radians).

Fig. 4 shows curves plotted against θ, with the θ axis labeled in degrees instead of radians. The most obvious difference from those in Fig. 3 is that the peak heights of the propagated waves rise near $\theta = 180°$. This is reasonable, since the wave converges on the south pole from all directions; the resulting diffusion from a larger into a smaller area could be expected to elicit a stronger response there than near the equator, where the area covered by the advancing wave remains more constant. The other features of Fig. 4 are qualitatively similar to those of Fig. 3, including the important result that the calcium concentration advances around the egg as a wave, increasing from a very low value and then decreasing back to that low value.

REFERENCES

Chambers, E. L., 1980, Fertilization and cleavage of eggs of the sea urchin, *Lytechinus variegatus*, in calcium-free seawater, *Eur. J. Cell Biol.* **22**:476.

Dale, B., DeFelice, L. J., and Ehrenstein, G., 1985, Injection of a soluble sperm fraction into sea-urchin eggs triggers the cortical reaction. *Experientia* **41**:1068.

Epel, D., 1982, The cascade of events initiated by rises in cytosolic Ca^{+2} and pH following fertilization in sea urchin eggs, in *Ions, Cell Proliferation, and Cancer,* (A. L. Boynton, W. L. McKeehan, J. F. Whitfield, eds.), Academic Press, New York, pp. 327–339.

Gilkey, J. C., Jaffe, L. F., Ridgeway, E. B., and Reynolds, G. T., 1978, A free calcium wave traverses the activating egg of the medaka, *Oryzias latipes, J. Cell Biol.* **76:**448–466.

Poenie, M., Patton, C., and Epel, D., 1982, Use of calcium precipitants during fixation of sea urchin eggs for electron microscopy: Search for the calcium store, *J. Cell Biol.* **95:**161a.

Steinhardt, R. A., Zucker, R., and Schatten, G., 1977, Intracellular calcium release at fertilization in the sea urchin egg, *Dev. Biol.***58:**185–196.

Whitaker, M., and Irvine, R. F., 1984, Inositol 1,4,5-trisphosphate microinjection activates sea urchin eggs, *Nature* **312:**636–639.

Index

Acetylcholine (ACh), 31
Acetylcholine esterase, 31, 117–118
Acetylcholine receptor (AChR)
  channel forming subunits, 82
  correlation analysis, 24
  current–voltage relationship, 73–83
  del Castillo–Katz model, 23
  density, 394
  desensitization, 169
  diffusion coefficient, 394–395
  α-helix model, 391
  ion permeation model, 54–83
  minimal functional unit, 262, 397
  monoclonal antibodies, 262, 388, 392
  reconstitution, 261–262
  relaxation measurements, 25
  structure, 269, 385–392
  see also Postsynaptic channels
Acrosome reaction, 291–293, 300, 302
Adenylate cyclase, 327
  activator, 338
Admittance, 41–43
β-adrenergic receptor, 340
Afterhyperpolarization
  amplitude, 159
  in myotubes, 364, 367–373, 378–381

Afterhyperpolarization (*cont.*)
  in nerve terminal, 151, 153
  origin, 157
Agonist
  definition, 25
  mixtures, 30
Allosteric site, 30
Amiloride
  derivatives, 402, 405–407
  inhibition of Na entry, 404–407, 409–412
  structure, 402
Amiloride-sensitive Na channel
  conductance, 413–414
  flux ratio, 407–409
  inhibition by amiloride, 404–407, 409–412
  interfacial potentials, 412
  modulation, 401, 417
  physiological role, 207, 401–402
  reconstitution, 412–417
  selectivity, 403–404, 413–414
  sensitivity to amiloride, 403
  surface charge density, 412
  turnover number, 401
cAMP, 327–330, 338–348

cAMP (cont.)
  membrane-permeable derivatives, 327
  pharmacological effect, 327
cAMP-dependent protein kinase, 338, 340–345
  catalytic subunit, 341
  reaction, 341
cAMP-dependent protein phosphorilation, 327, 329–330, 338, 345–347
Anomalous rectifier, 276
Antidiuretic hormone (ADH), 210
  role in K secretion, 215, 218
Antiport process, 403
Apamin, 364, 372
Apical membrane, 207
  depolarization, 217
  resistance, 218
  in sodium transport, 401–417
Area elasticity
  bilayers, 190
  defined, 189
  red blood cells, 190
ATP release
  ACh-induced, 168, 170
  Ca-dependence, 168
  granular nature, 170–173
Atropine, 170

Batrachotoxin (BTX), 93, 262; *see also* Na channel
Birefringence, 143
Blocking mechanism, 23
  open state lifetime, 27–28
  see also Channel blockers
Bungarotoxin, 184, 389

Ca-activated K channel [K(Ca)]
  apamin sensitive, 372
  Ba blockade, 214–215, 319–320
  barrier model, 317, 321
  blockade, 314–320, 375–377
  Ca activation, 212–213, 279, 308–310
  in β-cell, 354–356, 361
  conductance, 212, 279, 308, 312–313, 375
  ion selectivity, 216, 313–314
  in K secretion, 208
  in kidney, 208–218
  kinetic models, 308–310
  modulation, 215, 329, 331, 363–364, 367–381

Ca-activated K channel [K (Ca)] (cont.)
  multiion, 317–318, 321
  neurotrophic effect, 375–381
  physiological role, 104
  in plants, 196
  selectivity filter, 314
  surface charge effect, 267, 310–311, 313
  in T-tubules, 104
  voltage dependence, 211–212, 309–310, 375
Ca channel
  in ATP release, 173
  in -cell, 354, 356–357
  classes, 103
  conductance, 279
  nitrendipine binding, 103
  physiological role, 104
  in plants, 196
  in SR, 104–195
Ca ion
  in egg fertilization, 421–429
  intracellular regulator, 6, 226, 354–361
  pump, 102, 106, 108
  release, 104–105
  tubular depletion, 240
Ca currents
  in GH3 cell, 245, 249–252
  inactivation, 237–241, 251–253
  in invertebrate muscle, 235–241
  in pacemaking, 253
  ion selectivity, 252
  tail current, 250
  time constant, 239, 249, 252
  types, 249–253
  voltage dependence, 253
Caffeine, 338
Capacitance
  measurements, 40–43
  voltage-dependent, 43
Capacitative current, 38–39, 43; *see also* Gating currents
Cardiac muscle, 93–94
Catecholamine release, 172–174
A6 cell, 413–414
α-cell, 353
  ATP release, 164, 174–176
  bursting pattern model, 356–361
  channels, 354–356
  electrical activity, 244, 353–361
  secretion of insulin, 217, 244, 359, 361
δ-cell
Cell permeabilization, 164, 176

Index

Cell-to-cell channel
  physiological role, 335–337
  regulation by Ca, 337
  regulation by cAMP-dependent phosphorilation, 338–345
  tyrosine phosphorilation, 344–347
  SV-40 virus, 348
Channel
  affinity for toxins, 90
  chord conductance, 72–73
  definition, 1–2, 37
  gating, 307–308
  metabolic regulation
  one ion, 75
  in oocytes, 97
  physiological role, 1
  reconstitution, 258–269
  rate constant calculation, 9–15
  rate of ion movement, 128
  selectivity filter, 2
Channel blocker
  AChR, 27–28, 169, 184
  amiloride-sensitive Na channel, 401–416
  anomalous rectifier, 276
  Ca channel, 96, 154, 156–158, 173, 292–294
  definition, 25
  K(Ca) channel, 208, 213–215, 316–320, 364, 372
  K channel, 39, 196, 294, 300
  light-activated channel, 231
  Na channel, 39, 248–249, 279, 282–284
  in sea urchin sperm, 292–294
  SR K channel, 133–134
Chemically gated channel, 89
Cholesterol, 106, 113, 116
Chromaffin cell
  AChR desensitization, 169
  ATP content, 171, 175
  ATP release, 163–164, 168–177
  catecholamine release, 172–173
Cl channel, 127
  dimeric structure, 267–268
  in plants, 196
  protochannel, 267
  reconstitution, 267–268, 297
  in sea urchin sperm, 297
Cortical collecting tubule, 207–208
Curare, 27, 169
Current–voltage relationship
  polynomial fitting, 73
Cytochalasins, 190

Cytoskeleton, 190, 192, 347

D-600, 292, 294, 300
Debye length, 266
Diacylglycerol, 423, 425–426
Diazepan, 30
  action, 32–33
Dibutyryl cAMP, 338
Dichroism, 143–144
Dielectric current, 39
Dihydropyridines, 96, 293
Distal tubule, 217
cDNA, recombinant techniques, 389–393, 397
Double layer theory, 266
Dwell time distribution
  close and open states, 21, 27
  rate constant, 19
Dye
  application, 146
  quantum efficiency, 146
  types, 144
  voltage dependence, 144–146

Eadie–Hofstee plot, 71–72, 80
Eadie–Scatchard plot, 90
Electric dipole, 39, 42
Electronic potential, detection, 146
Energy barrier, ion channel closing and opening, 38
Energy profile
  asymmetric, 68–71
  in determining permeability parameters, 53
  fluctuating, 53–64, 67–72, 74–83
  flux coupling, 70
  fractional distances, 59
  static, 64–67
  voltage dependence, 59
Equivalent circuit, 44; *see also* Gating currents
Excitability-inducing material (EIM), 275
Excitation–contraction coupling, 101–102, 120
  role of channels, 102, 104–105
Eyring frequency factor, 59, 63

Fluctuating barrier, 53–64, 67–72, 74–83
  multiionlike behavior, 71
  see also Energy profile
Fluctuation rate constant, 60
Fluid mosaic model, 258, 394

Fluorescence probe, 130
Fluorescence quenching, 129–136
Flux coupling, *see* Energy profile
Forskolin, 210
  role in K secretion, 215
  channel modulation, 215–216, 218
Fourier transform, 41
  algorithm, 43
Freeze–thawing technique, 130
Frequency of translocation, 62
Fusion method, 127, 270

Gap junction, 337, 339
  permeability, 339–340, 342, 348
Gating currents
  capacitative currents, 38–39, 43
  equivalent circuit, 44
  measurements, 39–45
  methods to measure, 44–45
  rising phase, 48
  in single channels, 38
  time constant, 47
Gating kinetics
  definition, 1
  as a Markov process, 17–20
  as a stoschastic process, 17–24
Gating particle, 42, 50
  activating and inactivating interaction, 49
Gating subunit, 18
GH3 cell
  action potential, 244
  electrical activity, 244
  ionic currents, 247–253
  secretion, 244–245
  whole-cell patch-clamp, 245–247
Glucagon, 353
Gonadotropin, 340
Gramicidin, 84, 131
Growth hormone, 244

Hamilton R-factor, 84–85
Hexamethonium, 169
Hodgkin and Huxley model, 18, 24
  m^3 formulation, 47

Inactivating gate, 50
Inactivation
  Ca channel, 237–241, 251–253
  Na channel, 49

Inhibitory postsynaptic channel
  GABA response, 30–31
Inositol triphosphate, 102, 245, 423
Insulin release, 174, 244; *see also* β-cell
Inward rectifier channel, 328
Ion permeation model, 53–84
Ion translocation, rate constants 54–55, 59
Islet of Langerhans, *see* β-cell

Kinetics
  bursting, 28
  multistep, 186
  states and substates, 5–15, 19–33
  see also Channel, Gating kinetics
Knock-off, 317

Laplace law, 189
Laplace transform, 34
Light-activated channel
  conductance, 229, 231
  internal blocker, 231
  internal transmitter, 228, 231
  in rods, 224
  ion selectivity, 226
  role of Ca, 226, 228
  types, 227–228
  voltage dependence, 226, 229
Light scattering
  ion fluxes measured, 136–139
Lipid bilayers, 1
  made from monolayers, 262–263, 277–278, 331
  in pipettes, 278, 296, 331
  planar, 258–260, 277–278
Lipid motion, 394
Liposome, 258–260
  ion fluxes, 128
Local anesthetics, 93
Luciferin–luciferase reaction
  in ATP release, 163–168
  chemical requirements, 164–166

Markov process, 17–18, 20, *see also* Gating kinetics
Matrix method
  in channel kinetics, 20–23
  in ion permeation, 54–58
Mechanical transduction, 181
Mechano sensor, 181
Melittin, 138

Mg-ATPase, in plants, 195
Michaelis–Menten behavior, 71, 264
Microscopic reversivility, 56, 60–62
Multiion behavior, 71, 317–318, 321
Muscimol, 30
Muscle surface membrane
　isolation, 112–113
　markers, 113, 117–120
　ouabain binding, 116

Na–Ca exchanger, 403, 405
Na channel
　activation kinetics, 48–49
　affinity for TTX, 93–94, 369–370
　BTX-activated, 264, 276, 279–287
　Ca blockade, 285–287
　in cardiac muscle, 93–94
　closed states, 46–48
　currents, 46
　density, 97
　dipole moment, 393
　distribution in skeletal muscle, 94–98, 102, 117
　gating, 45–50
　inactivation, 49
　ion permeation, 283–285
　ion selectivity, 262, 281
　kinetic model, 50–51
　molecular structure, 39–274, 393
　molecular weight, 98, 393
　pharmacology, 90–96
　reconstitution studies, 103, 261–264, 271–276, 279–288
　sensitivity to STX, 282–284
　subunits, 263
　voltage sensor, 393
Na currents
　in GH3 cells, 248–249
　tail currents, 248
　types, 97
Na–proton exchanger
　amiloride inhibition 404–405
　half maximal saturation, 403–404
　regulation, 404
　selectivity, 403–404
　sensitivity to amiloride, 403
Neurohypophysis
　action potential, 147–159
　afterhyperpolarization, 151
　axons, 150
　Ca channel, 148, 153–159

Neurohypophysis (*cont.*)
　K channel, 148
　Na channel, 148, 154–159
　secretion, 147
　secretory ending, 150
Neuromuscular junction, 30
Neurotransmitter, in channel modulation, 326, 328–329
Nifedipine, 173
Nigericin, 292
Nisoldipine, 293–294, 300
Nitredipine, 112
　binding to T-tubules, 108, 116
　receptor, 103
Nonyltriethylammonium (C_9), 318–319
nystatin, 408

One-ion channel, 75
Oocytes, *see* Channel
Optical methods, 141–160, 163–177
　comparison with electrical measurements, 142–146
Optical recording, action potential, 147–159
Ouabain binding, 116

Paramecium, Ca currents, 240
Parathyroid hormone, 244
Partioned matrix, 20
Pentobarbital, 30
Phophatidylinositol-4,5-biphosphate (PIP_2), 423, 425
Phosphodiesterase inhibitor, 327, 330, 338
Phospholipase C, 425
Photoreceptors, 221–223, 225–228
　current, 323
　see also, Light activated channel Rods
Plant protoplast channels, 199–204
　conductance, 201
　ion selectivity, 204
　K channel, 201–202, 204
　voltage dependence, 200–201
Posterior pituitary, 150; *see also* Neurohypophysis
Postsynaptic channels
　agonist interaction, 25
　antagonists, 25
　blockers, 25, 27–28
　kinetic models, 24–30
Postsynaptic current, reconstruction, 30–33

Postsynaptic response, 28
Power spectrum, 43
Prostaglandin E_1, 340
Proton gradient
  in plants, 195
  synthesis of ATP, 243
Pump rate, 259

Quantum efficiency, 146
Quaternary ammonium ions, 318–319
Quencher molecule, 129
QX-222, 28

Rhodopsin, 221–223, 225, 229
Relaxation, macroscopic conductance, 19
Rods
  morphology, 223
  in vertebrates, 223–225
Rous sarcoma virus, *src* gene, 345–348

Sarcoplasmic reticulum (SR)
  Ca channel, 104–105
  Ca release, 102
  heavy, 106
  ionic channels, 103
  light, 104
  membrane composition, 106–107
  membrane isolation, 106–107, 120
Saxitoxin, 91, 94
  binding, 274–276
  structure, 92
Schwann cell, 48
Scorpion toxins, 91, 93, 95
Sea urchin egg
  fertilization, 421–423
  model for fertilization, 423–429
Sea urchin sperm
  activation, 291
  channel reconstitution, 294–300
  channels, 292–294
  Cl channel, 297
  entry factor, 423
  K channels, 297
  receptors, 300–302
Sensory cells
  adaptation, 222
Sensory stimulus, in photoreceptors, 221–222, 228
Series resistance, 48
  determination, 146
Serotonin, 328–329; *see also* channel modulation

Shunt pathway, in epithelia, 408
Squid giant axon, 142–144
  action potential, 144
SR K channel, 104
  ion fluxes, 129–139
  ion selectivity, 130
  K concentration dependence, 264
  K permeability, 135
  model, 265
  surface charge effect, 266–267
Steady state probability, 7–8, 11
Stern–Volmer relationship, 129–131
Stochastic process, 17–24
Strech-activated channel
  ion selectivity, 185
  effect of presure, 187
  multistep kinetics, 186
  pH effect, 191–192
  single channel activity, 183–184
Synaptic integration, model, 33
Synaptic potential, detection, 146
Synaptosomes, 275

Tetraethylammonium (TEA), 39, 196, 318–319, 370–372
Thyrotropin-releasing hormone (TRH), 244
Time constant, 5
Tonoplast, 196
Torpedo electroplax, 137, 139, 264, 267–268
α-toxin, 385
Transephitelial voltage, 408
Transfer function measurement, 40–42
Transmitter
  action, 25
  endogenous, 30, 32
Transverse tubule (T-tubule), 94–97, 102–104
  in charge movement, 102
  composition, 107, 108
  depolarization, 101, 103
  incorporation into bilayers, 103
  markers, 112
  membrane isolation, 107
TTX, 39, 46, 91–97, 154, 156–159, 274, 369–370
  derivatives, 95–97
  structure, 92
Tyrosine phosphorilation, in cell-to-cell channel, 344–347
Tyrosyl kinase, 347–348

Index

Uranyl ions, 412

Verapamil, 282
Veratridine, 93, 274–275
Vinculin, 347

Voltage gating process, definition, 37
Voltage sensor, 38, 49, 393; *see also* Gating currents

Whole-cell patch clamp, 198